CAMBRIDGE LIBRARY COLLECTION

Books of enduring scholarly value

Mathematics

From its pre-historic roots in simple counting to the algorithms powering modern desktop computers, from the genius of Archimedes to the genius of Einstein, advances in mathematical understanding and numerical techniques have been directly responsible for creating the modern world as we know it. This series will provide a library of the most influential publications and writers on mathematics in its broadest sense. As such, it will show not only the deep roots from which modern science and technology have grown, but also the astonishing breadth of application of mathematical techniques in the humanities and social sciences, and in everyday life.

A Treatise on Differential Equations

The need to support his family meant that George Boole (1815–64) was a largely self-educated mathematician. Widely recognised for his ability, he became the first professor of mathematics at Cork. Boole belonged to the British school of algebra, which held what now seems to modern mathematicians to be an excessive belief in the power of symbolism. However, in Boole's hands symbolic algebra became a source of novel and lasting mathematics. Also reissued in this series, his masterpiece was *An Investigation of the Laws of Thought* (1854), and his two later works *A Treatise on Differential Equations* (1859) and *A Treatise on the Calculus of Finite Differences* (1860) exercised an influence which can still be traced in many modern treatments of differential equations and numerical analysis. The beautiful and mysterious formulae that Boole obtained are among the direct ancestors of the theories of distributions and of operator algebras.

Cambridge University Press has long been a pioneer in the reissuing of out-of-print titles from its own backlist, producing digital reprints of books that are still sought after by scholars and students but could not be reprinted economically using traditional technology. The Cambridge Library Collection extends this activity to a wider range of books which are still of importance to researchers and professionals, either for the source material they contain, or as landmarks in the history of their academic discipline.

Drawing from the world-renowned collections in the Cambridge University Library and other partner libraries, and guided by the advice of experts in each subject area, Cambridge University Press is using state-of-the-art scanning machines in its own Printing House to capture the content of each book selected for inclusion. The files are processed to give a consistently clear, crisp image, and the books finished to the high quality standard for which the Press is recognised around the world. The latest print-on-demand technology ensures that the books will remain available indefinitely, and that orders for single or multiple copies can quickly be supplied.

The Cambridge Library Collection brings back to life books of enduring scholarly value (including out-of-copyright works originally issued by other publishers) across a wide range of disciplines in the humanities and social sciences and in science and technology.

A Treatise on Differential Equations

George Boole

CAMBRIDGE
UNIVERSITY PRESS

University Printing House, Cambridge, CB2 8BS, United Kingdom

Cambridge University Press is part of the University of Cambridge.
It furthers the University's mission by disseminating knowledge in the pursuit of education, learning and research at the highest international levels of excellence.

www.cambridge.org
Information on this title: www.cambridge.org/9781108067928

This edition first published 1859
This digitally printed version 2014

ISBN 978-1-108-06792-8 Paperback

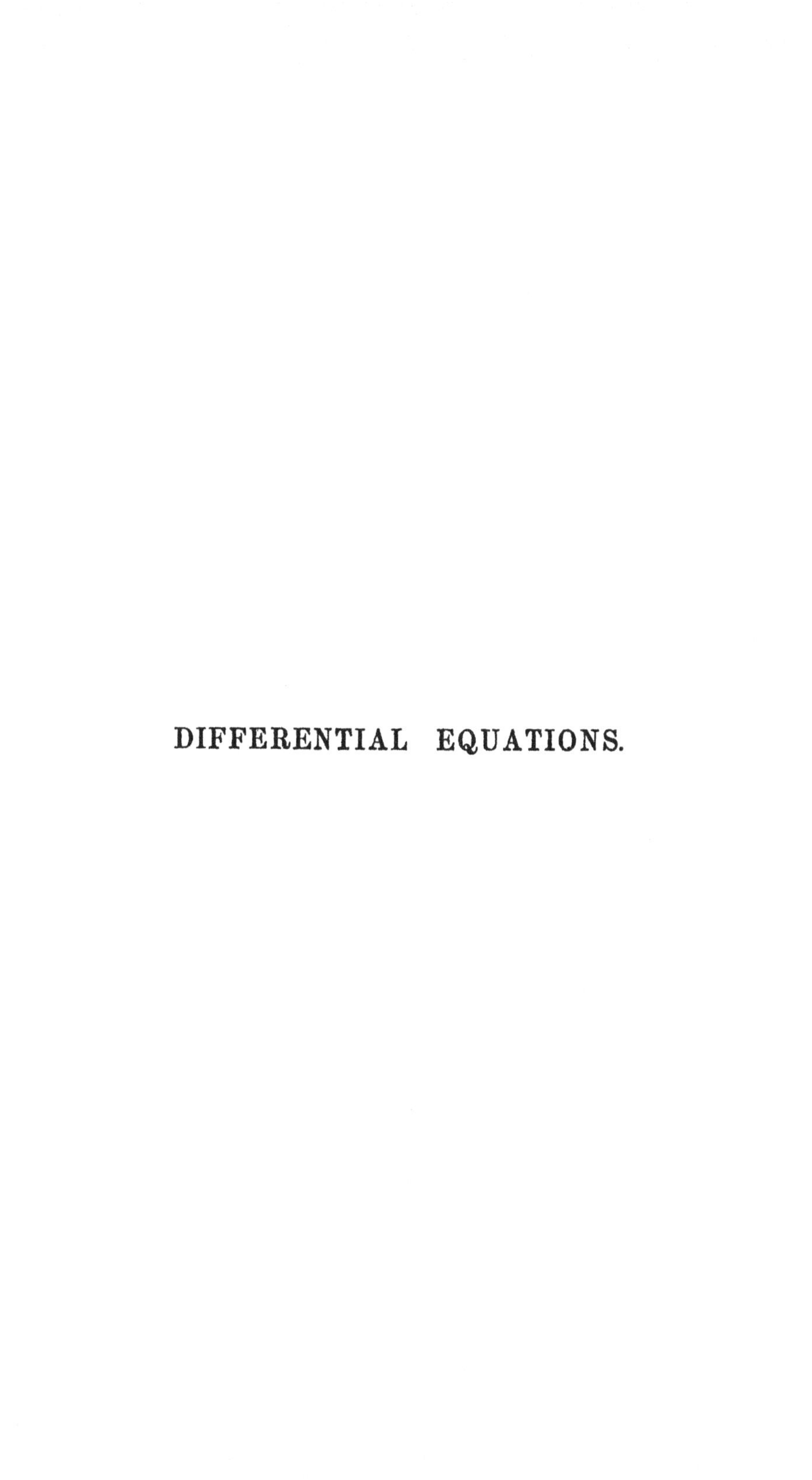

DIFFERENTIAL EQUATIONS.

CAMBRIDGE:
PRINTED BY C. J. CLAY, M.A. AT THE UNIVERSITY PRESS.

A TREATISE

ON

DIFFERENTIAL EQUATIONS.

BY

GEORGE BOOLE, F.R.S.

PROFESSOR OF MATHEMATICS IN THE QUEEN'S UNIVERSITY, IRELAND,
HONORARY MEMBER OF THE CAMBRIDGE PHILOSOPHICAL SOCIETY.

Cambridge:
MACMILLAN AND CO.
AND 23, HENRIETTA STREET, COVENT GARDEN, LONDON.
1859.

PREFACE.

I HAVE endeavoured, in the following treatise, to convey as complete an account of the present state of knowledge on the subject of Differential Equations, as was consistent with the idea of a work intended, primarily, for elementary instruction. It was my object, first of all, to meet the wants of those who had no previous acquaintance with the subject, but I also desired not quite to disappoint others who might seek for more advanced information. These distinct, but not inconsistent aims determined the plan of composition. The earlier sections of each chapter contain that kind of matter which has usually been thought suitable for the beginner, while the latter ones are devoted either to an account of recent discovery, or to the discussion of such deeper questions of principle as are likely to present themselves to the reflective student in connexion with the methods and processes of his previous course. An appendix to the table of contents will shew what portions of the work are regarded as sufficient for the less complete, but still not unconnected study of the subject.

The principles which I have kept in view in carrying out the above design, are the following:

1st, In the exposition of methods I have adhered as closely as possible to the historical order of their development.

I presume that few who have paid any attention to the history of the Mathematical Analysis, will doubt that it has been developed in a certain order, or that that order has been, to a great extent, necessary—being determined, either by steps of logical deduction, or by the successive introduction of new ideas and conceptions, when the time for their evolution had arrived. And these are causes which operate in perfect harmony. Each new scientific conception gives occasion to new applications of deductive reasoning; but those applications may be only possible through the methods and the processes which belong to an earlier stage.

Thus, to take an illustration from the subject of the following work,—the solution of ordinary simultaneous differential equations properly precedes that of linear partial differential equations of the first order; and this, again, properly precedes that of partial differential equations of the first order which are not linear. And in this natural order were the theories of these subjects developed. Again, there exist large and very important classes of differential equations the solution of which depends on some process of successive reduction. Now such reduction seems to have been effected at first by a repeated change of variables; afterwards, and with greater generality, by a combination of such transformations with others involving differentiation; last of all, and with greatest generality, by symbolical methods. I think it necessary to direct attention to instances like these, because the indications which they afford appear to me to have been, in some works of great ability, overlooked, and because I wish to explain my motives for departing from the precedent thus set.

Now there is this reason for grounding the order of exposition upon the historical sequence of discovery, that by so doing we are most likely to present each new form of truth to the mind, precisely at that stage at which the mind is most fitted to receive it, or even, like that of the discoverer, to go forth to meet it. Of the many forms of false culture, a premature converse with abstractions is perhaps the most likely to prove fatal to the growth of a masculine vigour of intellect.

In accordance with the above principles I have reserved the exposition, and, with one unimportant exception, the application of symbolical methods to the end of the work. The propriety of this course appears to me to be confirmed by an examination of the actual processes to which symbolical methods, as applied to differential equations, lead. Generally speaking, these methods present the solution of the proposed equation as dependent upon the performance of certain inverse operations. I have endeavoured to shew in Chap. XVI., that the expressions by which these inverse operations are symbolized are in reality a species of interrogations, admitting of answers, legitimate, but differing in species and character according to the nature of the transformations to which the expressions from which they are derived have been subjected. The solutions thus obtained may be particular or general,—they may be defective, wholly or partially, or complete or redundant, in those elements of a solution which are termed arbitrary. If defective, the question arises how the defect is to be supplied; if redundant, the more difficult question whether the redundancy is real or apparent, and in either case how it is to be dealt with, must be considered. And here the necessity of some prior acquaintance with the things themselves, rather than with the symbolic forms of their expression, must become apparent. The most accomplished in the use of symbols must sometimes throw aside his abstractions and resort to homelier methods for trial and verification—not doubting, in so doing, the truth which lies at the bottom of his symbolism, but distrusting his own powers.

The question of the true value and proper place of symbolical methods is undoubtedly of great importance. Their convenient simplicity—their condensed power—must ever constitute their first claim upon attention. I believe however that, in order to form a just estimate, we must consider them in another aspect, viz. as in some sort the visible manifestation of truths relating to the intimate and vital connexion of language with thought—truths of which it may be presumed that we do not yet see the entire scheme and connexion. But, while this consideration vindicates to them a high position, it seems to me clearly to define that position. As discussions about words can never remove the difficulties that exist in things, so no skill in the use of those aids to thought which language furnishes can relieve us from the necessity of a prior and more direct study of the things which are the subjects of our reasonings. And the more exact, and the more complete, that study of things has been, the more likely shall we be to employ with advantage all instrumental aids and appliances.

But although I have, for the reasons above mentioned, treated of symbolical methods only in the latter chapters of the work, I trust that the exposition of them which is there given will repay the attention of the student. I have endeavoured to supply what appeared to me to be serious defects in their logic, and I have collected under them a large number of equations, nearly all of which are important,—from their connexion with physical science or for other reasons.

2ndly, I have endeavoured, more perhaps than it has been usual to do, to found the methods of solution of differential equations upon the study of the modes of their formation. In principle, this course is justified by a consideration of the real nature of inverse processes, the laws of which must be ultimately derived from those of the direct processes to which they stand related; in point of expediency it is recommended by the greater simplicity, and even in some instances by the greater generality, of the demonstrations to which it leads. I would refer particularly to the demonstration of Monge's method for the solution of partial differential equations of the second order given in Chap. XV.

With respect to the sources from which information has been drawn, it is proper to mention that, on questions relating to the theory of differential equations, my obligations are greatest to Lagrange, Jacobi, Cauchy, and, of living writers, to Professor De Morgan. For methods and examples, a very large number of memoirs English and foreign have been consulted: these are, for the most part, acknowledged. At the same time it is right to add that, in almost every part of the work, I found it necessary to engage more or less in original investigation, and especially in those parts which relate to Riccati's equation, to integrating factors, to singular solutions, to the inverse problems of Geometry and Optics, to partial differential equations both of the first and second order, and, as has already been intimated, to symbolical methods. The demonstrations scattered through the work are also many of them new, at least in form.

In recent years much light has been thrown on certain classes of differential equations by the researches of Jacobi on the Calculus of Variations, and of the same great analyst, with Sir W. R. Hamilton and others, on Theoretical Dynamics. I have thought it more accordant with the design of an elementary treatise to endeavour to prepare the way for this order of inquiries than to enter systematically upon them. This object has been kept in view in the writing of various portions of the following work, and more particularly of that which relates to partial differential equations of the first order.

GEORGE BOOLE.

QUEEN'S COLLEGE, CORK,
February, 1859.

CONTENTS.

PAGE

CHAPTER V.

CHAPTER VI.

CHAPTER VII.

CHAPTER VIII.

PAGE

CHAPTER IX.

CHAPTER X.

CHAPTER XI.

CHAPTER XII.

CHAPTER XIII.

The following portions of the work are recommended to beginners.

Chap. I. Arts. 1—8, 11. Chap. II. Arts. 1—11. Chap. III. Chap. IV. Arts. 1—9. Chap. V. Arts. 1—4. Chap. VI. Arts. 1—10. Chap. VII. Arts. 1—8. Chap. VIII. Arts. 1—7. Chap. IX. Arts. 1, 3—12. Chap. X. Arts. 1—3. Chap. XI. Arts. 1—7. Chap. XII. Arts. 1—8. Chap. XIII. Chap. XIV. Arts. 1—12. Chap. XV. Arts. 1—8. Chap. XVI. Arts. 1—8. Chap. XVII. Arts. 1—12. Chap. XVIII. Arts. 1—8.

ERRATA.

Page 2, line 8, *for* this *read* that
" 15, " 25, for $n - 2 + 1$ read $n - r + 1$
" 30, " 6, *supply* $= 0$
" 36, " 11, *for* $\because$ *read* $\therefore$
" 80, " 3, *omit* concluding
" — " 4, *for* 5 read 6
" 85, " 9, *for* XXXIV. *read* XXIV.
" 119, " 13, *for* found *read* formed
" 128, last line, for x^4 read x^n
" 247, line 20, *for* (5) and (6) *read* (4) and (5)
" — " 22, *for* (4) *read* (6)
" 277, last line, for u read μ
" 280, line 20, *for* 3 *read* 4
" 290, " 1, for $\frac{d^2 y}{dx}$ read $\frac{d^2 y}{dx^2}$
" 294, " 2, *supply* compared with Chap. I. Art. 5.
" 252, last line, for 0 read s

DIFFERENTIAL EQUATIONS.

CHAPTER I.

OF THE NATURE AND ORIGIN OF DIFFERENTIAL EQUATIONS.

1. WHAT is meant by a differential equation?

To answer this question we must revert to the fundamental conceptions of the Differential Calculus.

The Differential Calculus contemplates quantity as subject to variation; and variation as capable of being measured. In comparing any two variable quantities x and y connected by a known relation, e. g. the ordinate and abscissa of a given curve, it defines the rate of variation of the one, y, as referred to that of the other, x, by means of the fundamental conception of a limit; it expresses that ratio by a differential coefficient $\frac{dy}{dx}$; and of that differential coefficient it shews how to determine the varying magnitude or value. Or, again, considering $\frac{dy}{dx}$ as a new variable, it seeks to determine the rate of its variation as referred to the same fixed standard, the variation of x, by means of a second differential coefficient $\frac{d^2y}{dx^2}$, and so on. But in all its applications, as well as in its theory and its processes, the primitive relation between the variables x and y is supposed to be *known*.

In the Integral Calculus, on the other hand, it is the relation among the primitive variables, x, y, &c. which is *sought*. In that branch of the Integral Calculus with which the student is supposed to be already familiar, the differential coefficient $\frac{dy}{dx}$ being given in terms of the independent variable x, it is proposed to determine the most general relation between y and x. Expressing the *given* relation in the form

$$\frac{dy}{dx} = \phi(x) \text{ ....(1),}$$

the relation *sought* is exhibited in the form

$$y = \int \phi(x)\, dx + c,$$

where the symbol $\int$ denotes a certain process of integration, the study of the various forms and conditions of which is, in a peculiar sense, the object of this part of the Integral Calculus.

In (1) we have a particular example of an equation in the expression of which a differential coefficient is involved. But instead of having as in that example $\frac{dy}{dx}$ expressed in terms of x, we might have that differential coefficient expressed in terms of y, or in terms of x and y. Or we might have an equation in which differential coefficients of a higher order, $\frac{d^2y}{dx^2}$, $\frac{d^3y}{dx^3}$, &c., were involved, with or without the primitive variables. All these including (1) are examples of differential equations. The *essential* character consists in the presence of differential coefficients.

The equations

$$\frac{d^2y}{dx^2} + 2\frac{dy}{dx} = 0 \text{ .... (2),}$$

$$x^2\frac{d^2y}{dx^2} + x\frac{dy}{dx} + y = \sin x \text{ .... (3),}$$

are seen to be differential equations, the latter of which contains, while the former does not contain, the primitive variables.

And thus we are led to the following definition.

DEF. *A differential equation is an expressed relation involving differential coefficients, with or without the primitive variables from which those differential coefficients are derived.*

That which gives to the study of differential equations its peculiar value, is the circumstance that many of the most important conceptions of Geometry and Mechanics can only be realized in thought by means of the fundamental conception of the limit. When such is the case, the only adequate expression of those conceptions in language is through the medium of differential coefficients,—the only adequate expression of the truths and relations of which they are the subjects is in the form of differential equations.

Species, Order and Degree.

2. The species of differential equations are determined either by the mode in which differential coefficients enter into their composition, or by the nature of the differential coefficients themselves. We may thus distinguish two great primary classes of differential equations, viz.:

1st. Ordinary differential equations, or those in which all the differential coefficients involved have reference to a single independent variable.

2ndly. Partial differential equations, characterized by the presence of *partial* differential coefficients, and therefore indicating the existence of two or more independent variables with respect to which those differential coefficients have been formed.

Thus an equation such as (2) or (3), involving no other differential coefficients than $\frac{dy}{dx}, \frac{d^2y}{dx^2}$, &c. is an ordinary differential equation, in which x is the independent, y the dependent variable. An equation involving $\frac{dz}{dx}$ and $\frac{dz}{dy}$ would, on the contrary, be a partial differential equation, having z for its dependent, x and y for its independent variables. The equation $x\frac{dz}{dx} + y\frac{dz}{dy} = z$ is a partial differential equation.

The present chapter will be chiefly devoted to the consideration of that class of ordinary differential equations in which there exists a single independent variable x, a single dependent variable y, and one or more of the differential coefficients of y taken with respect to x; the presence of the last element only, viz. the differential coefficient, being essential (Art. 1).

The two following equations, in addition to those already given, will exemplify some of the chief varieties of the species under consideration:

$$\frac{y - x\dfrac{dy}{dx}}{\sqrt{\left\{1+\left(\dfrac{dy}{dx}\right)^2\right\}}} = C \text{ .... (4)},$$

$$\frac{\left\{1+\left(\dfrac{dy}{dx}\right)^2\right\}^{\frac{3}{2}}}{\dfrac{d^2y}{dx^2}} = mx \text{ .... (5)}.$$

In (4) the independent variable x, the dependent variable y, and the differential coefficient $\frac{dy}{dx}$ are all involved; but, while in the previous examples $\frac{dy}{dx}$ appears only in the first degree, in the present one it appears in the second degree and under a radical sign. In (5) we meet with the second differential coefficient $\frac{d^2y}{dx^2}$ in addition to the first differential coefficient $\frac{dy}{dx}$ and the independent variable x.

The typical or general form of a differential equation of the species just described is

$$f\left(x, y, \frac{dy}{dx}, \frac{d^2y}{dx^2}, \dots \frac{d^ny}{dx^n}\right) = 0 \text{ .... (6)},$$

with the condition, already referred to, that one at least of the differential coefficients must explicitly present itself. All the above equations may at once be referred to the typical form by transposition of their second member.

3. Differential Equations are ranked in *order* and *degree* according to the following principles.

1st. The order of a differential equation is the same as the order of the highest differential coefficient which it contains.

2ndly. The degree of a differential equation is the same as the degree to which the differential coefficient which marks its order is raised, that coefficient being supposed to enter into the equation in a rational form.

Thus the equation

$$\left(\frac{dy}{dx}\right)^2 + a\frac{dy}{dx} = b,$$

is of the first order and of the second degree.

The equation

$$\frac{d^2y}{dx^2} + a\frac{dy}{dx} + by^2 = c,$$

is of the second order and of the first degree.

The equation

$$\frac{dy}{dx} = \sqrt{\left(y - x\frac{dy}{dx}\right)} \text{ .... } (7),$$

reduced to the rational form

$$(1 - x^2)\left(\frac{dy}{dx}\right)^2 + 2xy\frac{dy}{dx} - 1 = 0 \text{ .... } (8),$$

is seen to be of the first order and second degree.

The ground of the preference which is to be given to rational forms in the expression and in the classification of differential equations is, that a rational form is at the same time the most *general* form of which an equation is susceptible. Thus (8) includes both the equations which would be formed by giving different signs to the radical in (7).

The typical form of an ordinary differential equation of the first order is evidently

$$f\left(x, y, \frac{dy}{dx}\right) = 0 \text{ ....} (9).$$

4. When a differential equation is capable of being expressed in the form

$$\frac{d^ny}{dx^n} + X_1 \frac{d^{n-1}y}{dx^{n-1}} + X_2 \frac{d^{n-2}y}{dx^2} \dots + X_n y = X \dots\dots (10),$$

in which the coefficients X_1, $X_2 \dots X_n$ and the second member X are either constant quantities or functions of the independent variable x only, the equation is said to be linear. Equations (1), (2) and (3) are thus seen to be linear, but (4) and (5) are not linear. If we refer (3), after dividing both members by x^2, to the general form (10), we have

$$n = 2, \; X_1 = \frac{1}{x}, \; X_2 = \frac{1}{x^2}, \; X = \frac{\sin x}{x^2}.$$

When the coefficients X_1, X_2, &c. in the first member of a linear differential equation referred to the above general type are constant quantities, the equation is defined as a linear differential equation with constant coefficients. When those coefficients are not all constant it is defined as a linear differential equation with variable coefficients. The distinction is illustrated in the following examples:

$$\frac{d^3y}{dx^3} - 2\frac{d^2y}{dx^2} + 5\frac{dy}{dx} - 8y = \sin x,$$

$$(1 - x^2)\frac{d^2y}{dx^2} - x\frac{dy}{dx} + 4y = \cos x,$$

the former of which is a linear differential equation with constant coefficients, while the latter would be described as a linear differential equation with variable coefficients.

Meaning of the terms 'general solution,' 'complete primitive.'

5. In all differential equations there is, as has been seen, an implied reference to some relation among variable quantities dependent and independent; such reference being established through the medium of differential coefficients. Now the chief object of the study of differential equations is to enable us to

determine whenever it is possible, and in the most general manner which is possible, such implied relation among the primitive variables. That relation, when discovered, is, by the adoption of a term primarily applicable to the mode or process of its discovery, called the *solution* of the equation.

Thus if the given equation be

$$x\frac{dy}{dx}+y=\cos x \quad\text{.................... (11)},$$

the following process of solution may be adopted. Multiplying by dx, we have

$$xdy+ydx=\cos x dx,$$

and integrating, since each member is an exact differential,

$$xy=\sin x+c \quad\text{...................... (12)}.$$

The result is termed the solution, or, still more definitely, the *general* solution of the equation. It involves an arbitrary constant, c, by giving particular values to which a series of particular solutions is obtained. The equations

$$xy=\sin x,$$

$$xy=\sin x+1,$$

are particular solutions of the given differential equation.

The term solution is still employed, even when the integration necessary in order to obtain in a finite and explicit form the relation between the variables cannot be effected. Thus if we had the differential equation,

$$x\frac{dy}{dx}-y-xe^x=0 \quad\text{..................... (13)},$$

we should thence derive in succession

$$\frac{xdy-ydx}{x^2}=\frac{e^x dx}{x};$$

$$\frac{y}{x}=\int\frac{e^x dx}{x}+c \quad\text{..................... (14)},$$

and the last result is called the solution of the given equation, although it involves an integration which cannot be performed in finite terms.

The relation among the variables which constitutes the general solution of a differential equation, as above described, is also termed its *complete primitive*. The relation (14) involving the arbitrary constant c is virtually the complete primitive of the differential equation (13). It will be observed that the terms 'general solution' and 'complete primitive,' though applied to a common object, have relation to distinct processes and to a distinct order of thought. In the strict application of the former term we contemplate the differential equation as prior in the order of thought, and the explicit relation among the variables as thence deduced by a process of solution; while in the strict use of the latter term the order both of thought and of process is reversed.

Genesis of Differential Equations.

6. The theory of the genesis of differential equations from their primitives is to a certain extent explained in treatises on the Differential Calculus, but there are some points of great importance relating to the connexion of differential equations thus derived, not only with their primitive, but with each other, which need a distinct elucidation.

Suppose that the complete primitive expresses a relation between x, y and an arbitrary constant c. Differentiating on the supposition that x is the independent variable, we obtain a new equation which must involve $\frac{dy}{dx}$, and which may involve any or all of the quantities x, y and c. If it do not involve c, it will constitute the differential equation of the first order corresponding to the given primitive. If it involve c, then the elimination of c between it and the primitive will lead to the differential equation in question.

Thus if the complete primitive be

$$y = cx \text{ .... } (1),$$

we have on differentiation,

$$\frac{dy}{dx} = c \quad \text{.............................. (2),}$$

and, eliminating the constant c,

$$y = x\frac{dy}{dx} \quad \text{........................ (3),}$$

the differential equation of the first order of which (1) is the complete primitive.

That primitive might have been so prepared as to lead to the same final equation by mere differentiation. Thus, reducing the primitive to the form

$$\frac{y}{x} = c,$$

we have on differentiating and clearing the results of fractions,

$$x\frac{dy}{dx} - y = 0,$$

which agrees with (3). And generally, if a primitive involving an arbitrary constant c be reduced to the form $\phi(x, y) = c$, the corresponding differential equation will be obtained by mere differentiation and removal of irrelevant factors, i. e. of factors which do not contain $\frac{dy}{dx}$, and do not therefore affect the relation in which $\frac{dy}{dx}$ stands to x and y. For it is in that relation, as already intimated, Art. 2, that the *essential* character of the differential equation consists.

It is to be observed that when the differentiation of a primitive involving an arbitrary constant c does not alone cause that constant to disappear, the result to which it leads is still a differential equation, only not that differential equation of which the equation given constitutes the *complete* primitive. Thus, while the complete primitive of (3) is (1), that of (2) is $y = cx + c'$, c' being now the arbitrary constant,—arbitrary as being independent of anything contained in the differential equation. Indeed when we consider $\frac{dy}{dx} = c$ as the differential equation, the constant c, as entering into its complete primitive,

$$y = cx + c',$$

is not arbitrary, the value which it bears in the primitive being determined by that which it bears in the differential equation.

As another illustration of the same theory, the equation $y = ce^{ax}$ as complete primitive gives rise to the differential equation of the first order

$$\frac{dy}{dx} - ay = 0,$$

while the equation *immediately* derived from it by differentiation, viz. $\frac{dy}{dx} = cae^{ax}$, has for its complete primitive $y = ce^{ax} + c'$. To the last mentioned differential equation, $y = ce^{ax}$ stands in the relation of a particular primitive.

Second and Higher Orders.

7. It is shewn in the previous section that from an equation containing x and y with an arbitrary constant c, we can by differentiation, and elimination (if necessary) of that constant, obtain the differential equation of the first order, of which the given equation constitutes the complete primitive.

In like manner an equation connecting x, y, and two arbitrary constants being given, if we differentiate twice, and eliminate, should they not have already disappeared, the arbitrary constants, we shall arrive at a differential equation of the second order free from both the constants in question, and of which the given equation constitutes the complete primitive.

Thus, if we take as the primitive equation

$$y = ax^2 + bx \qquad (4),$$

we find on differentiation

$$\frac{dy}{dx} = 2ax + b \quad\quad (5),$$

and, eliminating b between these equations,

$$y = x\frac{dy}{dx} - ax^2 \quad\quad (6),$$

a differential equation of the first order free from the constant b. Differentiating this equation we have

$$x\frac{d^2y}{dx^2} = 2ax,$$

and, eliminating a between the last two equations,

$$x^2\frac{d^2y}{dx^2} - 2x\frac{dy}{dx} + 2y = 0 \quad\quad (7),$$

a differential equation of the second order free from both a and b.

In the above example the constant b was eliminated after the first differentiation, and the constant a after the second. But the same final result would have been arrived at if the order of the eliminations had been reversed. Thus, if a be eliminated between (4) and (5), we shall have

$$x\frac{dy}{dx} + bx - 2y = 0,$$

a differential equation of the first order, different in form from (6), and involving b instead of a. But on differentiating this equation and eliminating b, we shall arrive at the same final equation of the second order (7).

And generally *the order in which the constants are eliminated does not affect the form of the final differential equation.*

Now a little consideration will shew that this is necessarily the case. We are to remember that the generality which the primitive derives from the presence of its arbitrary constants consists only in this, that it is thus made to stand as the representative of an infinite number of particular equations, in each of which these constants receive particular and definite values. If in any one of the equations thus particularized we further give to x a definite value, definite values will also result for y, $\frac{dy}{dx}$, $\frac{d^2y}{dx^2}$, &c. Thus to a given abscissa of a given curve, i.e. of a curve determined as to its *species* by the form of its equation, and as to its *elements* by the values of the constants in that equation, correspond only definite values of the ordinate y determining the corresponding points of the curve, definite values of $\frac{dy}{dx}$ determining the inclination of the tangents at such points to the axis of x, and definite values of $\frac{d^2y}{dx^2}$ determining, in conjunction with the former, the measure of curvature at the same points. In other words, the species of the curve as defined by an equation of the form $\phi(x, y, a, b) = 0$ being fixed, the values of y, $\frac{dy}{dx}$, $\frac{d^2y}{dx^2}$ have a fixed dependence on those of a, b and x.

And hence the equation $\phi(x, y, a, b) = 0$ being given, any processes of differentiation, elimination, &c. applied thereto can only serve, either 1st, to bring out or manifest the dependence above referred to, or 2ndly, to modify the accidental form of its expression; but in no sense to *create* such dependence or affect its real nature. Now this dependence of y, $\frac{dy}{dx}$, $\frac{d^2y}{dx^2}$ upon a, b, and x, involves the existence of three equations among six quantities. Therefore the elimination which thus becomes possible of two of those quantities, a, b, must leave a single final relation between the remaining four, x, y, $\frac{dy}{dx}$, $\frac{d^2y}{dx^2}$. And this is the differential equation in question.

As another example, let us eliminate the arbitrary constants c and c' from the equation

$$y = ce^{ax} + c'e^{bx} \quad \text{.........................(8).}$$

Differentiating we have

$$\frac{dy}{dx} = ace^{ax} + bc'e^{bx} \qquad (9).$$

To eliminate c subtract from this equation the primitive (8) multiplied by a; we have

$$\frac{dy}{dx} - ay = (b - a)\, c'e^{bx} \qquad (10).$$

Again, differentiating

$$\frac{d^2y}{dx^2} - a\,\frac{dy}{dx} = b\,(b - a)\, c'e^{bx},$$

and (to eliminate c') subtracting from this the previous equation multiplied by b, we have

$$\frac{d^2y}{dx^2} - (a + b)\,\frac{dy}{dx} + aby = 0 \qquad (11),$$

the differential equation of the second order required.

If we had first eliminated c' we should in the place of (10) have obtained the equation

$$\frac{dy}{dx} - by = (a - b)\, ce^{ax} \qquad (12).$$

Differentiating this and eliminating c we again obtain the same final result (11).

That result is a differential equation of the second order, and (8), involving both the arbitrary constants c and c, is its complete primitive. The intermediate equations (10) and (12), each of which contains one of the arbitrary constants, and from each of which, by the elimination of that constant, the final differential equation may be derived, are its *first integrals*. As the term primitive has reference to the direct processes of differentiation, &c. by which a differential equation is formed, the term integral has reference to the inverse process of integration by which we reascend from a differential equation to its primitive. Considered with reference to these processes the primitive is sometimes termed the *final integral*.

It has been shewn that the order of succession in which arbitrary constants are eliminated is indifferent. It may be added, and upon the same ground, that the elimination may be simultaneous. If we write the primitive (8) in the form

$$y - ce^{ax} - c'e^{bx} = 0,$$

and differentiate it twice, we have

$$\frac{dy}{dx} - ace^{ax} - bc'e^{bx} = 0,$$

$$\frac{d^2y}{dx^2} - a^2ce^{ax} - b^2c'e^{bx} = 0,$$

and, from the above system of three equations eliminating the constants c and c' by the method of cross-multiplication, we again arrive at the final differential equation of the second order (11).

8. The above examples prepare us for the general statement of the theory of the genesis of differential equations. Let $F(x, y, c_1, c_2, \ldots c_n) = 0$ be a primitive equation between x and y involving n arbitrary constants $c_1, c_2, \ldots c_n$. Differentiating with respect to x, and regarding y as a function of x, we obtain directly, or by elimination of c_1, an equation of the first order of the form

$$\phi_1\left(x, y, \frac{dy}{dx}, c_2, c_3, \ldots c_n\right) = 0.$$

Differentiating this equation with respect to x, and regarding y and $\frac{dy}{dx}$ as functions of that quantity, we obtain directly, or by elimination of c_2, an equation of the second order of the general form

$$\phi_2\left(x, y, \frac{dy}{dx}, \frac{d^2y}{dx^2}, c_3, c_4, \ldots c_n\right) = 0.$$

Continuing the process, we arrive at a final result of the form

$$\phi_n\left(x, y, \frac{dy}{dx}, \frac{d^2y}{dx^2} \cdots \frac{d^ny}{dx^n}\right) = 0.$$

Now this is the type of an ordinary differential equation of the nth order, (6), Art. 2.

As, in the above process of differentiation and elimination, we might have begun by eliminating any other of the constants instead of c_1, it follows that to a primitive containing n arbitrary constants there belong n differential equations of the first order, each involving $n-1$ arbitrary constants. But as those differential equations are all formed by mere processes of elimination from *two* equations, viz. from the primitive and its first derived equation, *two* only of them are independent. Again, as the differential equations of the second order are formed by eliminating two of the constants $c_1, c_2, \ldots c_n$, and as from n constants, $n \cdot \frac{n-1}{2}$ combinations of two constants can be selected, it is seen that there will exist $n\frac{n-1}{2}$ differential equations of the second order, each containing $n-2$ arbitrary constants. Of these equations three only will however be independent, the whole system being derived actually or virtually from the primitive and its first and second derived equations;—actually if we differentiate twice before eliminating; virtually if each differentiation is followed by the elimination of a constant.

This process of deduction continued leads to the following general theorems, viz.:

1st. *To a given primitive involving x, y, and n arbitrary constants belong* $\frac{n(n-1)(n-2)\ldots(n-2+1)}{1 \cdot 2 \ldots r}$ *differential equations of the rth order (r being any whole number less than n), each involving $n-r$ arbitrary constants, but of those equations $r+1$ only will be independent.*

2nd. *There will exist one differential equation of the nth order free from arbitrary constants.*

The converse of the latter truth, viz. that a differential equation of the nth order implies the existence of a complete primitive involving n arbitrary constants, will be established in a future page.

Criterion of derivation from a common primitive.

9. It is established in Art. 7, 1st, that from a primitive equation involving two arbitrary constants arise two differential equations of the first order, each involving one of those constants; 2ndly, that each of these differential equations of the first order gives rise to the same differential equation of the second order, of which the original equation constitutes the complete primitive or final integral.

The second of the properties above noted constitutes a criterion by which it may be determined whether two differential equations of the first order, each involving an arbitrary constant, originate from the same primitive. We must differentiate each equation, and then eliminate its arbitrary constant. If the two results agree as differential equations of the second order, i. e. if they give the same value of $\frac{d^2y}{dx^2}$ as a function of x, y, and $\frac{dy}{dx}$, the differential equations of the first order must have originated in the same primitive. Furthermore, that primitive will be obtained by eliminating $\frac{dy}{dx}$ between the two differential equations given.

Ex. The differential equations of the first order

$$y\frac{dy}{dx} - ax = 0 \quad\text{.......................}\quad (1),$$

$$y^2 - xy\frac{dy}{dx} = b \quad\text{......................}\quad (2),$$

are both derived from the same primitive. Each of them leads on differentiation and elimination of its arbitrary constant to the differential equation of the second order,

$$xy\frac{d^2y}{dx^2} + x\left(\frac{dy}{dx}\right)^2 - y\frac{dy}{dx} = 0 \quad\text{..............}(3).$$

The primitive, found by eliminating $\frac{dy}{dx}$ from the given equations, is

$$y^2 - ax^2 = b \quad\text{.................... (4),}$$

a and b being arbitrary constants.

10. The differential equations of the first order which constitute the first integrals (Art. 7) of a differential equation of the second order (as, in the above example, (1) and (2) are first integrals of (3)), may by algebraic solution be reduced to the forms

$$\phi\left(x, y, \frac{dy}{dx}\right) = a \quad\text{.................. (5).}$$

$$\psi\left(x, y, \frac{dy}{dx}\right) = b \quad\text{.................. (6).}$$

Now a function of the arbitrary constants a and b, as $\Phi(a, b)$, is itself an arbitrary constant, and may be represented by C. Hence any equation of the form

$$\Phi\left\{\phi\left(x, y, \frac{dy}{dx}\right), \ \psi\left(x, y, \frac{dy}{dx}\right)\right\} = C \text{......(7)}$$

would, equally with (5) and (6), constitute a first integral of the supposed equation of the second order. It is evident that (7) is the general type of all such first integrals.

Thus the type of the first integrals of (3) would be

$$\Phi\left(\frac{y}{x}\frac{dy}{dx}, \ y^2 - xy\frac{dy}{dx}\right) = C.$$

But any two first integrals included under this type and independent of each other would lead us, as is obvious, to the same final integral (4), either under its actual or under an equivalent form.

While therefore, viewed as an independent system, the first integrals of a differential equation of the second order are but two, it is formally more correct to regard them as infinite in number, but as so related that any two of them which are independent contain by implication all the rest.

Such considerations are easily extended to differential equations of the higher orders.

Geometrical illustrations.

11. Geometry, by its peculiar conceptions of direction, tangency, and curvature, all developed out of the primary conception of the limit, Art. 1, throws much light on the nature of differential equations.

As the simplest illustration let the equation of a straight line

$$y = ax + b \qquad (1)$$

be taken as the complete primitive, a and b being arbitrary constants.

Differentiating, we have

$$\frac{dy}{dx} = a \qquad (2).$$

Eliminating a, we find

$$y - x\frac{dy}{dx} = b \qquad (3),$$

and again differentiating

$$\frac{d^2y}{dx^2} = 0 \qquad (4).$$

Of these equations, (4), which is free from arbitrary constants, is the general differential equation of the second order of a straight line; and (2) and (3), each of which contains one of the original arbitrary constants, are the two differential equations of the first order. Moreover, each of these differential equations expresses some general property of the straight line—(2), that its inclination to the axis is uniform; (3), that any vertical intercept between the straight line and a parallel through the origin will be of constant length; (4), that a straight line is nowhere either convex or concave;—and this property, which does not involve, in the same definite manner as the others do, the considerations of distance and of angular magnitude, is evidently the most absolute of the three.

The equation of the circle is

$$(x-a)^2+(y-b)^2=r^2 \dots\dots\dots\dots\dots\dots (5),$$

and if we regard a and b as arbitrary constants the corresponding differential equation of the second order will be

$$\frac{\left\{1+\left(\frac{dy}{dx}\right)^2\right\}^{\frac{3}{2}}}{\frac{d^2y}{dx^2}}=r \dots\dots\dots\dots\dots\dots (6),$$

expressing the property that the radius of curvature is invariable and equal to r.

If we proceed to another differentiation, we find

$$\left\{1+\left(\frac{dy}{dx}\right)^2\right\}\frac{d^3y}{dx^3}-3\frac{dy}{dx}\left(\frac{d^2y}{dx^2}\right)^2=0 \dots\dots\dots\dots (7),$$

which is the general differential equation of a circle free from arbitrary constants. And the geometrical property which this equation also expresses is the invariability of the radius of curvature, but the expression is of a more absolute character than that of the previous equation (6). For in that equation we may attribute to r a definite value, and then it ceases to be the differential equation of all circles, and pertains to that particular circle only whose radius is r. The equation (7) admits of no such limitation.

Monge has deduced the general differential equation of lines of the second order expressed by the algebraic equation

$$ax^2+bxy+cy^2+ex+fy=1.$$

It is

$$9\left(\frac{d^2y}{dx^2}\right)^2\frac{d^5y}{dx^5}-45\frac{d^2y}{dx^2}\frac{d^3y}{dx^3}\frac{d^4y}{dx^4}+40\left(\frac{d^3y}{dx^3}\right)^3=0.$$

But here our powers of geometrical interpretation fail, and results such as this can scarcely be otherwise useful than as a registry of integrable forms.

From the above examples it will be evident that the higher the order of the differential equation obtained by elimination of the determining constants from the equation of a curve, the higher and more absolute is the property which that differential equation expresses.

We reserve to a future chapter the consideration of the genesis of partial differential equations as well as of ordinary differential equations involving more than two variables.

EXERCISES.

1. Distinguish the following differential equations according to species, order, and degree, and take account of any peculiarities dependent upon their coefficients.

(1) $$\frac{dy}{dx}-x^2y=ax^5.$$

(2) $$\frac{d^2y}{dx^2}+\frac{2}{x}\frac{dy}{dx}-a^2y=0.$$

(3) $$y=x+x\frac{dy}{dx}+\left(\frac{dy}{dx}\right)^2.$$

(4) $$x\frac{dz}{dx}-y\frac{dz}{dy}=\frac{x^2}{y}.$$

(5) $$\frac{d^2u}{dx^2}+\frac{d^2u}{dy^2}+\frac{d^2u}{dz^2}=0.$$

2. Explain the term 'complete primitive,' and form the differential equations of the first order of which the following are the complete primitives, c being regarded as the arbitrary constant, viz.:

(1) $$y = cx + \sqrt{(1 + c^2)}.$$

(2) $$y = (x + c)\, e^{ax}.$$

(3) $$y = ce^{\tan^{-1}x} + \tan^{-1} x - 1.$$

(4) $$y = (cx + \log x + 1)^{-1}.$$

(5) $$y^2 - 2cx - c^2 = 0.$$

(6) $$y = cx + \phi\,(c).$$

3. Form the differential equations of the second order of which the following are the complete primitives, c and c' being regarded as arbitrary constants.

(1) $$y = c \cos mx + c' \sin mx.$$

(2) $$y = c \cos\,(mx + c').$$

(3) $$y = x \log \frac{c + c'x}{x}.$$

(4) $$y = c \sin nx + c' \cos nx + \frac{x \sin mx}{2m}.$$

4. State the criterion by which it may be determined whether differential equations are derived from a common primitive.

5. Are the differential equations

$$y\frac{dy}{dx} - ax = 0, \quad y^2\left\{1 - \left(\frac{dy}{dx}\right)^2\right\} = b,$$

derived from a common primitive involving a and b as arbitrary constants? If so, determine that primitive.

6. Apply a similar analysis to each of the following pairs of equations, in which p stands for $\frac{dy}{dx}$, viz.

(1) $$x + \frac{p}{\sqrt{(1 + p^2)}} = a, \quad \text{and } y - \frac{1}{\sqrt{(1 + p^2)}} = b.$$

(2) $$y - xp = a\,(y^2 + p), \quad \text{and } y - xp = b\,(1 + x^2 p).$$

7. How many first, second, third, &c. integrals, belong to the general differential equation of lines of the second order given in Art. 11, and how many of each order are independent?

8. From the equation $(y-b)^2 = 4m(x-a)$ assumed as the primitive, deduce 1st the differential equations of the first order involving a and b as their respective arbitrary constants; 2dly the general functional expression for all differential equations of the first order derivable from the same primitive.

9. Of what primitive involving two arbitrary constants would the functional equation

$$\Phi(y-2px, px) = C,$$

represent all possible differential equations of the first order?

10. How many independent differential equations of all orders are derivable from a given primitive involving x, y, and n arbitrary constants?

CHAPTER II.

ON DIFFERENTIAL EQUATIONS OF THE FIRST ORDER AND DEGREE BETWEEN TWO VARIABLES.

1. THE differential equations of which we shall treat in this chapter may be represented under the general form

$$M+N\frac{dy}{dx}=0,$$

M and N being functions of the variables x and y.

In this mode of representation x is regarded as the independent variable and y as the dependent variable.

We may, however, regard y as the independent and x as the dependent variable, on which supposition the form of the typical equation will be

$$M\frac{dx}{dy}+N=0.$$

For as any primitive equation between x and y enables us theoretically to determine either y as a function of x, or x as a function of y, it is indifferent which of the two variables we suppose independent.

It is usual, however, to treat the equation under the symmetrical form

$$Mdx+Ndy=0.$$

But it is to be remembered that this form is unmeaning except as a representative of one or other of the forms given above. In the present chapter we shall first inquire into the significance of the equation

$$M+N\frac{dy}{dx}=0;$$

we shall in the second place examine certain cases in which it admits of finite solution; and lastly deduce certain inferences from the form of its general solution in a series.

2. *The equation* $M + N\frac{dy}{dx} = 0$ *always implies the existence of a primitive relation between* x *and* y *of the form*

$$f(x, y) = c,$$

in which c *is an arbitrary constant.*

Let us first consider what is the *immediate* signification of the equation

$$M + N\frac{dy}{dx} = 0 \quad\text{..........................}\quad (1).$$

We know that if Δx represent any finite increment of x, and Δy the corresponding finite increment of y, $\frac{dy}{dx}$ will represent the *limit* to which the ratio $\frac{\Delta y}{\Delta x}$ approaches as Δx approaches to 0.

Let us then first examine the interpretation of the equation

$$M + N\frac{\Delta y}{\Delta x} = 0 \quad\text{..........................}\quad (2).$$

We have $\frac{\Delta y}{\Delta x} = -\frac{M}{N}$. The second member of this equation being a function of x and y, since M and N are functions of those variables, we may write

$$\frac{\Delta y}{\Delta x} = \phi(x, y) \quad\text{..........................}\quad (3),$$

the form of $\phi(x, y)$ being known when M and N are given.

Now if we assign to x any series of values, it is possible to assign a corresponding series of values of y, any one of which being fixed arbitrarily all the others will be determined by (3).

Thus let $x_0, x_1, x_2 \dots$ be the series of arbitrary values of x, and y_0 an arbitrary value of y corresponding to x_0 as the value of x, then, representing by Δx_0 the increment of x_0, i. e. the value which being added to x_0 converts it into x_1, we have by (3)

$$\Delta y_0 = \phi(x_0, y_0)\,\Delta x_0;$$

$$\therefore\ y_0 + \Delta y_0 = y_0 + \phi(x_0, y_0)\,\Delta x_0.$$

But as Δy_0 represents the increment of y_0 corresponding to Δx_0 as the increment of x_0 it is evident that $y_0 + \Delta y_0$ will be the value of y corresponding to $x_0 + \Delta x_0$ as the value of x. Representing then this value of y by y_1 we shall have

$$\begin{aligned} y_1 &= y_0 + \phi(x_0, y_0)\,\Delta x_0 \\ &= y_0 + \phi(x_0, y_0)\,(x_1 - x_0) \dots\dots\dots\dots\dots\dots (4). \end{aligned}$$

In like manner we shall find

$$y_2 = y_1 + \phi(x_1, y_1)\,(x_2 - x_1) \dots\dots\dots\dots\dots (5),$$

but, y_1 being already determined by (4), y_2 is determined, and, continuing the operation, a series of values of y will be determined, only one of which is arbitrary, while all the others are assigned in terms of that arbitrary value and of the known values of x.

If, for example, we have the particular equation

$$\Delta y = (x + y)\,\Delta x,$$

and assign to x the series of values 0, 1, 2, 3, 4, &c., and at the same time assume that when x is equal to 0, y is equal to 1, we shall have the two following corresponding series of values, viz.

$$x_0 = 0,\quad x_1 = 1,\quad x_2 = 2,\quad x_3 = 3,\quad x_4 = 4,\ \&c.$$

$$y_0 = 1,\quad y_1 = 2,\quad y_2 = 5,\quad y_3 = 12,\quad y_4 = 27,\ \&c.$$

By assigning a different value to y_0, or by assuming arbitrarily the value of some other term of the series y_0, y_1, y_2, &c. we should find another set of values of those quantities corresponding to the given values of x. But, in every such set, the values of all the terms but one will be determined by a law.

Now if the intervals between the successive values of x are diminished, while their number is proportionately increased, each of the corresponding sets of values of x and y will more and more approach to the state of continuous magnitude. And, in the limit, to every conceivable value of x will correspond a value of y, determined in subjection to a continuous law—to a law however which permits us to assign one of the values of y arbitrarily. The analytical expression of that law will be the solution of the differential equation given.

3. To illustrate the same doctrine geometrically, if x and y represent rectangular co-ordinates, any system such as the above would represent a series of points of which the abscissæ having been assumed arbitrarily, the corresponding values of y, except one, are determined by a continuous law. In the limit, that series of points would approximate to a curve the *species* of which as dependent upon the form of its equation would be determined by a law, but an *element* of which, represented by a constant in that equation, would be left arbitrary, so as to permit us to draw the curve through a given point.

The form of the analytical solution thus indicated is

$$f(x, y) = c \quad\quad (6).$$

The genesis of differential equations of the first order and degree from equations of this description has already been explained in Chap. I. Art. 6. It is evident that, as c is arbitrary, such a value may be assigned to it as to make a given value of y correspond to a given value of x. If those corresponding values are x_0, y_0, we have only to assume

$$f(x_0, y_0) = c \quad\quad (7),$$

whence c is determined. But c being once determined, all the values of y depend upon those of x, in obedience to the law expressed by (6).

Certain cases in which the equation $Mdx + Ndy = 0$ admits of finite solution.

4. *The equation* $\mathrm{Mdx} + \mathrm{Ndy} = 0$ *can always be solved when the variables in* M *and* N *admit of being separated; i.e. when the equation can be reduced to the form*

$$\mathrm{Xdx} + \mathrm{Ydy} = 0 \qquad (8),$$

in which X *is a function of* x *alone, and* Y *a function of* y *alone.*

To solve the equation in its reduced form (8), it is only necessary to integrate the two terms separately, and to equate the result to an arbitrary constant. Thus the solution will be

$$\int Xdx + \int Ydy = C \qquad (9).$$

On differentiating this result the arbitrary constant C disappears, and (8) is reproduced.

Thus the solution of the equation

$$xdx + ydy = 0$$

will be

$$\frac{x^2 + y^2}{2} = C,$$

or, since C is arbitrary,

$$x^2 + y^2 = C.$$

The solution of the equation

$$\frac{dx}{1+x} + \frac{dy}{1+y} = 0,$$

will in like manner be

$$\log(1+x) + \log(1+y) = c;$$

a result which may be simplified in the following manner. We have

$$\log(1+x)(1+y) = c;$$

$$\therefore (1+x)(1+y) = \epsilon^c.$$

But *a function of an arbitrary constant is itself an arbitrary constant.* Hence we may write as the solution

$$(1+x)(1+y) = C.$$

Indeed it frequently happens that solutions which present themselves in a transcendental form admit of being reduced to an algebraic form.

Thus also the solution of the equation

$$\frac{dx}{\sqrt{(1-x^2)}}+\frac{dy}{\sqrt{(1-y^2)}}=0 \text{ .... } (10)$$

being

$$\sin^{-1}x+\sin^{-1}y=c,$$

we shall have, on taking the sine of both members of the equation and replacing $\sin c$ by C,

$$x\sqrt{(1-y^2)}+y\sqrt{(1-x^2)}=C \text{ .... } (11),$$

which is algebraic.

5. Different modes of integration will also give rise to solutions which at first sight appear to be discordant. The discordance however will be only apparent. Thus if we express the equation last solved in the form

$$\frac{-dx}{\sqrt{(1-x^2)}}+\frac{-dy}{\sqrt{(1-y^2)}}=0,$$

and integrate by means of the formula

$$\int\frac{-dx}{\sqrt{(1-x^2)}}=\cos^{-1}x+\text{const.},$$

we shall have

$$\cos^{-1}x+\cos^{-1}y=C_1$$

and, taking the cosine of both members,

$$xy-\sqrt{\{(1-x^2)(1-y^2)\}}=\cos C_1 \text{ .... } (12).$$

The last result may however be reduced to the form

$$x\sqrt{(1-y^2)}+y\sqrt{(1-x^2)}=\sin C_1 \text{ .... } (13),$$

which, as $\sin C_1$ is arbitrary, agrees with the previous result, (11).

The constants C and C_1 are seen to be connected by a relation $C = \sin C_1$, which is independent of the variables x and y.

And in general the test of the accordance of two solutions of a differential equation, each involving an arbitrary constant, is, that on *eliminating one of the variables, the other variable will disappear also, and a relation between the arbitrary constants alone result.*

6. It sometimes happens that the variables may be separated by multiplying or dividing the equation by a factor. Thus the equation

$$\frac{xdx}{1+y} - \frac{ydy}{1+x} = 0,$$

becomes on multiplying by $(1+x)(1+y)$,

$$x(1+x)\,dx - y(1+y)\,dy = 0,$$

in which the variables are separated. Integration then gives

$$\frac{x^2}{2} + \frac{x^3}{3} - \frac{y^2}{2} - \frac{y^3}{3} = C.$$

The most general form of equations in which the variables can be separated by the process above mentioned is

$$X_1 Y_1 dx + X_2 Y_2 dy = 0 \quad \ldots\ldots\ldots\ldots\ldots\ldots (14),$$

in which X_1 and X_2 are functions of x only, and Y_1 and Y_2 functions of y only. On dividing the above equation by $Y_1 X_2$, or, which amounts to the same thing, multiplying it by the factor $\frac{1}{Y_1 X_2}$, we have

$$\frac{X_1}{X_2}dx + \frac{Y_2}{Y_1}dy = 0 \quad \ldots\ldots\ldots\ldots\ldots\ldots\ldots (15),$$

in which the variables are separated.

Ex. The equation $x\sqrt{(1+y^2)}\,dx + y\sqrt{(1+x^2)}\,dy = 0$ is thus reduced to

$$\frac{xdx}{\sqrt{(1+x^2)}} + \frac{ydy}{\sqrt{(1+y^2)}} = 0,$$

and has for its complete integral

$$\sqrt{(1+x^2)} + \sqrt{(1+y^2)} = c.$$

7. It frequently happens that the variables in the equation $Mdx + Ndy = 0$ admit of being separated after a preliminary transformation.

Ex. If in the equation $(x - y^2)\,dx + 2xydy = 0$ we assume $y = \sqrt{(xz)}$, we find

$$dy = \frac{zdx + xdz}{2\sqrt{(xz)}}.$$

Substituting these expressions for y and dy in the given equation, we have

$$xdx + x^2dz = 0,$$

whence
$$\frac{dx}{x} + dz = 0.$$

Integrating and replacing z by its value $\frac{y^2}{x}$, we find

$$\log x + \frac{y^2}{x} = C,$$

for the complete primitive.

It will be remarked that the transformation employed in the above example is not a very obvious one. It would scarcely be suggested by the form of the differential equation itself. And in the present state of analysis, it would be impossible to lay down any general direction on the subject. There are however certain classes of differential equations in which the nature of the required transformation can be determined. Among them a foremost place is due to homogeneous equations.

Homogeneous Equations.

8. The differential equation $Mdx + Ndy = 0$ is said to be homogeneous when M and N are homogeneous functions of x and y, and are of the same degree.

Thus the equation

$$\{y + \sqrt{(x^2 + y^2)}\} \, dx - x dy = 0,$$

is a homogeneous equation, M and N being here of the first degree.

To integrate a homogeneous equation it suffices to assume $y = vx$. In the transformed equation the variables x and v will then admit of separation.

Thus in the above example we should find

$$\{vx + x\sqrt{(1 + v^2)}\} \, dx - x \, (v dx + x dv) = 0,$$

whence dividing by x

$$\sqrt{(1 + v^2)} \, dx - x dv = 0,$$

from which result

$$\frac{dx}{x} - \frac{dv}{\sqrt{(1 + v^2)}} = 0\,;$$

$$\log x - \log \{v + \sqrt{(1 + v^2)}\} = C.$$

Replacing v by $\dfrac{y}{x}$, we have

$$\log x - \log \left\{\frac{y + \sqrt{(x^2 + y^2)}}{x}\right\} = C,$$

for the complete primitive.

As in Art. 5, the above solution admits of a simpler expression. Freed from transcendents and radicals, it gives

$$x^2 = 2Cy + C^2,$$

C being an arbitrary constant.

To demonstrate the above method generally, let us suppose that M and N are homogeneous functions of x and y of the nth degree. We may then, in accordance with the known type of homogeneous functions, write

$$M = x^n \phi \left(\frac{y}{x}\right), \quad N = x^n \psi \left(\frac{y}{x}\right),$$

so that the equation $Mdx + Ndy = 0$ becomes on substitution and division by the common factor x^n,

$$\phi\left(\frac{y}{x}\right) dx + \psi\left(\frac{y}{x}\right) dy = 0 \text{ .... } (16).$$

Now assuming $y = vx$, we have

$$\frac{y}{x} = v, \quad dy = vdx + xdv,$$

and the above equation becomes

$$\phi(v)\, dx + \psi(v)\,(vdx + xdv) = 0.$$

Or, $$\{\phi(v) + v\psi(v)\}\, dx + \psi(v)\, xdv = 0.$$

Therefore

$$\frac{dx}{x} + \frac{\psi(v)\, dv}{\phi(v) + v\psi(v)} = 0 \text{ .... } (17),$$

whence on integrating

$$\log x + \int \frac{\psi(v)\, dv}{\phi(v) + v\psi(v)} = C \text{ .... } (18).$$

It is obvious from the symmetry of the relation between x and y that we might equally employ the transformation $\frac{x}{y} = v$ and regard v and y as the new variables. What is essential in the method is the substitution, in place of the original variables x and y, of a new system of variables, consisting of one variable of the old system, and of the ratio which is borne to it by the other variable of that system.

Ex. It is required to integrate the equation

$$\{x - \sqrt{(xy)} - y\}\, dx + \sqrt{(xy)}\, dy = 0,$$

by the direct application of (18). Here, $n = 1$,

$$M = x - \sqrt{(xy)} - y = x\,\{1 - \sqrt{(v)} - v\},$$

$$N = \sqrt{(xy)} = x\sqrt{(v)}.$$

Thus we have

$$\phi(v) = 1 - v^{\frac{1}{2}} - v,$$
$$\psi(v) = v^{\frac{1}{2}},$$

and (18) gives

$$\log x + \int \frac{v^{\frac{1}{2}}\,dv}{1 - v^{\frac{1}{2}} - v + v^{\frac{3}{2}}} = C.$$

To effect the integration in the second term, let $v = t^2$.

Then $\displaystyle\int \frac{v^{\frac{1}{2}}\,dv}{1 - v^{\frac{1}{2}} - v + v^{\frac{3}{2}}} = \int \frac{2t^2\,dt}{1 - t - t^2 + t^3}$

$$= \frac{1}{1-t} + \tfrac{3}{2}\log(1-t) + \tfrac{1}{2}\log(1+t)$$

$$= \frac{1}{1-t} + \log(1-t) + \tfrac{1}{2}\log(1-t^2).$$

Hence finally, replacing t by $\dfrac{y^{\frac{1}{2}}}{x^{\frac{1}{2}}}$,

$$\frac{x^{\frac{1}{2}}}{x^{\frac{1}{2}} - y^{\frac{1}{2}}} + \log(x^{\frac{1}{2}} - y^{\frac{1}{2}}) + \tfrac{1}{2}\log(x - y) = C.$$

9. The equation $(ax + by + c)\,dx + (ex + fy + g)\,dy = 0$ may be made homogeneous by assuming

$$x = x' - \alpha, \quad y = y' - \beta \quad\text{..................(19)},$$

and properly determining the constants α and β.

The proposed transformations give

$$dx = dx', \qquad dy = dy'.$$

Substituting in the given equation, we have

$$(ax' + by' - a\alpha - b\beta + c)\,dx' + (ex' + fy' - e\alpha - f\beta + g)\,dy' = 0,$$

which, if we determine α and β so as to satisfy the conditions

$$\left.\begin{aligned} a\alpha + b\beta &= c \\ e\alpha + f\beta &= g \end{aligned}\right\} \quad\text{.....................}\ (20)$$

becomes

$$(ax'+by')\,dx'+(ex'+fy')\,dy'=0,$$

and is now homogeneous.

To solve this equation we must now assume, in accordance with the rule for homogeneous equations, $y'=vx'$. The transformed equation will be

$$\{a+(b+e)\,v+fv^2\}\,dx'+(e+fv)\,x'dv=0.$$

Whence $$\frac{dx'}{x'}+\frac{(e+fv)\,dv}{a+(b+e)\,v+fv^2}=0,$$

and it now only remains to integrate the rational fractions and substitute for x' and y' the respective values

$$x'=x+\frac{fc-bg}{af-be},\qquad y'=y+\frac{ag-ec}{af-be},$$

obtained from (19) and (20).

The equation $(ax+by+c)\,dx+(ex+fy+g)\,dy=0$ may also be rendered homogeneous by assuming

$$ax+by+c=x',$$

$$ex+fy+g=y'.$$

These equations give

$$adx+bdy=dx',$$

$$edx+fdy=dy',$$

and the values of x, y, dx, and dy obtained from the above systems, substituted in the given equation will render it homogeneous.

Both solutions fail if $af-be=0$. But in this case, since $f=\dfrac{eb}{a}$, the original equation assumes the form

$$(ax+by+c)\,dx+\frac{e}{a}\left(ax+by+\frac{ag}{e}\right)dy=0,$$

and the variables will be separated if we assume $ax + by = z$, and then regard either z and x or z and y as the new system of variables.

10. The linear differential equation of the first order and degree

$$\frac{dy}{dx} + Py = Q \quad\text{.....................(21)},$$

P and Q being functions of x, admits of being solved. When $Q = 0$ the solution is obtained by separating the variables; and when Q is not equal to 0, a solution may be founded upon that of the previous and simpler case.

It must be observed that the linear equation (21), when reduced to the form

$$(Py - Q)\,dx + dy = 0,$$

falls under the general type, $Mdx + Ndy = 0$.

1st, When $Q = 0$, we have

$$\frac{dy}{dx} + Py = 0.$$

Dividing by y, in order to separate the variables

$$\frac{dy}{y} = -Pdx.$$

Therefore, $\log y = -\int Pdx + c$, which gives

$$y = \epsilon^{-\int Pdx + c}$$

$$= C\epsilon^{-\int Pdx} \quad\text{.....................(22)},$$

C being an arbitrary constant substituted for ϵ^c. It has been already observed that a function of an arbitrary constant is itself an arbitrary constant.

2ndly, To solve the linear equation (21) when Q is not equal to 0, let us assign to the solution the general form (22) above obtained, but suppose C to be no longer a constant but a new variable quantity—an unknown function of x, which must be determined in accordance with the new conditions to which the solution must be subject.

Substituting then the above expression for y in (21), and observing that, since C is now variable, we have

$$\frac{dy}{dx}=\frac{d}{dx}(C\epsilon^{-\int Pdx})=\frac{dC}{dx}\epsilon^{-\int Pdx}-CP\epsilon^{-\int Pdx},$$

there results

$$\frac{dC}{dx}\epsilon^{-\int Pdx}=Q.$$

Hence $$\frac{dC}{dx}=\epsilon^{\int Pdx}Q.$$

$$C=\int\epsilon^{\int Pdx}Qdx+c,$$

c being an arbitrary constant. Substituting this *generalized* value of C in (22), we have finally

$$y=\epsilon^{-\int Pdx}\left(\int\epsilon^{\int Pdx}Qdx+c\right)\ \dots\dots\dots\dots\dots\ (23),$$

the solution required.

It will be observed that if $Q=0$, the above solution is reduced to the form (22) before obtained.

The method of generalizing a solution above exemplified is called the method of the *variation of parameters*, the term parameter, by an extension of its use in the conic sections, being applied to denote the arbitrary constants of the solution of a differential equation. It is only, however, in certain cases that this method is successful. It is always legitimate to endeavour to adapt a solution to wider conditions by a transformation, which, like the above, only introduces a new variable instead of an old one, or a new and adequate system of variables in the room of a former system. But it is not always that the equations thus obtained are, as in the above example, easier of solution than those of which they take the place.

Ex. 1. Given $\dfrac{dy}{dx} - \dfrac{2y}{x+1} = (x+1)^3$.

Here $P = \dfrac{-2}{x+1}$, $Q = (x+1)^3$.

Hence $\int P dx = -2 \log(x+1)$, $\epsilon^{\int P dx} = (x+1)^{-2}$.

$$\int \epsilon^{\int P dx} Q dx = \int (x+1)\, dx = \frac{(x+1)^2}{2} + C.$$

Therefore $$y = (x+1)^2 \left\{ \frac{(x+1)^2}{2} + C \right\}.$$

Ex. 2. Given $\dfrac{dy}{dx} - \dfrac{ny}{x+1} = \epsilon^x (x+1)^n$.

Here we find $$\int P dx = -n \log(x+1),$$

$$\epsilon^{\int P dx} = (x+1)^{-n},$$

$$\int \epsilon^{\int P dx} Q dx = \int \epsilon^x dx = \epsilon^x.$$

Therefore $$y = (x+1)^n (\epsilon^x + c).$$

11. Equations of the form

$$\frac{dy}{dx} + Py = Qy^n,$$

P and Q being functions of x, are reducible to a linear form. For, dividing by y^n, we have

$$y^{-n} \frac{dy}{dx} + P y^{1-n} = Q.$$

Now let $y^{1-n} = z$, then

$$(1-n)\, y^{-n} \frac{dy}{dx} = \frac{dz}{dx},$$

whence $$y^{-n} \frac{dy}{dx} = \frac{1}{1-n} \frac{dz}{dx},$$

so that the equation becomes

$$\frac{1}{1-n}\frac{dz}{dx}+Pz=Q,$$

or $$\frac{dz}{dx}+(1-n)Pz=(1-n)Q,$$

which is linear.

Ex. Given $\dfrac{dy}{dx}+\dfrac{y}{x+1}=\dfrac{-(x+1)^3y^3}{2}$.

Here, dividing by y^3, we have

$$y^{-3}\frac{dy}{dx}+\frac{y^{-2}}{x+1}=-\frac{(x+1)^3}{2},$$

and, assuming $y^{-2}=z$,

$$-\tfrac{1}{2}\frac{dz}{dx}+\frac{z}{x+1}=-\frac{(x+1)^3}{2},$$

or $$\frac{dz}{dx}-2\frac{z}{x+1}=(x+1)^3.$$

The solution of this equation, which is identical in form with that of Ex. 1, is

$$z=\frac{(x+1)^4}{2}+c(x+1)^2,$$

whence $$y=\left\{\frac{(x+1)^4}{2}+c(x+1)^2\right\}^{-\frac{1}{2}}.$$

General solution by development.

12. In the earlier portion of this chapter it was established, by considerations founded upon the nature and interpretation of the equation

$$Mdx+Ndy=0,$$

that it implied the existence of a primitive equation between x, y, and an arbitrary constant. The examples of finite solution which have been given above, illustrate this truth. But a further and more complete illustration is afforded by the presence of an arbitrary constant in the general integral of the equation, as developed in the form of a series by Taylor's theorem. This mode of solution we now proceed to exhibit.

From the given equation we have

$$\frac{dy}{dx} = -\frac{M}{N},$$

the second member of which, being a function of x and y, may be replaced by $f_1(x, y)$. Thus we may write

$$\frac{dy}{dx} = f_1(x, y) \text{ .... } (24).$$

And differentiating this equation

$$\frac{d^2y}{dx^2} = \frac{df_1(x, y)}{dx} + \frac{df_1(x, y)}{dy}\frac{dy}{dx}$$

$$= \frac{df_1(x, y)}{dx} + \frac{df_1(x, y)}{dy} f_1(x, y),$$

the second member of which, being a function of x and y, may be represented by $f_2(x, y)$. Thus we have, as a consequence of (24),

$$\frac{d^2y}{dx^2} = f_2(x, y) \text{ .... } (25).$$

Repeating on this equation the above process of differentiation and substitution, we have

$$\frac{d^3y}{dx^3} = f_3(x, y) \text{ .... } (26),$$

wherein

$$f_3(x, y) = \frac{df_2(x, y)}{dx} + \frac{df_2(x, y)}{dy} f_1(x, y).$$

And, continuing thus to repeat the same operation, we obtain a series of equations determining the successive differential coefficients of y, in the form

$$\frac{d^n y}{dx^n} = f_n(x, y) \text{ .... } (27),$$

the dependence of $f_n(x, y)$ upon $f_{n-1}(x, y)$, and hence ultimately upon $f_1(x, y)$, being determined by the general equation

$$f_n(x, y) = \frac{df_{n-1}(x, y)}{dx} + \frac{df_{n-1}(x, y)}{dy} f_1(x, y) \text{ ..... } (28).$$

Hence M and N being given, the expressions for

$$\frac{dy}{dx}, \frac{d^2y}{dx^2} \cdots$$

are implicitly given also.

Now $\frac{dy}{dx}$, $\frac{d^2y}{dx^2}$, &c. determine the coefficients of the several terms *after the first* in the development of y in ascending powers of x, by Taylor's theorem, or more generally in ascending powers of $x - x_0$, where x_0 is a particular value of x. Leaving that first term arbitrary, the development is thus seen to be possible, and the result, while constituting the general integral of the given differential equation, shews that that integral involves an arbitrary constant.

Actually to obtain the development, let $\phi(x)$ represent the general value of y, and let y_0 be the particular value of y corresponding to some particular and definite value, x_0, of the variable x. Then, writing $\phi(x)$ in the form

$$\phi(x_0 + x - x_0),$$

we have, by Taylor's theorem,

$$y = \phi(x_0) + \phi'(x_0)(x - x_0) + \phi''(x_0)\frac{(x - x_0)^2}{1.2} + \&\text{c.} \ldots (29).$$

But $\phi(x_0)$ is what y becomes when $x = x_0$. Hence $\phi(x_0) = y_0$. Again, $\phi'(x_0)$ is what $\frac{d\phi(x)}{dx}$, i.e. $\frac{dy}{dx}$, becomes when $x = x_0$. Hence $\phi'(x_0) = f_1(x_0, y_0)$ by (24). In like manner $\phi''(x_0)$ is what $\dfrac{d^2y}{dx^2}$ becomes when $x = x_0$, and is therefore equal to $f_2(x_0, y_0)$. Determining thus the successive coefficients of (29), we have finally

$$y = y_0 + f_1(x_0, y_0)(x - x_0) + f_2(x_0, y_0)\frac{(x - x_0)^2}{1.2} + \&c. \ldots (30),$$

which is the general integral.

If we assume $x_0 = 0$, and represent the corresponding value of y by c, we have

$$y = c + f_1(0, c)\,x + f_2(0, c)\frac{x^2}{1.2} + \&c. \ldots\ldots (31).$$

Should however any of the coefficients in this development become infinite we must revert to the previous form, and give to x_0 such a value as will render the coefficients finite, and therefore justify the application of Taylor's theorem.

Virtually the integral (30) involves like (31) only one arbitrary constant. For in applying it we are supposed to give to x_0 a *definite* value, and this being done the corresponding arbitrary value of y_0 constitutes the single arbitrary constant of the solution.

EXERCISES.

1. Integrate the differential equations:

(1) $(1 + x)\,ydx + (1 - y)\,xdy = 0.$

(2) $(y^2 + xy^2)\,dx + (x^2 - yx^2)\,dy = 0.$

(3) $xy\,(1 + x^2)\,dy - (1 + y^2)\,dx = 0.$

(4) $(1 + y^2)\,dx - \{y + \sqrt{(1 + y^2)}\}\,(1 + x^2)^{\frac{3}{2}}\,dx = 0.$

(5) $\sin x \cos v\, lx - \cos x \sin ydy = 0.$

(6) $\sec^2 x \tan ydx + \sec^2 y \tan xdy = 0.$

2. Different processes of solution present the primitive of a differential equation under the following different forms, viz.

$$\tan^{-1}(x+y)+\tan^{-1}(x-y)=c,$$
$$y^2-x^2+1=2Cx.$$

Are these results accordant?

3. Integrate the homogeneous equations:

(1) $(y-x)\,dy+ydx=0.$

(2) $\{2\sqrt{(xy)}-x\}\,dy+ydx=0.$

(3) $xdy-ydx-\sqrt{(x^2+y^2)}\,dx=0.$

(4) $\left(x-y\cos\frac{y}{x}\right)dx+x\cos\frac{y}{x}\,dy=0.$

(5) $(8y+10x)\,dx+(5y+7x)\,dy=0.$

4. Integrate the equations:

(1) $(2x-y+1)\,dx+(2y-x-1)\,dy=0.$

(2) $(3y-7x+7)\,dx+(7y-3x+3)\,dy=0;$

the former as an exact differential equation, the latter by reduction to a homogeneous form.

5. Explain what is meant by variation of parameters, and, having integrated the equation $x\frac{dy}{dx}-ay=0$, deduce by that method the solution of the equation $x\frac{dy}{dx}-ay=x+1.$

6. Integrate, by the direct application of (23) the linear equations,

(1) $$\frac{dy}{dx}+\frac{x}{1+x^2}y=\frac{1}{2x(1+x^2)}.$$

(2) $$x(1-x^2)\frac{dy}{dx}+(2x^2-1)\,y=ax^3.$$

(3) $$\frac{dy}{dx}+\frac{y}{(1-x^2)^{\frac{3}{2}}}=\frac{2+\sqrt{(1-x^2)}}{(1-x^2)^2}.$$

(4) $$\frac{dy}{dx} + y\cos x = \frac{\sin 2x}{2}.$$

(5) $$(1+x^2)\frac{dy}{dx} + y = \tan^{-1}x.$$

6. Shew that the solution of the general linear equation $\frac{dy}{dx} + Py = Q$ may be expressed in the form

$$y = \frac{Q}{P} - \epsilon^{-\int P dx}\left(C + \int \epsilon^{+\int P dx}\, d\frac{Q}{P}\right).$$

7. Shew that, $\phi(x)$ being any function of x, the solution of the linear equation

$$\frac{dy}{dx} - y\phi'(x) = \phi(x)\phi'(x),$$

will be $y = c\epsilon^{\phi(x)} - \phi(x) - 1$.

8. Shew that if in the linear equation $\frac{dy}{dx} + Py = Q$ we represent $\frac{dy}{dx}$ by p, and then, differentiating and eliminating y, form a differential equation between p and x, that equation will also be linear.

9. Integrate the differential equations:

(1) $$(1-x^2)\frac{dz}{dx} - xz = axz^2.$$

(2) $$z^2\frac{dz}{dx} - az^3 = x + 1.$$

(3) $$\frac{dz}{dx} + 2xz = 2ax^3z^3.$$

(4) $$\frac{dz}{dx} + z\cos x = z^n \sin 2x.$$

(5) $$x\frac{dy}{dx} + y = y^2 \log x.$$

CHAPTER III.

EXACT DIFFERENTIAL EQUATIONS OF THE FIRST DEGREE.

1. THE conditions described in the previous chapter under which the equation $Mdx + Ndy = 0$ is integrable by the separation of the variables, embrace but a small number of the cases in which a solution expressible in finite terms exists. Analysts have therefore engaged in the more fundamental inquiry of which the following are the objects, viz.

1st, To ascertain under what conditions the equation

$$Mdx + Ndy = 0$$

is derived by immediate differentiation from a primitive of the form $f(x, y) = c$, and how, when those conditions are satisfied, the primitive may be found.

2ndly, To ascertain whether, when those conditions are not satisfied, it is possible to discover a factor by which the equation $Mdx + Ndy = 0$ being multiplied, its first number will become an exact differential.

These inquiries will form the subject of this and the following chapter.

Theorem. *The one necessary and sufficient condition under which the first member of the equation $Mdx + Ndy = 0$ is an exact differential is*

$$\frac{dM}{dy} = \frac{dN}{dx} \quad \text{.......................} (1).$$

Let it be considered in the first place what is meant by the supposition that $Mdx + Ndy$ is an exact differential. It is that M and N are partial differential coefficients with respect to x and y,—that there exists some function V, such that we may have

$$\frac{dV}{dx} = M \quad\text{.......................(2)}$$

$$\frac{dV}{dy} = N \quad\text{....................... (3).}$$

Any relation between M and N which we can derive independently of the form of V from the above equations will be a *necessary* condition of $Mdx + Ndy$ being an exact differential. And conversely, any relation between M and N which suffices to enable us to discover a function V actually satisfying the above equations (2), (3), will be a *sufficient* condition of $Mdx + Ndy$ being an exact differential. And if the same condition should present itself in both cases, it will be both *necessary and sufficient.*

Differentiating (2) with respect to y, and (3) with respect to x, we have

$$\frac{d^2V}{dydx} = \frac{dM}{dy}, \quad \frac{d^2V}{dxdy} = \frac{dN}{dx} \quad\text{............. (4).}$$

But the first members of these equations being, by a known theorem of the Differential Calculus, equal, we have

$$\frac{dM}{dy} = \frac{dN}{dx} \quad\text{....................... (5).}$$

This, therefore, is a *necessary* condition of $Mdx + Ndy$ being an exact differential. It is also, as will next be shewn, a sufficient condition.

In the first place the function V, if such exist, must satisfy the equation (2).

Integrating this equation relatively to x alone (since the differentiation in $\frac{dV}{dx}$ is relative to x alone), we have

$$V = \int Mdx + C \quad\text{.................... (6),}$$

C being a quantity which is constant relatively to x, so that $\frac{dC}{dx}=0$. Hence, though C does not vary with x, it may vary with y, and there is nothing to limit the manner of its variation. It is therefore an *arbitrary* function of y, and we may write

$$V=\int Mdx+\phi(y) \text{ .... } (7).$$

This is the most general form of V as a function of x and y, which satisfies the equation (2).

In the second place V must satisfy the equation (3). Substituting in that equation the value of V given in (7), we have

$$\frac{d\int Mdx}{dy}+\frac{d\phi(y)}{dy}=N.$$

Therefore $$\frac{d\phi(y)}{dy}=N-\frac{d\int Mdx}{dy}.$$

Whence $$\phi(y)=\int\left(N-\frac{d\int Mdx}{dy}\right)dy+C \text{ .... } (8),$$

C being simply an arbitrary constant, since, as the constant of integration it cannot contain y, and as part of the expression for $\phi(y)$ it cannot contain x.

Now the integration in the second member is theoretically possible (though its expression in finite terms may not be possible) if the coefficient of dy, viz. $N-\frac{d\int Mdx}{dy}$, is a function of y only, i.e. if its differential coefficient with respect to x is 0. Expressing this condition, we have

$$\frac{dN}{dx}-\frac{d}{dx}\cdot\frac{d\int Mdx}{dy}=0 \text{ .... } (9).$$

But $$\frac{d}{dx}\cdot\frac{d\int Mdx}{dy}=\frac{d}{dy}\frac{d\int Mdx}{dx}$$

$$=\frac{dM}{dy}.$$

Thus the condition (9) becomes

$$\frac{dN}{dx} - \frac{dM}{dy} = 0 \text{........................(10)}.$$

This then is a *sufficient*, as it has before been shewn to be a *necessary* condition of $Mdx + Ndy$ being an exact differential.

The substitution in (7) of the value of $\phi(y)$ found in (8) gives

$$V = \int Mdx + \int \left(N - \frac{d\int Mdx}{dy}\right) dy + C \text{.........(11)}.$$

Finally, supposing still the condition (10) satisfied, the solution of the equation $Mdx + Ndy = 0$ will be

$$\int Mdx + \int \left(N - \frac{d\int Mdx}{dy}\right) dy = C \text{......... (12)}.$$

2. The practical rule to which the above investigation leads is the following.

To solve the equation $Mdx + Ndy = 0$ when its first member is an exact differential. integrate Mdx with respect to x, regarding y as constant, and adding, instead of an arbitrary constant, an arbitrary function of y, which must afterwards be determined by the condition that the differential coefficient of the sum with respect to y shall be equal to N. Then that sum equated to an arbitrary constant will be the solution required.

Ex. 1. Given $(x^2 - 4xy - 2y^2)\,dx + (y^2 - 4xy - 2x^2)\,dy = 0$.

Here $M = x^2 - 4xy - 2y^2$ and $N = y^2 - 4xy - 2x^2$, whence

$$\frac{dM}{dy} = \frac{dN}{dx} = -4x - 4y,$$

and the first member of the given equation is an exact differential.

Now $$\int M dx = \frac{x^3}{3} - 2x^2y - 2y^2x + \phi(y) \text{ ....(1)},$$

the arbitrary function $\phi(y)$ occupying, according to the Rule, the place of the constant of integration. To determine $\phi(y)$, we have

$$\frac{d}{dy}\left\{\frac{x^3}{3} - 2x^2y - 2y^2x + \phi(y)\right\} = y^2 - 4xy - 2x^2.$$

Whence $$\frac{d\phi(y)}{dy} = y^2,$$

$$\phi(y) = \frac{y^3}{3}.$$

Substituting this value in the second member of (1), and equating the result to an arbitrary constant, we have

$$\frac{x^3}{3} - 2x^2y - 2y^2x + \frac{y^3}{3} = C,$$

the solution required.

Ex. 2. Given $$\frac{dx}{\sqrt{(x^2+y^2)}} + \left\{1 - \frac{x}{\sqrt{(x^2+y^2)}}\right\}\frac{dy}{y} = 0.$$

Here $M = \dfrac{1}{\sqrt{(x^2+y^2)}}$, $N = \dfrac{1}{y} - \dfrac{x}{y\sqrt{(x^2+y^2)}}$.

Hence we find

$$\frac{dM}{dy} = \frac{-y}{(x^2+y^2)^{\frac{3}{2}}} = \frac{dN}{dx}.$$

To obtain the complete integral we will on this occasion employ directly the general form of solution (12). We have

$$\int M dx = \log\{x + \sqrt{(x^2+y^2)}\},$$

$$\frac{d}{dy}\int M dx = \frac{1}{y} - \frac{x}{y\sqrt{(x^2+y^2)}}.$$

Hence $N - \dfrac{d}{dy}\int M dx = 0$, so that (12) gives simply

$$\log\{x + \sqrt{(x^2+y^2)}\} = c.$$

Substituting $\log C$ for c, and then freeing the equation from logarithmic signs and from radicals, we have

$$y^2 = C^2 - 2Cx.$$

3. When the criterion (1) is satisfied, we may often simplify the subsequent process of solution by noting that the complementary function of y which remains to be added after the integration with respect to x has been performed, can only be derived from terms in N which do not contain x. If then ν represent the aggregate of such terms, the complementary function will be $\int \nu dy$.

Thus in Ex. 1, the only term in N which does not involve x being y^2, the complementary function of y is $\int y^2 dy$ or $\frac{y^3}{3}$.

Lastly, we may in many cases dispense with the application of the criterion, or greatly simplify its application, by attending to the two following principles, viz.

1st, If $Mdx + Ndy$ can be divided into two portions, one of which is manifestly an exact differential, it suffices to ascertain whether the other is such.

2ndly, If $Mdx+Ndy$, or that portion of it which, according to the above principle, it may suffice to examine, can be resolved into two factors, one of which is manifestly the exact differential of a function of x and y, which we will represent by u, then when the other factor is expressible as a function of u, we shall have an expression of the form $f(u)du$ which is necessarily an exact differential. The converse truth, that when one factor is du the other factor must be of the form $f(u)$ in order to make the product an exact differential, will be established in the following chapter.

Ex. Given $\left\{x + \frac{1}{\sqrt{(y^2 - x^2)}}\right\} dx + \left\{y - \frac{x}{y\sqrt{(y^2 - x^2)}}\right\} dy = 0.$

This equation may be expressed in the form

$$xdx + ydy + \frac{ydx - xdy}{y\sqrt{(y^2 - x^2)}} = 0.$$

Now, $xdx + ydy$ being an exact differential, it suffices to examine whether the term $\dfrac{ydx - xdy}{y\sqrt{(y^2 - x^2)}}$ is such also.

This term may be expressed in the form of the product

$$\frac{y}{\sqrt{(y^2 - x^2)}} \times \frac{ydx - xdy}{y^2},$$

the second factor of which is the differential of $\dfrac{x}{y}$. If we make $\dfrac{x}{y} = u$ the product assumes the form $\dfrac{du}{\sqrt{1 - u^2)}}$, which is the differential of $\sin^{-1} u$.

The complete primitive is therefore

$$\frac{x^2 + y^2}{2} + \sin^{-1}\frac{x}{y} = c.$$

EXERCISES.

1. $(x^3 + 3xy^2)\,dx + (y^3 + 3x^2y)\,dy = 0.$

2. $\left(1 + \dfrac{y^2}{x^2}\right) dx - 2\dfrac{y}{x}\,dy = 0.$

3. $\dfrac{2xdx}{y^3} + \left(\dfrac{1}{y^2} - \dfrac{3x^2}{y^4}\right) dy = 0.$

4. $xdx + ydy + \dfrac{xdy - ydx}{x^2 + y^2} = 0.$

5. $\left(1 + e^{\frac{x}{y}}\right) dx + e^{\frac{x}{y}}\left(1 - \dfrac{x}{y}\right) dy = 0.$

6. $e^x\,(x^2 + y^2 + 2x)\,dx + 2ye^x dy = 0.$

7. $\{n \cos(nx + my) - m \sin(mx + ny)\}\,dx$
$+ \{m \cos(nx + my) - n \sin(mx + ny)\}\,dy = 0.$

8. Shew, without applying the criterion, that the following are exact differentials, viz.

1st, $$\frac{xdx+ydy}{(1+x^2+y^2)^{\frac{1}{2}}}+\frac{ydx-xdy}{x^2+y^2}=0.$$

2ndly, $$\frac{xdx+ydy}{(x^2+y^2)^{\frac{1}{2}}(1-x^2-y^2)^{\frac{1}{2}}}+\left\{\frac{1}{y\sqrt{(y^2-x^2)}}+\frac{e^{\frac{x}{y}}}{y^2}\right\}(ydx-xdy).$$

9. Integrate the above equations.

10. Integrate the equation $\dfrac{x^a dy-ayx^{a-1}\,dx}{by^2-cx^{2a}}+x^{a-1}dx=0$,

distinguishing between the different cases which present themselves according, 1st, as b and c are of the same or of opposite signs; 2ndly, as a is equal to, or not equal to, 0.

11. Shew by the criterion that the expression

$$\frac{\phi\left(\frac{x}{y}\right)dx+\psi\left(\frac{x}{y}\right)dy}{x\phi\left(\frac{x}{y}\right)+y\psi\left(\frac{x}{y}\right)}$$

is generally an exact differential and exhibit the functional forms which $\dfrac{dM}{dy}$ and $\dfrac{dN}{dx}$ assume.

CHAPTER IV.

ON THE INTEGRATING FACTORS OF THE DIFFERENTIAL EQUATION $Mdx + Ndy = 0$.

1. THE first member of the equation $Mdx + Ndy = 0$ not being necessarily an exact differential, analysts have sought to render it such by multiplying the equation by a properly determined factor.

Thus the first member of the equation

$$(1 + y^2)\, dx + xy\, dy$$

is not an exact differential, since it does not satisfy the condition $\frac{dM}{dy} = \frac{dN}{dx}$, but it becomes an exact differential if the equation be multiplied by $2x$, and its integration, which then becomes possible, leads to the primitive equation

$$x^2 (1 + y^2) = c.$$

The multiplier $2x$ is termed an integrating factor.

We propose in this chapter, after establishing a certain preliminary theorem, 1st to demonstrate that integrating factors of the equation $Mdx + Ndy$ always exist; 2ndly, to investigate some of their properties and relations, and to shew how in certain cases they may be discovered. To complete this subject we shall, in the following chapter, investigate a partial differential equation, upon the solution of which their general determination depends, and shall examine some of the conditions under which the solution of that equation is possible.

2. THEOREM. *If V and v are any functions of x and y, the expression Vdv will be an exact differential, only when V is expressible as a function of v alone, involving x and y only through its involving v.*

Let us transform Vdv to an expression in which x and v are regarded as the primary variables instead of x and y.

The transformation is a *possible* one, for if, v being a function of x and y, we write

$$v = \psi(x, y) \text{ ....} (1),$$

it follows that y is implicitly a function of x and v, and therefore V, which is by hypothesis a function of x and y, is also implicitly a function of x and v. The actual expression for V in terms of x and v would be found by determining y in terms of x and v from (1), and substituting the value in the primary expression for V. Suppose the result to be

$$V = \chi(x, v),$$

then

$$Vdv = \chi(x, v)\,dv.$$

If we compare the second member of this equation with the expression for the *exact* differential of a function of x and v, viz. $Mdx + Ndv$, in which M and N satisfy the condition

$$\frac{dM}{dv} = \frac{dN}{dx},$$

(1) Chap. III., we find $M = 0$, $N = \chi(x, v)$, whence the above condition gives

$$0 = \frac{d\chi(x, v)}{dx}.$$

This equation implies that $\chi(x, v)$, that is V, considered originally as a function of x and v, must, under the actual conditions, be considered as a function of v only, since its differential coefficient with respect to x vanishes. Hence the theorem is established.

There is another form of demonstration which has the advantage of connecting the theorem to be proved with another of equal importance.

Since v is by hypothesis a function of x and y, we have

$$Vdv = V\frac{dv}{dx}dx + V\frac{dv}{dy}dy \quad\ldots\ldots\ldots\ldots\ldots\ldots (2).$$

In order that the second member of this equation may be an exact differential, it is necessary and sufficient that we have *identically*

$$\frac{d}{dy}\left(V\frac{dv}{dx}\right) = \frac{d}{dx}\left(V\frac{dv}{dy}\right),$$

which, on effecting the differentiations, becomes

$$\frac{dV}{dy}\frac{dv}{dx} - \frac{dV}{dx}\frac{dv}{dy} = 0 \quad\ldots\ldots\ldots\ldots\ldots\ldots (3).$$

Now reasoning as before, we see that V, whatever may be its constitution as a function of x and y, is reducible to the form $\chi(x, v)$. If in that expression we substitute for v its value $\psi(x, y)$, V is again reduced to a function of x and y. Thus instead of contemplating V as *directly* a function of x and y, we may consider it as mediately such through the system

$$V = \chi(x, v), \quad v = \psi(x, y).$$

Hence according to the rules of implicit differentiation,

$$\frac{dV}{dx} = \frac{d\chi(x, v)}{dx} + \frac{d\chi(x, v)}{dv}\frac{dv}{dx},$$

$$\frac{dV}{dy} = \frac{d\chi(x, v)}{dv}\frac{dv}{dy}.$$

Substituting these values in (3) there results simply

$$\frac{d\chi(x, v)}{dx}\frac{dv}{dy} = 0.$$

Now the constitution of v is either such that v contains y in its expression $\psi(x, y)$, or that it reduces to a mere function of x. If v contains y, then $\frac{dv}{dy}$ does not identically vanish, and therefore we must have

$$\frac{d\chi(x, v)}{dx} = 0.$$

This implies, as before, that V is a function of v alone. But if v is a mere function of x, then reciprocally x is merely a function of v, and therefore $\chi(x, v)$, or V, is merely a function of v.

Wherefore always, in order that Vdv may be an exact differential, V must be expressible as a function of v alone.

3. The above demonstration establishes also the following allied theorem.

THEOREM. If V and v are two functions of x and y, which satisfy the equation

$$\frac{dV}{dy}\frac{dv}{dx} - \frac{dV}{dx}\frac{dv}{dy} = 0 \ldots\ldots\ldots\ldots\ldots\ldots (4),$$

then V is expressible as a function of v only.

We are now prepared to enter upon the proper subject of this chapter.

4. To every differential equation of the form

$$Mdx + Ndy = 0,$$

pertain an infinite number of integrating factors, all of which are included under a single functional expression.

It has been shewn, Chap. II. Art. 2, that the above equation always implies the existence of a complete primitive of the form

$$\psi(x, y) = c \ldots\ldots\ldots\ldots\ldots\ldots\ldots\ldots (5).$$

Differentiating the last equation, we have

$$\frac{d\psi(x, y)}{dx} + \frac{d\psi(x, y)}{dy}\frac{dy}{dx} = 0 \ldots\ldots\ldots\ldots\ldots (6).$$

The value of $\frac{dy}{dx}$ determined as a function of x and y from this equation must be the same as the value of $\frac{dy}{dx}$ furnished by the given differential equation expressed in the form

$$M + N\frac{dy}{dx} = 0.$$

Hence eliminating $\frac{dy}{dx}$ between these equations we have

$$\frac{\frac{d\psi(x, y)}{dx}}{M} = \frac{\frac{d\psi(x, y)}{dy}}{N} \text{.....................(7).}$$

Let μ be the value of each of these ratios, then

$$\frac{d\psi(x, y)}{dx} = \mu M, \quad \frac{d\psi(x, y)}{dy} = \mu N.$$

As μM and μN are therefore the partial differential coefficients with respect to x and y of the same function $\psi(x, y)$, the expression $\mu M dx + \mu N dy$ will be an exact differential. Thus $Mdx + Ndy$ is always susceptible of being made an exact differential by a factor μ.

5. The form of the complete primitive is however without gain or loss of generality susceptible of variation: Thus the primitive $x^2(1+y^2) = c$, Art. 1, might, without becoming more or less general, be presented in the forms

$$\sin\{x^2(1+y^2)\} = c_1, \quad \log\{x^2(1+y^2)\} = c_2,$$

or in the functional form $f\{x^2(1+y^2)\} = c$, where c, c_1, c_2 are arbitrary constants. And these variations in the form of the primitive indicate corresponding variations in the form of the integrating factor, a special determination of which has already been given Art. 1.

6. To investigate the general form under which all such special determinations are included, let us suppose μ to be a particular integrating factor of $Mdx + Ndy$, and let

$\mu Mdx + \mu Ndy$ be the exact differential of a function $\psi(x, y)$. Then representing for the present $\psi(x, y)$ by v, we have

$$\mu Mdx + \mu Ndy = dv.$$

Multiply this equation by $f(v)$, an arbitrary function of v; such being, by the theorem just demonstrated, the general form of a factor which will render the second member an exact differential. We have

$$\mu f(v)(Mdx + Ndy) = f(v)\,dv.$$

Now the second member of this equation being an exact differential the first is so also. As moreover the first member of the above equation can only become an exact differential simultaneously with the second, the factor $\mu f(v)$ is the general form of a factor which renders $Mdx + Ndy$ an exact differential.

We may express the above result in the following theorem.

If μ be an integrating factor of the equation $Mdx + Ndy = 0$, and if $v = c$ be the complete primitive obtained by multiplying the equation by that factor and integrating, then $\mu f(v)$ will be the typical form of all the integrating factors of the equation.

Furthermore, $f(v)$ being an *arbitrary* function of v, the number of such factors is infinite.

Ex. The equation

$$(x^3y - 2y^4)\,dx + (y^3x - 2x^4)\,dy = 0,$$

becomes integrable on multiplying it by the factor $\left(\frac{1}{xy}\right)^3$, the actual solution thus obtained being

$$\frac{x}{y^2} + \frac{y}{x^2} = c.$$

Hence the general form of the integrating factor of the equation is

$$\frac{1}{x^3y^3} f\left(\frac{x}{y^2} + \frac{y}{x^2}\right).$$

From the typical form of the integrating factor of the equation $Mdx + Ndy = 0$, it follows that if we know two particular integrating factors of the equation, the solution may be inferred without integration.

For μ being one of the factors given, the other must be of the form $\mu f(v)$. If we determine their ratio by division and equate the result to an arbitrary constant we shall have

$$f(v) = c,$$

which, from what has been said at the commencement of this Article, is a form of the complete primitive.

It has been observed, Art. 1, that the discovery of an integrating factor of the differential equation $Mdx + Ndy = 0$ generally depends on the solution of another differential equation, but there are some cases in which it presents itself on inspection. The equation

$$(xy^2 + y)\,dx - xdy = 0,$$

becomes integrable on being multiplied by the factor $\frac{1}{y^2}$, and this factor is at once suggested if we place the equation in the form

$$y^2xdx + ydx - xdy = 0.$$

We could thus, also by inspection, assign the integrating factors of any equation of the form

$$y^2dx + \phi(x)\,(ydx - xdy) = 0,$$

and many other forms will readily suggest themselves. The following analysis will however lead to results of greater generality and importance.

Special Determinations of Integrating Factors.

7. Whatever may be the constitution of the functions M and N we have identically

$$Mdx + Ndy = \frac{1}{2}\left\{(Mx+Ny)\left(\frac{dx}{x} + \frac{dy}{y}\right) + (Mx-Ny)\left(\frac{dx}{x} - \frac{dy}{y}\right)\right\},$$

But $$\frac{dx}{x} + \frac{dy}{y} = d\log(xy),\quad \frac{dx}{x} - \frac{dy}{y} = d\log\left(\frac{x}{y}\right).$$

Hence,

$$Mdx + Ndy = \frac{1}{2}\left\{(Mx + Ny)\, d\log xy + (Mx - Ny)\, d\log\frac{x}{y}\right\} \quad (1).$$

The functions $Mx + Ny$ and $Mx - Ny$ appear in the second member of this equation as the coefficients of exact differentials. And upon the nature and relations of these functions the inquiry will now depend.

Whatever may be the constitution of M and N some one, and only one, of the following cases will present itself. Either the functions $Mx + Ny$ and $Mx - Ny$ will be *both* identically equal to 0, or *one* of them will be so and not the other, or *neither* of them will be identically equal to 0. These cases we will separately consider.

1st. The case of $Mx + Ny$ and $Mx - Ny$ being both identically equal to 0 may be dismissed, as it would involve the supposition that M and N are each identically equal to 0. This is seen by addition and subtraction of the equations

$$Mx + Ny = 0,$$

$$Mx - Ny = 0.$$

2ndly. Suppose that one of the functions $Mx + Ny$ is identically 0 and not the other, and first let $Mx + Ny$ be identically 0, then (1) becomes

$$Mdx + Ndy = \tfrac{1}{2}(Mx - Ny)\, d\log\frac{x}{y};$$

whence dividing by $Mx - Ny$,

$$\frac{Mdx + Ndy}{Mx - Ny} = \tfrac{1}{2} d\log\frac{x}{y} \quad \text{..................} (2).$$

Now the second member being an exact differential the first member is also one. In this case then $Mdx + Ndy$ is made an exact differential by the factor $\frac{1}{Mx - Ny}$. By parallel reasoning it follows that if $Mx - Ny$ is identically equal to 0 and not $Mx + Ny$, an integrating factor of $Mdx + Ndy$ will be $\frac{1}{Mx + Ny}$.

And thus we are led to the following theorem.

THEOREM. *If one only of the functions $Mx + Ny$ and $Mx - Ny$ is identically equal to 0, the reciprocal of the other function will be an integrating factor of the equation*

3rdly. Let neither of the functions $Mx + Ny$ and $Mx - Ny$ be identically equal to 0. Then first dividing the fundamental equation (1) by $Mx + Ny$, we have

$$\frac{Mdx + Ndy}{Mx + Ny} = \tfrac{1}{2} d \log xy + \tfrac{1}{2} \frac{Mx - Ny}{Mx + Ny} d \log \frac{x}{y} \ldots (3).$$

Now by the theorem of Art. 2 the second member of the above equation becomes an exact differential (its first term being already such) if $\frac{Mx - Ny}{Mx + Ny}$ is a function of $\log \frac{x}{y}$; therefore if it is a function of $\frac{x}{y}$; therefore if it is a homogeneous function of x and y of the degree 0, for the typical form of such a function is $\phi\left(\frac{x}{y}\right)$; therefore, finally, if M and N are homogeneous functions of x and y of a common degree. For let M and N be homogeneous and of the nth degree. Then $Mx - Ny$ and $Mx + Ny$ are each of the degree $n + 1$, and $\frac{Mx - Ny}{Mx + Ny}$ of the degree 0. Thus M and N being homogeneous functions of the nth degree, the second member, and therefore the first member of (3), is an exact differential.

From this conclusion, combined with the previous one, we arrive at the following theorem.

THEOREM. *The equation $Mdx + Ndy = 0$ when homogeneous is made integrable by the factor $\frac{1}{Mx + Ny}$, unless $Mx + Ny$ is identically equal to 0, in which case $\frac{1}{Mx - Ny}$ is an integrating factor.*

Always then the homogeneous equation $Mdx + Ndy = 0$ is made integrable either by the factor $\frac{1}{Mx + Ny}$, or by the factor $\frac{1}{Mx - Ny}$.

In the second place, dividing the fundamental equation (1) by $Mx - Ny$, we have

$$\frac{Mdx + Ndy}{Mx - Ny} = \frac{1}{2}\left(\frac{Mx + Ny}{Mx - Ny}\, d\log xy + d\log\frac{x}{y}\right) \ldots\ldots (4),$$

of which the second member, and therefore also the first member, becomes an exact differential if $\frac{Mx + Ny}{Mx - Ny}$ is a function of $\log xy$; therefore if it is a function of xy; therefore, finally, if M and N are of the respective forms

$$M = F_1(xy)\, y, \quad N = F_2(xy)\, x;$$

since this supposition would give

$$\frac{Mx + Ny}{Mx - Ny} = \frac{F_1(xy) + F_2(xy)}{F_1(xy) - F_2(xy)},$$

of which the second member is a function of the product xy.

Hence the following theorem.

THEOREM. *The equation $Mdx + Ndy$ is made integrable by the factor $\frac{1}{Mx - Ny}$, when M and N are of the respective forms*

$$M = F_1(xy)\, y, \quad N = F_2(xy)\, x,$$

unless $Mx - Ny$ is identically equal to 0, in which case $\frac{1}{Mx + Ny}$ is an integrating factor.

Or the theorem might be thus expressed. *The equation*

$$F_1(xy)\, ydx + F_2(xy)\, xdy$$

is made integrable by the factor

$$\frac{1}{xy\{F_1(xy) - F_2(xy)\}},$$

unless we have identically $F_1(xy) - F_2(xy) = 0$, in which case

$$\frac{1}{xy\{F_1(xy) + F_2(xy)\}}$$

is an integrating factor.

We may, however, remark that, in the particular case in which $F_1(xy) - F_2(xy) = 0$, no factor is needed, as the differential equation may then be expressed in the form

$$F_1(xy)(ydx + xdy) = 0,$$

the first member being manifestly an exact differential.

8. The results of the above investigation may be summed up as follows.

If either of the functions $Mx + Ny$, $Mx - Ny$ is identically equal to 0, the reciprocal of the other function is an integrating factor of $Mdx + Ndy = 0$; but if neither of these functions is equal to 0, then $\frac{1}{Mx + Ny}$ is an integrating factor for the equation when homogeneous, and $\frac{1}{Mx - Ny}$ an integrating factor of the equation when susceptible of expression in the form

$$F_1(xy)\,ydx + F_2(xy)\,xdy = 0.$$

Ex. 1. Given $x^3dx + (3x^2y + 2y^3)\,dy = 0$.

This is a homogeneous equation, and its integrating factor according to the rule above given will be

$$\frac{1}{x^4 + 3x^2y^2 + 2y^4}.$$

Thus we have, as an *exact* differential equation,

$$\frac{x^3dx}{x^4 + 3x^2y^2 + 2y^4} + \frac{(3x^2y + 2y^3)dy}{x^4 + 3x^2y^2 + 2y^4} = 0 \text{ ....} (1).$$

Referring then to Art. 2. Chap. III, we have

$$\int M dx = \int \frac{x^3 dx}{x^4 + 3x^2y^2 + 2y^4}.$$

$$= \int \left(\frac{2x}{x^2 + 2y^2} - \frac{x}{x^2 + y^2}\right) dx$$

$$= \log \frac{x^2 + 2y^2}{\sqrt{(x^2 + y^2)}} + \phi(y).$$

Differentiating this expression with respect to y, and comparing the result with the corresponding term in (1), we find $\frac{d\phi(y)}{dy} = 0$, whence $\phi(y) = \text{const.}$, and we have

$$\log \frac{x^2 + 2y^2}{\sqrt{(x^2+y^2)}} = c,$$

$$\text{or} \quad x^2 + 2y^2 = C\sqrt{(x^2 + y^2)}$$

for the integral required.

Ex. 2. Given $(y + xy^2) dx + (x - yx^2) dy = 0$.

This equation may be expressed in the form

$$(1 + xy) y dx + (1 - xy) x dy = 0.$$

Hence its integrating factor, as given by the rule, will be

$$\frac{1}{Mx - Ny} = \frac{1}{(1 + xy) xy - (1 - xy) xy}$$

$$= \frac{1}{2x^2y^2}.$$

Rejecting the constant $\frac{1}{2}$, we have, on multiplying the given equation by $\frac{1}{x^2y^2}$,

$$\frac{1 + xy}{x^2y} dx + \frac{1 - xy}{xy^2} dy = 0.$$

Hence

$$\int M dx = \int \frac{dx}{x^2y} + \int \frac{dx}{x} = \log x - \frac{1}{xy} + \phi(y).$$

Now $Ndy = \frac{dy}{xy^2} - \frac{dy}{y}$. Hence the complementary function $\phi(y)$ will be $-\log y$. Thus we have

$$\log x - \log y - \frac{1}{xy} = C$$

for the integral required.

Ex. 3. Given $(x^2y^2 + xy^3)\, dx - (x^3y + x^2y^2)\, dy = 0$.

If we treat this as a homogeneous equation regardless of the implied conditions, we find

$$\frac{1}{Mx + Ny} = \frac{1}{0}.$$

The rule however shews that when $Mx + Ny$ is, as in the above example, identically equal to 0, $\frac{1}{Mx - Ny}$ represents an integrating factor, which in the above case will be

$$\frac{1}{2(x^3y^2 + x^2y^3)}.$$

The equation is thus reduced to

$$\frac{dx}{x} - \frac{dy}{y} = 0,$$

whence we find $y = cx$ as the complete integral.

9. From the theorems of the preceding article others of greater generality may be deduced by transformation. Thus, since the equation $F_1(xy)\, ydx + F_2(xy)\, xdy$ is made integrable by the factor $\frac{1}{xy\{F_1(xy) - F_2(xy)\}}$, it follows that the equation

$$F_1(uv)\, vdu + F_2(uv)\, udv = 0$$

is made integrable by the factor $\frac{1}{uv\{F_1(uv) - F_2(uv)\}}$, u and v being any functions of x and y. Hence expressing du in the form $\frac{du}{dx}dx + \frac{du}{dy}dy$, and dv in the form $\frac{dv}{dx}dx + \frac{dv}{dy}dy$, we see that the equation

$$\left\{F_1(uv)v\frac{du}{dx} + F_2(uv)u\frac{dv}{dx}\right\}dx + \left\{F_1(uv)v\frac{du}{dy} + F_2(uv)u\frac{dv}{dy}\right\}dy = 0$$

is made integrable by the factor $\frac{1}{uv\{F_1(uv) - F_2(uv)\}}$, whatever functions of x and y are represented by u and v. And, on giving particular forms to these functions, particular conditions of integration of the equation $Mdx + Ndy = 0$ present themselves.

10. An integrating factor for homogeneous equations may also be found by the following method, due to Professor Stokes, who first pointed out the necessity of the condition relative to the function $Mx + Ny$ (*Cambridge Mathematical Journal*, Vol. IV. p. 241. First Series).

Suppose M and N to be homogeneous functions of x and y of the degree n. Then we may write

$$M = x^n\phi(v), \quad N = x^n\psi(v) \text{ .... } (1),$$

where v stands for $\frac{y}{x}$.

Hence $Mdx + Ndy = x^n\phi(v)dx + x^n\psi(v)dy$............(2).

But $y = xv$, therefore $dy = xdv + vdx$. Substituting this value of dy in the second member, we have

$$Mdx + Ndy = x^n\{\phi(v) + v\psi(v)\}dx + x^{n+1}\psi(v)dv \text{ ... } (3).$$

Two cases here present themselves.

First, the constitution of the functions $\phi(v)$ and $\psi(v)$ may be such that $\phi(v) + v\psi(v)$ may be *identically* equal to 0. This will happen if $Mx + Ny$ is identically equal to 0, since by (1)

$$Mx + Ny = x^{n+1}\{\phi(v) + v\psi(v)\} \text{ .... } (4).$$

In this case the equation (3) reduces itself to

$$Mdx + Ndy = x^{n+1} \psi (v)\, dv,$$

$$\text{or} \qquad \frac{Mdx + Ndy}{x^{n+1}} = \psi (v)\, dv.$$

Now the second member being an exact differential the first is so also, and $Mdx + Ndy$ is therefore made integrable by the factor $\frac{1}{x^{n+1}}$.

Secondly, the constitution of $\phi (v)$ and $\psi (v)$ may be such that $\phi (v) + v\psi (v)$ is not identically equal to 0. And this happens when $Mx + Ny$ is not identically equal to 0.

In this case dividing both members of (3) by

$$x^{n+1} \{\phi(v) + v\, \psi (v)\},$$

we have

$$\frac{Mdx + Ndy}{x^{n+1} \{\phi (v) + v\psi (v)\}} = \frac{dx}{x} + \frac{\psi (v)\, dv}{\phi (v) + v\psi (v)}.$$

But the second member being an exact differential the first also is such. Now

$$\frac{Mdx + Ndy}{x^{n+1} \{\phi (v) + v\psi (v)\}} = \frac{Mdx + Ndy}{Mx + Ny} \text{ by (4).}$$

Here then $Mdx + Ndy$ is made integrable by the factor

$$1$$

Combining these results together, we see that the homogeneous equation $Mdx + Ndy = 0$ is made integrable by the factor $\frac{1}{Mx + Ny}$, unless the constitution of M and N is such as to make that factor infinite. In the latter case $\frac{1}{x^{n+1}}$ will be an integrating factor, n being the degree of M and N.

The form of the supplementary integrating factor as given by the above investigation is different from that before obtained. The results are however perfectly consistent.

For a more complete analysis of the problem which has for its object the discovery of the integrating factors of a homogeneous equation we must have recourse to the method of the next chapter.

EXERCISES.

1. Shew by the application of the theorem of Art. 3, that the expression $x^2y^2+x^2+y^2+2(xy-1)(x+y)$ is a function of x and y, only as being a function of $xy+x+y$.

2. A particular integrating factor of the equation

$$2xydx+(y^2-3x^2)\,dy=0 \text{ is } y^{-4}.$$

Prove this, and deduce another integrating factor by the formula established in Art. 7 for homogeneous equations.

3. Exhibit the general form under which all the integrating factors of the above equation are comprehended.

4. Deduce in like manner the functional expression for all the integrating factors of the equation

$$\frac{dx}{x}+\frac{dy}{y}+2\left(\frac{dx}{y}-\frac{dy}{x}\right)=0.$$

5. Obtain integrating factors for the homogeneous equations:

(1) $xdy-ydx=\sqrt{(x^2+y^2)}\,dx.$

(2) $(8y+10x)\,dx+(5y+7x)\,dy=0.$

(3) $(x^2+2xy-y^2)\,dx+(y^2+2xy-x^2)\,dy=0.$

(4) $y^2+(xy+x^2)\dfrac{dy}{dx}=0.$

(5) $\left(x\cos\frac{y}{x}+y\sin\frac{y}{x}\right)ydx+\left(x\cos\frac{y}{x}-y\sin\frac{y}{x}\right)xdy=0.$

5. Exhibit the corresponding integrals of the above equations.

6. The formula $\frac{1}{Mx+Ny}$ fails to give an integrating factor for the homogeneous equation $\frac{xdy-ydx}{x^2+y^2}=0$. What formula ought here to be employed and to what result does it lead?

7. Determine an integrating factor of each of the equations

(1) $(x^2y^2+xy)\,ydx+(x^2y^2-1)\,xdy=0.$

(2) $(x^3y^3+x^2y^2+xy+1)\,ydx+(x^3y^3-x^2y^2-xy+1)\,xdy=0.$

CHAPTER V.

ON THE GENERAL DETERMINATION OF THE INTEGRATING FACTORS OF THE EQUATION $Mdx + Ndy = 0$.

1. Prop. It is required to form a differential equation for determining in the most general manner the integrating factors of the equation $Mdx + Ndy = 0$.

Let μ be any integrating factor of the above equation, then since $\mu Mdx + \mu Ndy$ is by hypothesis an exact differential, we have by Art. 1, Chap. III.

$$\frac{d(\mu N)}{dx} = \frac{d(\mu M)}{dy}.$$

Hence

$$N\frac{d\mu}{dx} + \mu\frac{dN}{dx} = M\frac{d\mu}{dy} + \mu\frac{dM}{dy},$$

or, by transposition,

$$N\frac{d\mu}{dx} - M\frac{d\mu}{dy} = \left(\frac{dM}{dy} - \frac{dN}{dx}\right)\mu \text{..........} (1),$$

which is the equation required.

Now this equation involves the partial differential coefficients of μ taken with respect to x and y. It is therefore a *partial* differential equation. We have not the means of solving it *generally*, and it will hereafter appear that its general solution would demand a previous general solution of the differential equation $Mdx + Ndy = 0$, of which μ is the integrating factor. But there are many cases in which we can solve the equation under some restrictive condition or hypothesis, and the form of the solution obtained will always indicate when the supposed condition or hypothesis is legitimate.

The following are examples of such solutions.

2. Let μ be a function of one of the variables only, e.g. suppose $\mu = \phi(x)$, then since $\frac{d\mu}{dy} = 0$, we have from (1)

$$N\phi'(x) = \left(\frac{dM}{dy} - \frac{dN}{dx}\right)\phi(x).$$

Therefore

$$\frac{\phi'(x)}{\phi(x)} = \frac{\frac{dM}{dy} - \frac{dN}{dx}}{N},$$

Or

$$\frac{d}{dx}\log\phi(x) = \frac{\frac{dM}{dy} - \frac{dN}{dx}}{N}.$$

Now *if the second member of this equation is a function of* x the equation is integrable, and we have

$$\log\phi(x) = \int \frac{\frac{dM}{dy} - \frac{dN}{dx}}{N}\,dx.$$

Whence

$$\mu = \epsilon^{\int \frac{\frac{dM}{dy} - \frac{dN}{dx}}{N}\,dx} \text{..........................} (2).$$

We have seen that the hypothesis assumed as the basis of the above solution, viz. that the integrating factor μ is a function of x only, is legitimate when the constitution of the functions M and N is such that the expression

$$\left(\frac{dM}{dy} - \frac{dN}{dx}\right) \div N$$

is a function of x only. In this case (2) enables us to determine the value of μ.

In like manner the condition under which μ is a function of y only, is

$$\frac{\frac{dN}{dx} - \frac{dM}{dy}}{M} = \text{a function of } y \text{ only.......(3)},$$

and the value of μ, on this hypothesis, is

$$\mu = \epsilon^{\int \frac{\frac{dN}{dx} - \frac{dM}{dy}}{M} dy} \text{.....................} (4).$$

Ex. Let us inquire whether the equation

$$(3x^2 + 6xy + 3y^2) dx + (2x^2 + 3xy) dy = 0 \text{.........} (5),$$

admits of an integrating factor which is a function of x only.

Making $M = 3x^2 + 6xy + 3y^2$, $N = 2x^2 + 3xy$, we find

$$\frac{\frac{dM}{dy} - \frac{dN}{dx}}{N} = \frac{6x + 6y - (4x + 3y)}{2x^2 + 3xy} = \frac{1}{x},$$

and this result being a function of x alone, the determination of μ as a function of x alone is seen to be possible. From (2) we now find

$$\mu = \epsilon^{\int \frac{dx}{x}} = Cx,$$

C being an arbitrary constant.

Now multiplying (5) by Cx, we have

$$C\{(3x^3 + 6x^2y + 3xy^2) dx + (2x^3 + 3x^2y) dy\} = 0.$$

The first member of this equation remains a complete differential whatever value we assign to C. If we make $C = 1$, and integrate, we find

$$\frac{3x^4}{4} + 2x^3y + \frac{3x^2y^2}{2} = c,$$

the integral sought.

The student may obtain also the same result by solving (5) as a homogeneous equation.

The linear differential equation of the first order

$$\frac{dy}{dx} + Py - Q = 0 \text{............................} (6),$$

P and Q being functions of x, may be solved by the above method.

For, reducing it to the form

$$(Py - Q) dx + dy = 0 \text{...................} (7),$$

we have $M = Py - Q$, $N = 1$, whence

$$\frac{\frac{dM}{dy} - \frac{dN}{dx}}{N} = P,$$

which being a function of x we find from (2)

$$\mu = \epsilon^{\int Pdx}.$$

Multiplying (7) by the factor thus determined, we have

$$\epsilon^{\int Pdx}(Py - Q)\,dx + \epsilon^{\int Pdx}\,dy = 0,$$

the first member of which is now the exact differential of the function

$$\epsilon^{\int Pdx}y - \int \epsilon^{\int Pdx}\,Qdx.$$

Equating this expression to an arbitrary constant C, we find

$$y = \epsilon^{-\int Pdx}\{C + \int \epsilon^{\int Pdx}\,Qdx\} \quad\text{.................}(8),$$

which agrees with the result of Art. 10, Chap. II.

3. *Let it be required to determine the conditions under which the equation $Mdx + Ndy = 0$, can be made integrable by a factor μ which is a function of the product xy.*

Representing xy by v and making $\mu = \phi(v)$, the partial differential equation (1) becomes

$$N\phi'(v)\frac{dv}{dx} - M\phi'(v)\frac{dv}{dy} = \left(\frac{dM}{dy} - \frac{dN}{dx}\right)\phi(v),$$

whence, since $\frac{dv}{dx} = y$, $\frac{dv}{dy} = x$, we find

$$\frac{\phi'(v)}{\phi(v)} = \frac{\frac{dM}{dy} - \frac{dN}{dx}}{Ny - Mx} \quad\text{.........................}(9).$$

Thus the condition sought is that the second member of the

above equation be reducible to a function of v alone, i. e. of xy alone. And the corresponding value of μ is

$$\mu = \epsilon^{\int \frac{\frac{dM}{dy} - \frac{dN}{dx}}{Ny - Mx} dv} \dots\dots\dots\dots\dots\dots (10).$$

One case in which the above condition is satisfied is the following, viz.

$$F_1(xy)\, ydx + F_2(xy)\, xdy = 0 \dots\dots\dots\dots\dots\dots (11).$$

Making $M = F_1(v)\, y$, $N = F_2(v)\, x$, and observing that since

$$v = xy, \quad \frac{dv}{dx} = y, \quad \frac{dv}{dy} = x, \text{ we find}$$

$$\frac{\frac{dM}{dy} - \frac{dN}{dx}}{Ny - Mx} = \frac{F_1(v) + vF_1'(v) - F_2(v) - vF_2'(v)}{v\{F_2(v) - F_1(v)\}}$$

$$= -\frac{F_1(v) - F_2(v) + v\{F_1'(v) - F_2'(v)\}}{v\{F_1(v) - F_2(v)\}}$$

$$= -\frac{1}{v} - \frac{F_1'(v) - F_2'(v)}{F_1(v) - F_2(v)},$$

a function of v alone.

Multiplying by dv and integrating, we have

$$\int \frac{\frac{dM}{dy} - \frac{dN}{dx}}{Ny - Mx} dv = -\log v - \log\{F_1(v) - F_2(v)\}.$$

Hence,
$$\mu = \frac{1}{v\{F_1(v) - F_2(v)\}}$$

$$= \frac{1}{xy\{F_1(xy) - F_2(xy)\}}.$$

This accords with a result of Art. 7, Chap. IV.

Ex. 1. Thus the equation $(x^2y^2 + 1)\, ydx + (x^2y^2 - 1)\, xdy = 0$ becomes integrable on being multiplied by the factor $\frac{1}{2xy}$, which is found by substituting in the previous expression $x^2y^2 + 1$ for $F_1(xy)$, and $x^2y^2 - 1$ for $F_2(xy)$.

The final solution is

$$\frac{1}{4}x^2y^2+\frac{1}{2}\log xy=c.$$

Ex. 2. The equation

$$(2x^3y^2-y)\,dx+(2x^2y^3-x)\,dy=0,$$

does not fall under the type (11), but the values which it furnishes for M and N give

$$\frac{\frac{dM}{dy}-\frac{dN}{dx}}{Ny-Mx}=\frac{4x^3y-1-(4y^3x-1)}{2x^2y^4-xy-(2x^4y^2-xy)}$$

$$=-\frac{2}{xy}=-\frac{2}{v},$$

so that the condition of integrability by a factor of the form $f(xy)$ is satisfied. Hence

$$\mu=\epsilon^{\int\frac{-2dv}{v}}$$

$$=\frac{1}{v^2}=\frac{1}{x^2y^2}.$$

Multiplying the equation by this factor, and integrating, we find for the primitive

$$x^2+\frac{1}{xy}+y^2=C.$$

4. *It is required to investigate the conditions under which the equation $Mdx+Ndy=0$ can be made integrable by a factor μ which is a homogeneous function of x and y of the degree* 0.

As μ must be of the form $\phi\left(\frac{y}{x}\right)$ let us represent $\frac{y}{x}$ by v, and then assuming $\mu=\phi(v)$, and observing that

$$\frac{dv}{dx}=\frac{-y}{x^2},\quad \frac{dv}{dy}=\frac{1}{x},$$

the partial differential equation (1) becomes

$$-N\phi'(v)\frac{y}{x^2}-M\phi'(v)\frac{1}{x}=\left(\frac{dM}{dy}-\frac{dN}{dx}\right)\phi(v)\ldots\ldots\ldots\ldots(12),$$

whence
$$\frac{\phi'(v)}{\phi(v)}=\frac{x^2\left(\dfrac{dN}{dx}-\dfrac{dM}{dy}\right)}{Mx+Ny}.$$

Thus the condition sought is that the second member of the above equation should be a function of v, i.e. of $\frac{y}{x}$.

And the corresponding value of μ is

$$\mu=\epsilon^{\int\frac{x^2\left(\frac{dN}{dx}-\frac{dM}{dy}\right)}{Mx+Ny}dv}.$$

But since every function of $\frac{y}{x}$ is homogeneous and of the degree 0, with reference to the variables x and y, we may express the above results in the following theorem.

In order that the equation $Mdx+Ndy=0$ may be made integrable by a factor μ which is a homogeneous function of x and y of the degree 0, it is necessary and sufficient that the function

$$\frac{x^2\left(\dfrac{dN}{dx}-\dfrac{dM}{dy}\right)}{Mx+Ny}\ldots\ldots\ldots\ldots\ldots\ldots(13),$$

should be also homogeneous and of the degree 0. This condition being satisfied, the value of μ will be

$$\mu=\epsilon^{\int f(v)dv}\ldots\ldots\ldots\ldots\ldots\ldots(14),$$

where v stands for $\frac{y}{x}$, and $f(v)$ is what the function (13) is reduced to by this transformation.

The above investigation fails when the constitution of the functions M and N is such that we have *identically*

$$Mx+Ny=0.$$

An integrating factor for this case has already been found in the preceding chapter.

We proceed to notice some of the consequences of the above theorem.

It is evident that the condition which it involves will be satisfied when M and N are homogeneous functions of x and y. For, supposing them to be homogeneous and of the nth degree, the numerator and denominator of the fraction (13) will each be of the $n+1^{\text{th}}$ degree, and the fraction itself therefore of the degree 0, the condition required.

It is not however by homogeneous equations only that this condition is satisfied, and it is sometimes worth while to inquire into its applicability in other cases. Thus for the equation

$$\left(\frac{1}{y}+\sec\frac{y}{x}\right)dx-\frac{x}{y^2}dy=0$$

we should find the integrating factor $\cos\frac{y}{x}$.

5. *It is required to investigate the conditions under which the equation $Mdx+Ndy=0$ can be made integrable by a factor μ, which is a homogeneous function of the degree n.*

Assuming $\mu=x^n\phi\left(\frac{y}{x}\right)$, the partial differential equation (1) becomes

$$N\left\{nx^{n-1}\phi\left(\frac{y}{x}\right)-x^{n-2}y\phi'\left(\frac{y}{x}\right)\right\}-Mx^{n-1}\phi'\left(\frac{y}{x}\right)$$
$$=\left(\frac{dM}{dy}-\frac{dN}{dx}\right)x^n\phi\left(\frac{y}{x}\right).$$

Dividing by x^{n-2} and transposing, we get

$$(Mx+Ny)\,\phi'\left(\frac{y}{x}\right)=\left\{x^2\left(\frac{dN}{dx}-\frac{dM}{dy}\right)+nNx\right\}\phi\left(\frac{y}{x}\right),$$

whence

$$\frac{\phi'\left(\frac{y}{x}\right)}{\phi\left(\frac{y}{x}\right)}=\frac{x^2\left(\frac{dN}{dx}-\frac{dM}{dy}\right)+nNx}{Mx+Ny}.$$

Let $\frac{y}{x}=v$, and suppose the second member to assume the form $f(v)$; then, multiplying both sides by dv and integrating, we have

$$\log \phi(v) = \int f(v)\, dv.$$

$$\text{Hence}\quad \mu = x^n \phi(v) = x^n \epsilon^{\int f(v)dv}.$$

Thus we arrive at the following theorem.

THEOREM. *In order that the equation $Mdx + Ndy = 0$ may be made integrable by a factor μ, which is a homogeneous function of x and y of the nth degree, it is necessary, and it suffices, that on making $y = vx$ the function*

$$\frac{x^2\left(\frac{dN}{dx}-\frac{dM}{dy}\right)+nNx}{Mx+Ny} \quad\ldots\ldots\ldots\ldots\ldots\ldots (15),$$

should assume the form $f(v)$. This condition being satisfied, the expression for μ will be

$$\mu = x^n \epsilon^{\int f(v)dv} \quad\ldots\ldots\ldots\ldots\ldots\ldots\ldots\ldots (16).$$

wherein $v=\frac{y}{x}$.

It will be noted that the condition that (15) shall be a function of v, is the same as the condition that it shall be a homogeneous function of x and y of the degree 0.

The theorem fails when $Mx + Ny = 0$, a case already considered.

Ex. 1. Required to determine whether the equation

$$(2x^3 + 3x^2y + y^2 - y^3)\, dx + (2y^3 + 3xy^2 + x^2 - x^3)\, dy = 0$$

admits of an integrating factor which is a homogeneous function of x and y.

Here $M = 2x^3 + 3x^2y + y^2 - y^3$, $N = 2y^3 + 3xy^2 + x^2 - x^3$,

$$\frac{dM}{dy} = 3x^2 + 2y - 3y^2, \qquad \frac{dN}{dx} = 3y^2 + 2x - 3x^2,$$

$$\frac{dN}{dx} - \frac{dM}{dy} = 6y^2 - 6x^2 + 2x - 2y.$$

Hence, on substitution,

$$\frac{x^2\left(\frac{dN}{dx} - \frac{dM}{dy}\right) + nNx}{Mx + Ny}$$

$$= \frac{-(n+6)x^4 + (3n+6)x^2y^2 + 2nxy^3 + (n+2)x^3 - 2x^2y}{2x^4 + 2x^3y + 2xy^3 + 2y^4 + x^2y + xy^2}.$$

We are now to inquire whether there exists any value of n which reduces the second member of the above equation to a homogeneous function of x and y of the degree 0.

That member may be expressed in the form

$$\frac{-x}{x+y} \times \frac{(n+6)x^3 - (3x+6)xy^2 - 2ny^3 - (n+2)x^2 + 2xy}{2x^3 + 2y^3 + xy},$$

and it is now plain that if any value of n will answer the required condition, it must be one which will make the terms containing xy^2 and x^2 in the numerator of the second factor vanish. Making then $n = -2$, we have

$$\frac{-x}{x+y} \times \frac{4x^3 + 4y^3 + 2xy}{2x^3 + 2y^3 + xy} = \frac{-2x}{x+y}$$

$$= \frac{-2}{1+v}.$$

$$\text{Hence } \mu = x^{-2} \epsilon^{\int \frac{-2dv}{1+v}} = \frac{c}{x^2(1+v)^2}$$

$$= \frac{c}{(x+y)^2}.$$

Multiplying the given equation by this factor and integrating, we find as the primitive equation

$$\frac{x^3+xy+y^3}{x+y}=C.$$

In the case of homogeneous equations the condition involved in the general theorem will be satisfied *independently of the value of n*, the particular case in which $Mx+Ny=0$ excepted. It follows hence that with this exception we can find an integrating factor of any proposed degree for the homogeneous equation $Mdx+Ndy=0$.

Ex. 2. Required two integrating factors of the respective degrees 0 and 1 for the equation

$$(3x+2y)\,dx+xdy=0.$$

First making $M=3x+2y$, $N=x$, and $n=0$, we have

$$\frac{x^2\left(\frac{dN}{dx}-\frac{dM}{dy}\right)+nNx}{Mx+Ny}=\frac{-x}{3\,(x+y)}.$$

Hence $$f(v)=\frac{-1}{3\,(1+v)},$$

$$\mu=\epsilon^{\int f(v)dv}=c\,(v+1)^{-\frac{1}{3}}=c\left(\frac{x}{x+y}\right)^{\frac{1}{3}}.$$

Secondly, making $M=3x+2y$, $N=x$, $n=1$, we have

$$\frac{x^2\left(\frac{dN}{dx}-\frac{dM}{dy}\right)+Nx}{Mx+Ny}=0.$$

Hence $$f(v)=0,$$

$$\mu=x\,\epsilon^{\int f(v)dv}=c'x.$$

Thus replacing each of the constants c and c' by unity, the integrating factors in question are $\left(\frac{x}{x+y}\right)^{\frac{1}{3}}$ and x.

Multiplying by the second factor x and integrating, we find $x^3 + x^2 y = C$ for the primitive.

Again, if in illustration of the concluding observation of Art. 6, Chap. IV., we equate to an arbitrary constant the ratio of the second factor to the first, we have

$$x^{\frac{2}{3}} (x + y)^{\frac{1}{3}} = \text{constant},$$

which being equivalent to

$$x^2 (x + y) = \text{constant},$$

agrees with the previous solution.

Let us next examine the *general* results to which the theorem leads, when M and N are homogeneous and of the mth degree.

The general forms of M and N will be on putting v for $\frac{y}{x}$,

$$M = x^m \phi(v), \quad N = x^m \psi(v).$$

Hence, observing that

$$\frac{dN}{dx} = mx^{m-1} \psi(v) - x^{m-2} y\psi'(v),$$

$$\frac{dM}{dy} = x^{m-1} \phi'(v);$$

we have on substituting in the expression for $f(v)$, and dividing numerator and denominator of the result by x^{m+2},

$$f(v) = -\frac{(m+n)\psi(v) - v\psi'(v) - \phi'(v)}{\phi(v) + v\psi(v)} \quad \text{...............(17)}.$$

If we make n, the value of which may be chosen at pleasure, equal to $-m-1$, we have

$$f(v) = -\frac{\psi(v) + v\psi'(v) + \phi'(v)}{\phi(v) + v\psi(v)}.$$

Multiplying by dv and integrating,

$$\int f(v)\, dv = -\log\{\phi(v) + v\psi(v)\}.$$

Hence,

$$\mu = x^n \epsilon^{\int f(v)\,dv}$$

$$= \frac{C}{x^{m+1}\{\phi(v) + v\psi(v)\}}$$

$$= \frac{C}{Mx + Ny} \quad\text{.........................(18).}$$

And here again it results that the homogeneous equation $Mdx + Ndy = 0$, may be made integrable by the factor $\frac{1}{Mx + Ny}$, except in the particular case in which the constitution of M and N is such as to make $Mx + Ny = 0$. Moreover this theorem is seen to be only a particular consequence of the *general* theory of the integrating factors of homogeneous equations.

Resuming (17) which we may write in the form

$$f(v) = \frac{(m + n + 1)\,\psi(v) - \{\psi(v) + v\psi'(v) + \phi'(v)\}}{\phi(v) + v\psi(v)},$$

we have

$$\int f(v)\,dv = (m + n + 1)\int \frac{\psi(v)\,dv}{\phi(v) - v\psi(v)} - \log\{\phi(v) + v\psi(v)\},$$

by the substitution of which, combined with the previous reduction, the *general* value of μ becomes

$$\mu = \frac{x^{m+n+1}\,\epsilon^{(m+n+1)\int\frac{\psi(v)\,dv}{\phi(v)+v\psi(v)}}}{Mx + Ny} \quad\text{................(19),}$$

which is the general expression for an integrating factor of the nth degree, supposing n *not* equal to $-m-1$.

If we now equate to an arbitrary constant the ratio borne by the last value of μ to the previous one (18), we have

$$x^{m+n+1}\,\epsilon^{(m+n+1)\int\frac{\psi(v)\,dv}{\phi(v)+v\psi(v)}} = C,$$

which is readily reducible to

$$\log x + \int \frac{\psi(v)\, dv}{\phi(v) + v\psi(v)} = C \quad \ldots\ldots\ldots\ldots\ldots (20).$$

Now this is the very solution of the homogeneous equation $Mdx + Ndy = 0$, obtained by the direct assumption $y = vx$, in Art. 8, Chap. II.

We thus see that in the case of homogeneous equations the employment of integrating factors conducts us, but by a more lengthened route, to the same *final* integrals as the direct method of Chap. II. It is difficult to lay down any general rule as to the value of concurrent methods, but it would probably be not very remote from truth to say, that the *peculiar* advantage of the theory of integrating factors consists rather in its appropriateness for the investigation of conditions under which solution is possible, than in the actual processes of solution to which it leads.

6. The following application of the theorem is of a more general character.

The equation

$$P_1 dx + P_2 dy + Q(xdy - ydx) = 0 \quad \ldots\ldots\ldots\ldots (21),$$

where P_1 and P_2 are homogeneous functions of x and y of the degree p, and Q is a homogeneous function of x and y of the degree q, may be rendered integrable by a factor μ which is a homogeneous function of x and y of the degree $-q-2$.

Here $M = P_1 - Qy$, $N = P_2 + Qx$.

Hence $Mx + Ny = P_1 x + P_2 y$.

Thus the denominator of (15) is the same as if M and N were reduced to their first terms P_1 and P_2. And the numerator remains the same also: For the addition which the second terms of M and N, viz. $-Qy$ and Qx make thereto, is

$$x^2 \left\{\frac{d}{dx}(Qx) + \frac{d}{dy}(Qy)\right\} + nQx^2,$$

which, on effecting the differentiation, becomes

$$x^2 \left\{x \frac{dQ}{dx} + y \frac{dQ}{dy} + (n+2) Q\right\},$$

but Q being by hypothesis homogeneous of the qth degree, whence,

$$x \frac{dQ}{dx} + y \frac{dQ}{dy} = qQ,$$

the above expression reduces to

$$x^2 (q+n+2) Q,$$

and vanishes if n is made equal to $-q-2$. Thus (15) assumes the same form as if M and N were homogeneous of the degree p, and the condition of the theorem is satisfied.

7. All the applications which we have hitherto made of the partial differential equation (1) are of one type. The general problem which they exemplify is the following. Under what condition does the equation $Mdx + Ndy = 0$ admit of being made integrable by a factor of the form $\phi(v)$ where v is a known and definite function of x and y? Let us examine the general form of its solution.

On substituting $\phi(v)$ for μ in (1), we find

$$\frac{\phi'(v)}{\phi(v)} = \frac{\dfrac{dM}{dy} - \dfrac{dN}{dx}}{N\dfrac{dv}{dx} - M\dfrac{dv}{dy}} \quad \text{.....................(22).}$$

The condition sought then is that the second member of this equation should be a function of v. Representing that function by $f(v)$ the corresponding value of μ is

$$\mu = \epsilon^{\int f(v)dv} \quad \text{.............................(23).}$$

Any special case may be treated either independently as in the previous examples, or by directly referring it to the above general form.

6—2

Thus a direct reference to the above theorem shews that the condition which must be satisfied in order that the equation $Mdx + Ndy = 0$ may admit of an integrating factor of the form $\phi(x^2 + y)$ is that the function

$$\frac{\frac{dM}{dy} - \frac{dN}{dx}}{2Nx - M},$$

should be a function of $x^2 + y$. And the mode of determining this point would be to assume $x^2 + y = v$, and, thence deducing $y = v - x^2$, to substitute that value of y in the above function, and see whether the result assumed the form $f(v)$. The equation (23) would then give the value of μ. And this mode of procedure is general.

8. When by the discovery of an integrating factor the possibility of solving a differential equation has been established, there is no more valuable exercise than to endeavour to effect the same object by other means.

Let us take as an example the equation considered in Art. 6, viz.

$$P_1 dx + P_2 dy + Q(xdy - ydx) = 0 \quad \text{............(24)}.$$

P_1 and P_2 being homogeneous of the degree p, and Q homogeneous of the degree q.

Let $P_1 = x^p \phi\left(\frac{y}{x}\right)$, $P_2 = x^p \psi\left(\frac{y}{x}\right)$, $Q = x^q \chi\left(\frac{y}{x}\right)$, then making $\frac{y}{x} = v$, whence flow

$$dy = xdv + vdx$$

$$xdy - ydx = x^2 dv,$$

the given equation expressed in terms of the variables x and v, becomes

$$x^p \phi(v)\, dx + x^p \psi(v)\,(xdv + vdx) + x^q \chi(v) \times x^2 dv = 0,$$

and assumes on transposition and division the form

$$\frac{dx}{dv} + \frac{\psi(v)}{\phi(v) + v\psi(v)} x = - \frac{\chi(v)}{\phi(v) + v\psi(v)} \times x^{q-p+2} \text{......(25)}.$$

Now the reducibility of an equation of this form to a linear form has been established in Chap. II. Art. 11.

Under the general form (24) are virtually included some remarkable equations which have been made the subjects of distinct investigations.

Thus Jacobi has, by an analysis of a very peculiar character, solved the differential equation. (Crelle's *Journal*, Vol. XXIV.)

$$(A + A'x + A''y)(xdy - ydx) - (B + B'x + B''y)\,dy$$
$$+ (C + C'x + C''y)\,dx = 0 \text{.........(26)}.$$

If, however, we assume in that equation

$$x = \xi + \alpha, \quad y = \eta + \beta,$$

we can, by a proper determination of the constants α and β, reduce it to the form

$$(a\xi + a'\eta)(\xi d\eta - \eta d\xi) - (b\xi + b'\eta)\,d\eta + (c\xi + c'\eta)\,d\xi = 0,$$

which falls under (24). On effecting the substitution in question the equations for determining α and β will be found to be

$$\alpha(A + A'\alpha + A''\beta) - (B + B'\alpha + B''\beta) = 0,$$
$$-\beta(A + A'\alpha + A''\beta) + C + C'\alpha + C''\beta = 0.$$

The most convenient mode of solving these equations is to write them in the symmetrical form

$$\frac{B + B'\alpha + B''\beta}{\alpha} = \frac{C + C'\alpha + C''\beta}{\beta} = A + A'\alpha + A''\beta,$$

then, equating each of these expressions to λ, we find

$$A - \lambda + A'\alpha + A''\beta = 0,$$
$$B + (B' - \lambda)\alpha + B''\beta = 0,$$
$$C + C'\alpha + (C'' - \lambda)\beta = 0,$$

from which eliminating α and β we have the cubic equation

$$(A-\lambda)(B'-\lambda)(C''-\lambda)-B''C'(A-\lambda)-A''C(B'-\lambda)$$
$$-A'B(C'-\lambda)+A'B''C-A''BC'=0 \text{..............} (27).$$

If a value of λ be found from this equation, any two equations of the preceding system will give α and β.

9. The present chapter would be incomplete without some notice of a method which was largely employed by Euler.

That method consisted in assuming μ to be a function definite in form as respects the variable y, but involving unknown functions of x as the coefficients of the several powers of y.

After the substitution of this form of μ in the partial differential equation (1), the result is arranged according to the powers of y, and the coefficients of those powers separately equated to 0. This gives a series of simultaneous differential equations for the determination of the unknown functions of x. But for the success of the method it is necessary that the primary assumption for μ should have been chosen with some special fitness to the object proposed. The following is an example.

Required the conditions under which the equation

$$Pydx+(y+Q)\,dy=0$$

admits of being made integrable by a factor of the form

$$\frac{1}{y^3+Ry^2+Sy},$$

P, Q, R and S being functions of x.

In the partial differential equation (1), making

$$M=Py, \quad N=y+Q, \quad \mu=\frac{1}{y^3+Ry^2+Sy},$$

clearing the result of fractions and arranging it according to the powers of y, we have

$$\left(2P+\frac{dQ}{dx}-\frac{dR}{dx}\right)y^3+\left(PR+R\frac{dQ}{dx}-Q\frac{dR}{dx}-\frac{dS}{dx}\right)y^2$$
$$+\left(S\frac{dQ}{dx}-Q\frac{dS}{dx}\right)y=0\ldots\ldots\ldots\ldots(28).$$

Whence, equating separately to 0 the coefficients of the different powers of y, we have the ternary system

$$2P+\frac{dQ}{dx}-\frac{dR}{dx}=0\ \ldots\ldots\ldots\ldots\ldots\ldots\ldots\ldots\ (29),$$

$$PR+R\frac{dQ}{dx}-Q\frac{dR}{dx}-\frac{dS}{dx}=0\ \ldots\ldots\ldots\ldots\ (30),$$

$$S\frac{dQ}{dx}-Q\frac{dS}{dx}=0\ \ldots\ldots\ldots\ldots\ldots\ldots\ldots\ldots\ (31).$$

The last equation gives $S=cQ$, c being an arbitrary constant. Substituting this value of S in the equation obtained by eliminating P from the first two equations of the system, we find

$$(2c-R)\,dQ+2\,QdR=RdR,$$

or, regarding therein R as the independent and Q as the dependent variable,

$$(2c-R)\frac{dQ}{dR}+2\,Q=R,$$

a linear equation of which the solution is

$$Q=(R-c)+c'\,(R-2c)^2.$$

Hence we have

$$S=c\,(R-c)+cc'\,(R-2c)^2,$$

and from the substitution of these values in the first equation of the ternary system,

$$P=-c'\,(R-2c)\frac{dR}{dx}.$$

These values of S, Q, and P, in which R is arbitrary, reduce the given differential equation to the form

$$\{(R-c)+c'(R-2c)^2+y\}\,dy-c'y\,(R-2c)\,dR=0 \;\ldots\; (32),$$

and present its integrating factor in the form

$$\frac{1}{y^3+Ry^2+\{c(R-c)+cc'(R-2c)^2\}y},$$

R being an arbitrary function of x.

For other examples the student is referred to Lacroix (*Traité du Calcul. Diff. et du Calcul. Int.* Vol. II. Cap. IV.) The results of this method are usually of a very complex character, while their generality is limited by the restrictions which must be imposed in order to render the system of reducing equations solvable. Thus Euler's equation above considered is virtually only a limited case of the general equation (21). If we assume

$$y+c=s, \quad R-2c=t,$$

it becomes

$$(s+t)\,ds+cc'tdt+c't\,(tds-sdt),$$

which evidently falls under that equation.

EXERCISES.

1. The following equations admit of integrating factors of the form $\phi(x)$, viz.

(1) $(x^2+y^2+2x)\,dx+2y\,dy=0.$

(2) $(x^2+y^2)\,dx-2xy\,dy=0.$

Determine these factors and integrate the equations.

2. The equation $2xy\,dx+(y^2-3x^2)\,dy=0$, has an integrating factor which is a function of y. Determine it, and integrate the equation.

3. Find those integrating factors of the equation

$$ydx+(2y-x)\,dy=0$$

which are homogeneous functions of x and y of the respective degrees 0 and -2, and from the consideration of those factors deduce the complete primitive of the equation.

4. For each of the following equations examine whether there exists an integrating factor μ satisfying the particular condition specified, and if so determine the factor, and integrate the equation.

(1) $y(x^2+y^2)\,dx + x(xdy - ydx) = 0$, μ a homogeneous function of the degree -3.

(2) $(y^3 + axy^2)\,dy - ay^3dx + (x+y)(xdy - ydx)$, μ as in the previous example.

(3) $(y+x)\,dy + ydx - xd\left(\dfrac{x}{y}\right) = 0$, μ homogeneous of the degree -1.

(4) $(x^2+y^2+1)\,dx - 2xydy = 0$, μ a function of $y^2 - x^2$.

(5) $(y - 3x^2y^3 - 2x^3)\,dx + (2y^3 + 3x^3y^2 - x)\,dy = 0$, μ a function of $x^3 + y$.

(6) $(x^2+x^2y+2xy-y^2-y^3)\,dx+(y^2+xy^2+2xy-x^2-x^3)\,dy=0$, μ a function of the product $(1+x)(1+y)$.

(7) $(3y^2 - x)\,dx + (2y^3 - 6xy)\,dy = 0$, μ a function of $x + y^2$.

5. The equation $y(x^2+y^2)\,dx + x(xdy - ydx) = 0$ has an integrating factor of the form $\epsilon^x \phi(x^2+y^2)$. Determine it, and, from the comparison of the result with that of (1) Ex. 4, deduce the complete primitive.

6. The linear equation $\dfrac{dy}{dx} + Py = Q$ having an integrating factor of the form $\epsilon^{\int Pdx}$, deduce a corresponding expression for an integrating factor of the equation

$$\frac{dy}{dx} + Py = Qy^n.$$

7. Prove that the equation

$$\frac{dy}{dx} + y^2 = \frac{dP}{dx} + P^2,$$

where P is any function of x, has an integrating factor of the form $\frac{\epsilon^{-2\int P dx}}{(y-P)^2}$. Lacroix, Tom. II. p. 278.

8. Deduce a similar expression for an integrating factor of the equation $\frac{dy}{dx}+y^2+\frac{dP}{dx}+P^2=0$. *Ib.*

9. Investigate the conditions under which the equation

$$\frac{dy}{dx}+\frac{P}{y}=Q,$$

where P and Q are functions of x, can be made integrable by a factor of the form $\frac{y}{\{y+f(x)\}^n}$, and determine the form of $f(x)$.

CHAPTER VI.

OF SOME REMARKABLE EQUATIONS OF THE FIRST ORDER AND DEGREE.

1. THERE are certain differential equations of the first order and degree, to which, in addition to their intrinsic claims upon our notice, some degree of historical interest belongs. Among such, a prominent place is due to two equations which, having been first discussed by the Italian mathematician Riccati and by Euler respectively, have from this circumstance derived their names. To these equations, and to some other allied forms, the present chapter will be devoted.

Riccati's equation is usually expressed in the form

$$\frac{du}{dx} + bu^2 = cx^m \quad \text{.......................} \quad (1).$$

But as both it and some other equations closely related to it and possessing a distinct interest, may, either immediately or after a slight reduction, be referred to the more general equation

$$x\frac{dy}{dx} - ay + by^2 = cx^n \quad \text{.....................} \quad (2),$$

the discussion of which happens to be much more easy than that of the special equations which are included under it, we shall consider this equation first.

To reduce Riccati's equation under the general form (2), it suffices to assume $u = \frac{y}{x}$. We find, as the result of this substitution in (1),

$$\frac{xdy}{dx} - y + by^2 = cx^{m+2} \quad \text{.....................} \quad (3),$$

which is seen to be a particular case of (2).

Of the equation $x\dfrac{dy}{dx} - ay + by^2 = cx^n$.

2. The discussion upon which we are entering may be divided into two parts. First, we shall shew that the equation is solvable when $n = 2a$. Secondly, we shall establish a series of transformations by which a corresponding series of other cases may be reduced to the above.

3. First. The equation $x\dfrac{dy}{dx} - ay + by^2 = cx^n$ is solvable when $n = 2a$.

For, assuming $y = x^a v$, we find on substitution

$$x^{a+1}\frac{dv}{dx} + bx^{2a}v^2 = cx^n,$$

whence, dividing by x^{2a}, we have

$$x^{1-a}\frac{dv}{dx} + bv^2 = cx^{n-2a}.$$

Now if $n = 2a$ the above becomes

$$x^{1-a}\frac{dv}{dx} + bv^2 = c,$$

whence

$$\frac{dv}{c - bv^2} = \frac{dx}{x^{1-a}},$$

an equation in which the variables are separated. If we restore to v its value $\dfrac{y}{x^a}$ and transpose, this becomes

$$\frac{x^a dy - ayx^{a-1}dx}{by^2 - cx^{2a}} + x^{a-1}dx = 0 \quad\text{..............(4)},$$

an exact differential equation.

4. Secondly. The solution of the equation

$$x\frac{dy}{dx} - ay + by^2 = cx^n,$$

is always reducible by transformation to the preceding case whenever $\frac{n \pm 2a}{2n} = i$, a positive integer.

For let $y = A + \frac{x^n}{y_1}$, y_1 being a new variable which is to replace y, and A a constant whose value is yet to be determined. On substitution and arrangement of the terms we have

$$-aA + bA^2 + (n - a + 2bA)\frac{x^n}{y_1} + b\frac{x^{2n}}{y_1^2} - \frac{x^{n+1}}{y_1^2}\frac{dy_1}{dx} = cx^n \ldots (5).$$

Now let $-aA + bA^2 = 0$, then $A = \frac{a}{b}$ or 0. These values of A we shall employ in succession.

5. First. If we assume $A = \frac{a}{b}$ the above equation becomes

$$(n + a)\frac{x^n}{y_1} + b\frac{x^{2n}}{y_1^2} - \frac{x^{n+1}}{y_1^2}\frac{dy_1}{dx} = cx^n.$$

Multiplying this equation by $\frac{y_1^2}{x^n}$ and transposing, we have

$$x\frac{dy_1}{dx} - (a + n)y_1 + cy_1^2 = bx^n \ldots\ldots\ldots\ldots\ldots (6).$$

Now this equation is of the same *form* as the given equation between y and x. The coefficients however differ, in that b and c have changed their places, and a has become $a + n$. And this transformation has been effected by the assumption

$$y = \frac{a}{b} + \frac{x^n}{y_1}.$$

Hence, if in the transformed equation (6) we make a second assumption

$$y_1 = \frac{a+n}{c} + \frac{x^n}{y_2},$$

we shall have as the result

$$x\frac{dy_2}{dx} - (a+2n)\,y_2 + by_2^2 = cx^n \ldots\ldots\ldots\ldots\ldots (7),$$

b and c again changing places, and $a+n$ becoming $a+2n$. And the result of i successive transformations of the same series will be to reduce the given equation either to the form

$$x\frac{dy_i}{dx} - (a+in)\,y_i + cy_i^2 = bx^n \ldots\ldots\ldots\ldots\ldots (8),$$

or to the form

$$x\frac{dy_i}{dx} - (a+in)\,y_i + by_i^2 = cx^n \ldots\ldots\ldots\ldots\ldots (9),$$

according as the integer i is odd or even.

Now by what has been established in Art. 3 the above equations will be integrable if we have

$$n = 2\,(a+in),$$

an equation which gives

$$\frac{n-2a}{2n} = i \ldots\ldots\ldots\ldots\ldots\ldots\ldots\ldots (10).$$

6. Secondly, If we assign to A its second value 0, (5) becomes

$$(n-a)\,\frac{x^n}{y_1} + b\,\frac{x^{2n}}{y_1^2} - \frac{x^{n+1}}{y_1^2}\,\frac{dy_1}{dx} = cx^n.$$

Or, multiplying by $\frac{y_1^2}{x^n}$ and transposing,

$$x\frac{dy_1}{dx} - (n-a)\,y_1 + cy_1^2 = bx^n \ldots\ldots\ldots\ldots (11).$$

Now this equation for y_1 differs from the equation (6) obtained for y_1 in the previous series of transformations only in that a in the coefficient of the second term has become $-a$. With this change only then that series of transformations may be adopted in the present instance. The change of a into $-a$ in the final condition (10) gives

$$\frac{n+2a}{2n} = i$$

as a new condition under which the equation in y is solvable. If $i = 1$ this gives $n = 2a$, the condition first arrived at, and upon which the subsequent researches were based.

Collecting these results together we see that *the equation* $x\frac{dy}{dx} - ay + by^2 = cx^n$ *is integrable whenever* $\frac{n \pm 2a}{2n}$ *is a positive integer.*

7. Let us now examine the form in which the solution is presented.

If $\frac{n-2a}{2n} = i$, which is the condition arrived at in Art. (5), we have the series of transformations

$$y = \frac{a}{b} + \frac{x^n}{y_1},$$

$$y_1 = \frac{a+n}{c} + \frac{x^n}{y_2},$$

$$y_2 = \frac{a+2n}{b} + \frac{x^n}{y_3},$$

and finally

$$y_{i-1} = \frac{a+(i-1)n}{k} + \frac{x^n}{y_i},$$

where $k = b$ or c, according as i is odd or even; and the effect of these transformations is to reduce the given equation to one or the other of the forms (8) and (9).

If in the above expression for y we substitute for y_1 its value in terms of y_2, in that result again, for y_2 its value in terms of y_3, and so on, we find

$$y = \frac{a}{b} + \cfrac{x^n}{\cfrac{a+n}{c} + \cfrac{x^n}{\cfrac{a+2n}{b} \dots\dots}} \quad \text{(A)},$$

the last denominator being $\frac{a+(i-1)n}{k} + \frac{x^n}{y_i}$. The value of y_i must then be determined by the solution of (8) or of (9), these equations being now susceptible of expression as exact differential equations in the forms

$$\frac{x^{a+in}dy_i - (a+in)y_i x^{a+in-1}dx}{cy_i^2 - bx^n} + x^{a+in-1}dx = 0 \dots\dots \text{(B)},$$

$$\frac{x^{a+in}dy_i - (a+in)y_i x^{a+in-1}dx}{by_1^2 - cx^n} + x^{a+in-1}dx = 0 \dots\dots \text{(C)}.$$

When therefore $\frac{n-2a}{2n} = i$ *a positive integer, the solution of the equation* $x\frac{dy}{dx} - ay + by^2 = cx^n$ *will be expressed in the form of a continued fraction by* (A), *the value of* y_i *in the last denominator being given by the solution of the exact differential equation* (B) *or* (C) *according as* i *is odd or even.*

Secondly, if $\frac{n+2a}{2n} = i$, which is the condition arrived at in Art. 6, we have the series of transformations

$$y = \frac{x^n}{y_1},$$

$$y_1 = \frac{n-a}{c} + \frac{x^n}{y_2}.$$

$$y_2 = \frac{2n-a}{b} + \frac{x^n}{y_3},$$

$$\text{.....................}$$

$$y_{i-1} = \frac{(i-1)n-a}{k} + \frac{x^n}{y_i} \text{ .... } (12),$$

where $k=b$ or c, according as i is odd or even. From these, eliminating, as before, the intermediate variables $y_1\,y_2 \ldots y_{i-1}$, we find

$$y = \cfrac{x^n}{\cfrac{n-a}{c} + \cfrac{x^n}{\cfrac{2n-a}{b} + \cfrac{x^n}{\cfrac{3n-a}{c}}}} \text{ .... } (D),$$

the last denominator being $\frac{(i-1)\,n-a}{k} + \frac{x^n}{y_i}$. In this case, however, the equation for y_i formed by changing a into $-a$ in B and C will be

$$\frac{x^{in-a}\,dy_i - (in-a)\,y_i x^{in-a-1}\,dx}{cy_i^2 - bx^n} + x^{in-a-1}\,dx = 0 \text{ ...... } (E),$$

or $$\frac{x^{in-a}\,dy_i - (in-a)\,y_i x^{in-a-1}\,dx}{by_i^2 - cx^n} + x^{in-a-1}\,dx = 0 \text{ ...... } (F),$$

as i is odd or even.

When therefore $\frac{n+2a}{2n} = i$ *a positive integer, the solution of* $x\frac{dy}{dx} - ay + by^2 = cx^n$ *is expressed by* (D), *the value of* y_i *in the last denominator being given by the exact differential equation* (E) *or* (F) *according as* i *is odd or even.*

Ex. Given $x\frac{dy}{dx} - y + y^2 = x^{\frac{2}{3}}$.

Here $n=\frac{2}{3}$, $a=1$, and as $\dfrac{n+2a}{2n}=2$ while $\dfrac{n-2a}{2n}=-1$, the formulæ (D) and (F) must be employed. Assuming therein $a=1$, $b=1$, $c=1$, $n=\frac{2}{3}$, $i=2$, we have

$$y=\frac{x^{\frac{2}{3}}}{-\frac{1}{3}+\dfrac{x^{\frac{2}{3}}}{y_2}}=\frac{3x^{\frac{2}{3}}y_2}{3x^{\frac{2}{3}}-y_2}\ \dots\dots\dots\dots\dots\dots\ (13),$$

y_2 being given by the exact differential equation

$$\frac{x^{\frac{1}{3}}dy_2-\frac{1}{3}y_2x^{-\frac{2}{3}}dx}{y_2^{\ 2}-x^{\frac{2}{3}}}+x^{-\frac{2}{3}}dx=0\ \dots\dots\dots\dots\dots\ (14),$$

from which we find

$$\tfrac{1}{2}\log\left(\frac{y_2-x^{\frac{1}{3}}}{y_2+x^{\frac{1}{3}}}\right)+3x^{\frac{1}{3}}=C\ \dots\dots\dots\dots\dots\ (15).$$

The elimination of y_2 between (13) and (15) gives

$$\log\frac{3yx^{\frac{1}{3}}-3x^{\frac{2}{3}}-y}{3yx^{\frac{1}{3}}+3x^{\frac{2}{3}}+y}+6x^{\frac{1}{3}}=C\dots\dots\dots\dots\dots(16),$$

which is the complete primitive.

Ex. 2. Given $\dfrac{du}{dx}+u^2=x^{-\frac{4}{3}}$.

This is an example of Riccati's equation. Assuming therefore $u=\dfrac{y}{x}$, we find $x\dfrac{dy}{dx}-y+y^2=x^{\frac{2}{3}}$, which is identical with the equation last considered. Substituting therefore in (18) ux for y, we find after reduction

$$\log\frac{3ux^{\frac{2}{3}}-3-ux^{\frac{1}{3}}}{3ux^{\frac{2}{3}}+3+ux^{\frac{1}{3}}}+6x^{\frac{1}{3}}=C\dots\dots\dots\dots(17),$$

General Observations.

8. The connexion between the two conditions for the solution of the equations $x\frac{dy}{dx} - ay + by^2 = cx^n$, implied by the double sign in the equation $\frac{n \pm 2a}{2n} = i$, may otherwise be established as follows.

If the differential equation be written in the form

$$x\frac{dy}{dx} + by\left(y - \frac{a}{b}\right) = cx^n \quad\text{..................} (18),$$

it becomes evident that it is symmetrical with respect to y and $y - \frac{a}{b}$. Assume then $y - \frac{a}{b}$ as a new variable in place of y, and writing $y - \frac{a}{b} = y'$, $y = y' + \frac{a}{b}$, the equation becomes

$$x\frac{dy'}{dx} + b\left(y' + \frac{a}{b}\right)y' = cx^n \quad\text{..............} (19),$$

or

$$x\frac{dy'}{dx} + ay' + by'^2 = cx^n \quad\text{..............} (20),$$

an equation which differs from the given equation only in that y has become y', and a has changed its sign. Hence the conditions $n = \frac{-2a}{2i-1}$ and $n = \frac{2a}{2i-1}$ are mutually dependent, and the value of y having been obtained for the former case, its value in the latter will be found by changing therein a into $-a$, and finally adding $\frac{a}{b}$.

It is here also to be noted that instead of beginning with an assumption of the form $y = A + \frac{x^n}{y_1}$ as in Art. 4, we might have commenced our reductions by the assumption $y = \frac{x^n}{B + y_1}$, the former of the above being proper for increasing by n, the latter for diminishing by n the quantity a. And as the first led *directly* to the solution (A), so would the second have led directly to the solution (D).

Lastly, it may be remarked that each of the above assumptions is only the *inverse* of the other. To increase the value of a by n we had to employ the assumption

$$y_{i-1} = A + \frac{x^n}{y_i},$$

which gives

$$y_i = \frac{x^n}{-A + y_{i-1}},$$

and this indicates the form of the assumption for the case in which a is to be diminished. Hence also by admitting negative as well as positive values of i, the two forms of solution might be replaced by a single one.

9. We have seen in Art. 1 that Riccati's equation

$$\frac{du}{dx} + bu^2 = cx^m$$

is reduced by the assumption $u = \frac{y}{x}$ to the form

$$x\frac{dy}{dx} - y + by^2 = cx^{m+2}.$$

Hence the condition for the solution of Riccati's equation, found by substituting in the final theorem of Art. 6, 1 for a and $m+2$ for n, will be

$$\frac{m+2 \pm 2}{2m+4} = i,$$

whence

$$m = -2 \pm \frac{2}{2i-1} \dots\dots\dots\dots\dots\dots (21),$$

i being a positive integer.

We may give to the expression for m another form, viz. $m=\frac{-4i}{2i\pm 1}$, i admitting of the value 0 together with positive integral values. In order to prove this, let it be observed that two values of m included in (21) are

$$m=\frac{-4i}{2i-1}, \text{ and } m=\frac{-4(i-1)}{2i-1}.$$

If in the second of these values we change $i-1$ into i, and therefore i into $i+1$, a change which merely involves that we interpret i as admitting of the value 0 as well as of positive integral values, we find

$$m=\frac{-4i}{2i+1} \quad\text{.........................} (22).$$

When $i=0$ this gives $m=0$, and as this value also results from the first of the expressions for m on making $i=0$, we are permitted in that formula also to regard i as admitting of the same range of values. Hence, combining the two formulæ in a single expression, we have

$$m=\frac{-4i}{2i\pm 1} \quad\text{.........................} (23),$$

i being 0, or a positive integer.

10. Riccati's equation may also be reduced, and it usually has been reduced, by a series of *double* transformations, of which the following will serve as an example.

The equation being $\frac{du}{dx}+bu^2=cx^m$, let $u=\frac{1}{bx}+\frac{1}{x^2u_1}$.

We have

$$\frac{du}{dx}=-\frac{1}{bx^2}-\frac{2}{x^3u_1}-\frac{1}{x^2u_1^2}\frac{du_1}{dx},$$

$$bu^2=\frac{1}{bx^2}+\frac{2}{x^3u_1}+\frac{b}{x^4u_1^2}.$$

Substituting these values in the given equation, we have

$$\frac{b}{x^4 u_1^{\,2}} - \frac{1}{x^2 u_1^{\,2}} \frac{du_1}{dx} = cx^m.$$

Whence,

$$x^2 \frac{du_1}{dx} + cx^{m+4} u_1^{\,2} - b = 0.$$

In this equation assume $x = z^{\frac{1}{m+3}}$, then

$$\frac{du_1}{dx} = \frac{du_1}{dz} \frac{dz}{dx} = (m+3)\, z^{\frac{m+2}{m+3}} \frac{du_1}{dz},$$

whence, after substitution and reduction,

$$\frac{du_1}{dz} + \frac{c}{m+3} u_1^{\,2} = \frac{b}{m+3} z^{-\frac{m+4}{m+3}} \text{ .... } (24),$$

an equation differing from the given equation, as to its coefficients and indices, in that b has been converted into $\frac{c}{m+3}$, c into $\frac{b}{m+3}$, and m into $-\frac{m+4}{m+3}$; but which is still of Riccati's form. The transformation, it will be observed, is a double one, as it affects the independent as well as the dependent variable.

Now if m be of the form $\frac{-4i}{2i-1}$, we find on substitution and reduction

$$-\frac{m+4}{m+3} = \frac{-4(i-1)}{2(i-1)-1}.$$

Hence, a second double transformation of the same nature as the last will reduce the differential equation to a form in which the index in the second member will become $-\frac{4(i-2)}{2(i-2)-1}$. And thus after a series of i transformations the index is reduced to 0, and the equation becomes solvable by separation of the variables.

To establish another condition of solution, assume in the given equation $u=\frac{1}{y}$, $x=z^{\frac{1}{m+1}}$, then, after substitution and reduction, we have

$$\frac{dy}{dz}+\frac{c}{m+1}y^2=\frac{b}{m+1}z^{-\frac{m}{m+1}},$$

which, by what has preceded, will be solvable if we have

$$-\frac{m}{m+1}=-\frac{4i}{2i-1}$$

$$\text{whence, } m=-\frac{4i}{2i+1}.$$

Combining these results it appears that Riccati's equation is integrable if $m=\frac{-4i}{2i\pm 1}$, i being 0 or a positive integer. This agrees with (23).

It is manifest from the complexity both of the transformations above described and of the results to which they lead, that Riccati's equation is, in its actual form, far less adapted for such transformations than the equation

$$x\frac{dy}{dx}-ay+by^2=cx^n,$$

to which it is so easily reduced.

11. Riccati's equation becomes linear on assuming

$$u=\frac{1}{bw}\frac{dw}{dx}.$$

The transformed equation is

$$\frac{d^2w}{dx^2}-bcx^m w=0 \quad \text{.................. (25).}$$

We shall consider it under this form in a subsequent chapter.

To Riccati's equation some others of greater generality may be reduced by a change of variables, e. g. the equation

$$\frac{du}{dx} + bx^m u^2 = cx^n \quad\text{.................}\quad (26),$$

by assuming $x^{m+1} = t$.

Euler's Equation.

It has already appeared that the solution of a differential equation may sometimes be freed from transcendents introduced by integration. An example of this has been afforded in the instance of the equation

$$\frac{dx}{\sqrt{(1-x^2)}} + \frac{dy}{\sqrt{(1-y^2)}} = 0,$$

(Chap. II.), the solution of which is capable of being exhibited in an algebraic form, although immediate integration introduces the transcendental functions $\sin^{-1}x$, $\sin^{-1}y$. The inquiry is here suggested whether in any other cases the direct integration may be evaded, an inquiry the more important as our means of integration are so limited. Euler succeeded in obtaining without direct integration the solution of the equation

$$\frac{dx}{\sqrt{(a+bx+cx^2+ex^3+fx^4)}} + \frac{dy}{\sqrt{(a+by+cy^2+ey^3+fy^4)}} = 0,$$

and of some related forms. The result belongs to the theory of the elliptic functions, and may be established independently by the methods which more peculiarly pertain to that theory. But the method by which Euler arrived at that result demands notice here.

12. To integrate the equation

$$\frac{dx}{\sqrt{(a+bx+cx^2+ex^3+fx^4)}} + \frac{dy}{\sqrt{(a+by+cy^2+ey^3+fy^4)}} = 0 \text{.......} (1).$$

Representing the polynomials $a + bx + cx^2 + ex^3 + fx^4$, and

$a+by+cy^2+ey^3+fy^4$ by X and Y respectively, we have to integrate

$$\frac{dx}{\sqrt{(X)}}+\frac{dy}{\sqrt{(Y)}}=0 \quad\text{..........................} (2).$$

The ordinary *solution* of this equation in the sense of Art. 5, Chap. I. would be

$$\int\frac{dx}{\sqrt{(X)}}+\int\frac{dy}{\sqrt{(Y)}}=C,$$

but it is our present object to obtain an algebraical relation between x and y without performing the integrations above implied.

$$\text{Let } \int\frac{dx}{\sqrt{(X)}}=t, \text{ then}$$

$$\frac{dx}{dt}=\sqrt{(X)},\ \frac{dy}{dt}=-\sqrt{(Y)} \quad\text{..................} (3).$$

Also let $x+y=p$, $x-y=q$. We shall endeavour to form a differential equation in which p and q are dependent variables, and t the independent variable.

From (3) we have

$$\frac{dp}{dt}=\sqrt{(X)}-\sqrt{(Y)} \quad\text{..............................} (4),$$

$$\frac{dq}{dt}=\sqrt{(X)}+\sqrt{(Y)} \quad\text{..............................} (5);$$

$$\therefore \frac{dp}{dt}\frac{dq}{dt}=X-Y$$

$$=bq+cpq+\tfrac{1}{4}eq(3p^2+q^2)+\tfrac{1}{2}fpq(p^2+q^2) \quad\text{...} (6),$$

since the transformations $x+y=p$, $x-y=q$ give

$$x^2-y^2=pq,$$

$$x^3-y^3=(x-y)(x^2+xy+y^2)=q\left(\frac{3p^2+q^2}{4}\right),$$

$$x^4 - y^4 = (x^2 - y^2)(x^2 + y^2) = pq\left(\frac{p^2 + q^2}{2}\right).$$

Again, from (3) we have

$$\frac{d^2x}{dt^2} = \frac{d\sqrt{(X)}}{dt} = \frac{dx}{dt}\,\frac{d\sqrt{(X)}}{dx} = \frac{1}{2}\,\frac{dX}{dx},$$

$$\frac{d^2y}{dt^2} \qquad\qquad = \frac{1}{2}\,\frac{dY}{dy};$$

whence by addition

$$\frac{d^2p}{dt^2} = \frac{1}{2}\left(\frac{dX}{dx} + \frac{dY}{dy}\right)$$

$$= b + cp + \frac{3}{4}e(p^2 + q^2) + \frac{1}{2}fp(p^2 + 3q^2) \text{ .... (7)},$$

on effecting the differentiations and transforming as before from x and y to p and q.

Multiplying (7) by q, and from the result subtracting (6), we have

$$q\frac{d^2p}{dt^2} - \frac{dp}{dt}\,\frac{dq}{dt} = \frac{eq^3}{2} + fpq^3$$

$$= \frac{q^3}{2}(e + 2fp).$$

Therefore

$$\frac{2}{q^2}\,\frac{d^2p}{dt^2} - \frac{2}{q^3}\,\frac{dp}{dt}\,\frac{dq}{dt} = e + 2fp.$$

Now multiplying both sides by $\frac{dp}{dt}$,

$$\frac{2\frac{dp}{dt}\,\frac{d^2p}{dt^2}}{q^2} - \left(\frac{dp}{dt}\right)^2 \times \frac{2\frac{dq}{dt}}{q^3} = (e + 2fp)\frac{dp}{dt} \text{ .... (8)},$$

from which, each member being an exact differential, we have on integration

$$\frac{1}{q^2}\left(\frac{dp}{dt}\right)^2 = ep + fp^2 + C,$$

C being an arbitrary constant.

Hence $$\frac{dp}{dt} = q\sqrt{(C + ep + fp^2)}.$$

Therefore by (4)

$$\sqrt{(X)} - \sqrt{(Y)} = (x-y)\sqrt{\{C + e(x+y) + f(x+y)^2\}} \ldots (9),$$

the integral required.

The student may apply the same series of transformation and reduction to the equation

$$\frac{dx}{\sqrt{(a + bx + cx^2 + ex^3 + fx^4)}} - \frac{dy}{\sqrt{(a + by + cy^2 + ey^3 + fy^4)}} = 0 \ldots\ldots\ldots\ldots (10).$$

The resulting integral will be

$$\sqrt{(X)} + \sqrt{(Y)} = (x-y)\sqrt{\{C + e(x+y) + f(x+y)^2\}} \ldots (11).$$

13. It will probably appear that there is something arbitrary in the mode in which, in the above investigation, the final differential equation (8) between p, q, and t, upon which the solution of the problem depends, is formed. The analysis which is subjoined may throw some light upon its real nature, and shew of what general theorem that equation constitutes an expression.

Prop. Whatever may be the form of the function $f(x)$, the following theorem of development holds good, viz.

$$\begin{aligned} f(y) - f(x) &= A_1\{f'(y) + f'(x)\}(y-x) \\ &+ A_3\{f'''(y) + f'''(x)\}(y-x)^3 \\ &+ A_5\{f^{\text{v}}(y) + f^{\text{v}}(x)\}(y-x)^5 + \text{\&c.} \ldots (12), \end{aligned}$$

wherein A_1, A_3, A_5, &c. are the coefficients of the successive powers of x, in the development of the function $\frac{\epsilon^x - 1}{\epsilon^x + 1}$ in a series of the form

$$A_1 x + A_3 x^3 + A_5 x^5 + \text{\&c.}$$

For let $y = x + h$, then, employing a well-known symbolical form of Taylor's theorem,

$$f(y) - f(x) = f(x+h) - f(x)$$

$$= (\epsilon^{h\frac{d}{dx}} - 1) f(x)$$

$$= \frac{\epsilon^{h\frac{d}{dx}} - 1}{\epsilon^{h\frac{d}{dx}} + 1} (\epsilon^{h\frac{d}{dx}} + 1) f(x)$$

$$= \frac{\epsilon^{h\frac{d}{dx}} - 1}{\epsilon^{h\frac{d}{dx}} + 1} \{f(x+h) + f(x)\}$$

$$= \left\{A_1 h \frac{d}{dx} + A_3 h^3 \left(\frac{d}{dx}\right)^3 + \&c.\right\} \{f(x+h) + f(x)\} \quad \text{.........(13)},$$

where A_1, A_3, &c. have the series of values above described. Hence, performing the differentiations and replacing $x + h$ by y, and h by $y - x$, we have

$$f(y) - f(x) = A_1 \{f'(y) + f'(x)\} (y - x)$$
$$+ A_3 \{f'''(y) + f'''(x)\} (y-x)^3 + \&c. \ldots (14),$$

which is the proposition in question.

The values of A_1, A_3, A_5, &c. may be expressed by means of Bernoulli's numbers, but they may also be calculated very simply by developing the exponentials in the fraction $\frac{\epsilon^x - 1}{\epsilon^x + 1}$, and then expanding the fraction itself by division. We readily find

$$A_1 = \frac{1}{2}, \quad A_3 = -\frac{1}{24}, \quad A_5 = \frac{1}{240}, \quad A_7 = \frac{-17}{40320}, \ \&c.$$

When $f(x)$ is a polynomial of the fourth degree, we have

$$f^{v}(x) = 0, \quad f^{vi}(x) = 0, \ \&c.,$$

and the theorem is reduced to the following, viz.:

$$f(y)-f(x)=\frac{1}{2}\{f'(y)+f'(x)\}(y-x)$$

$$-\frac{1}{24}\{f'''(y)+f'''(x)\}(y-x)^3 \text{.......} (15).$$

Now the differential equation (8) into whose origin we are inquiring is merely a transformation of the last theorem. We will on this occasion, and for the sake of variety, exemplify the above remark in the solution of the differential equation

$$\frac{dx}{\sqrt{\{f(x)\}}}=\frac{dy}{\sqrt{\{f(y)\}}} \text{....................} (16),$$

in which

$$f(x)=a+bx+cx^2+ex^3+fx^4 \text{...........} (17),$$

$$f(y)=a+by+cy^2+ey^3+fy^4 \text{...........} (18).$$

Representing either member of (16) by dt and assuming t as an independent variable, substitute the values hence determined for $f(x)$, $f'(x)$, $f'''(x)$, &c. in the theorem (15). There will result

$$\frac{dx}{dt}=\sqrt{\{f(x)\}},\quad \frac{dy}{dt}=\sqrt{\{f(y)\}}.$$

Hence $$f(x)=\left(\frac{dx}{dt}\right)^2,\quad f(y)=\left(\frac{dy}{dt}\right)^2,$$

$$f'(x)=\frac{d}{dx}\left(\frac{dx}{dt}\right)^2=\frac{dt}{dx}\,\frac{d}{dt}\left(\frac{dx}{dt}\right)^2$$

$$=2\frac{d^2x}{dt^2},$$

$$f'(y)=2\frac{d^2y}{dt^2}.$$

Lastly from (17) and (18)

$$f'''(x)=24fx+6e,$$

$$f'''(y)=24fy+6e,$$

by which substitution, (15) becomes

$$\left(\frac{dy}{dt}\right)^2 - \left(\frac{dx}{dt}\right)^2 = \left(\frac{d^2y}{dt^2} + \frac{d^2x}{dt^2}\right)(y-x) - \left(fx + fy + \frac{e}{2}\right)(y-x)^3.$$

Or, transposing

$$\left(\frac{dy}{dt}\right)^2 - \left(\frac{dx}{dt}\right)^2 + (x-y)\left(\frac{d^2x}{dt^2} + \frac{d^2y}{dt^2}\right) = \left(fx + fy + \frac{e}{2}\right)(y-x)^3 \ldots (19).$$

Now the very *form* of this equation suggests the transformation $x + y = p$, $x - y = q$, by which it becomes

$$-\frac{dp}{dt}\frac{dq}{dt} + q\frac{d^2p}{dt^2} = \left(fp + \frac{e}{2}\right)q^3,$$

whence multiplying by $\frac{2}{q^3}dp$ and integrating

$$\left(\frac{dp}{dt}\right)^2 \div q^2 = fp^2 + ep + C;$$

$$\therefore \left[\frac{\sqrt{\{f(x)\}} + \sqrt{\{f(y)\}}}{x-y}\right]^2 = f(x+y)^2 + e(x+y) + C \ldots (20),$$

the integral sought.

EXERCISES.

1. $$x\frac{dy}{dx} - ay + y^2 = x^{-2a}.$$

2. $$x\frac{dy}{dx} - ay + y^2 = x^{-\frac{2a}{3}}.$$

3. $$\frac{du}{dx} + u^2 = cx^{-\frac{4}{3}}.$$

4. $$\frac{du}{dx} + bu^2 = cx^{-4}.$$

5. $$\frac{du}{dx} - u^2 = 2x^{-\frac{8}{3}} dx.$$

6. Assuming the conditions for the solution of Riccati's equation, Art. 9, investigate those under which the equation $\frac{du}{dx} + 6x^m u^2 = cx^n$ is integrable.

7. Assuming the conditions, Art. 6, under which

$$x\frac{dy}{dx} - ay + by^2 = cx^n$$

is integrable in finite terms, investigate those under which the equation

$$x\frac{dy}{dx} + \alpha + \beta y + \gamma y^2 = \delta x^n,$$

is integrable in finite terms.

8. Transforming the equation $x\frac{dy}{dx} - ay + by^2 = cx^{2a}$, by assuming $x^a = t$, an integrating factor may be found by Art. 6, Chap. V.

9. The equation $\frac{du}{dx} + bu^2 = cx^m + \frac{h}{x^2}$, more general than Riccati's, is reducible to the form $x\frac{dy}{dx} - a'y + b'y^2 = c'x^n$, considered in Art. 3, by an assumption of the form $u = \frac{y - A}{x}$.

10. Hence investigate the conditions under which the former equation may be solved.

11. The same equation may be reduced to Riccati's form by an assumption of the form $y = Ax^{-1} + z\phi(x)$, followed by a transformation affecting only x.

12. Integrate the equation

$$\frac{dx}{\sqrt{(a + bx + cx^2 + ex^3 + fx^4)}} + \frac{dy}{\sqrt{(a + by + cy^2 + ey^3 + fy^4)}} = 0,$$

by the application of the theorem of Art. 13.

13. Deduce from that theorem the following expression for the value of a definite integral, viz.:

$$\int_a^b \phi(x)\,dx = \frac{\phi(a)+\phi(b)}{2}(y-x) - \frac{\phi''(a)+\phi''(b)}{24}(y-x)^3$$

$$+\frac{\phi^{iv}(a)+\phi^{iv}(b)}{240}(y-x)^5 - \&c.$$

CHAPTER VII.

ON DIFFERENTIAL EQUATIONS OF THE FIRST ORDER, BUT NOT OF THE FIRST DEGREE.

1. REFERRING to the general type of differential equations of the first order, viz.:

$$F\left(x, y, \frac{dy}{dx}\right) = 0,$$

we have now to consider those cases in which $\frac{dy}{dx}$ is so involved that the given equation cannot be reduced to the form

$$M + N\frac{dy}{dx} = 0,$$

already considered.

Freed from radicals the supposed equation will, however, present itself in the form

$$\left(\frac{dy}{dx}\right)^n + P_1\left(\frac{dy}{dx}\right)^{n-1} + P_2\left(\frac{dy}{dx}\right)^{n-2} \ldots + P_n = 0 \ldots\ldots\ldots\ldots (1),$$

where P_1, P_2, ... P_n are functions of x and y.

An obvious preparation for the solution of such an equation, is to resolve its first member, considered as algebraic with respect to the differential coefficient $\frac{dy}{dx}$, into its component factors of the first degree. If $p_1, p_2 \ldots p_n$ be the roots of (1) thus considered, we shall have

$$\left(\frac{dy}{dx} - p_1\right)\left(\frac{dy}{dx} - p_2\right) \cdots \left(\frac{dy}{dx} - p_n\right) = 0 \ldots\ldots\ldots\ldots\ldots (2),$$

$p_1, p_2 \dots p_n$ being supposed to be determined as known functions of x and y. And it is now manifest that any relation between x and y which makes either one or more than one of the factors of the first member to vanish, will be a solution of the equation, and that no relation between x and y not possessing this character will be such. Hence if we solve the separate equations

$$\frac{dy}{dx} - p_1 = 0, \quad \frac{dy}{dx} - p_2 = 0 \dots \frac{dy}{dx} - p_n = 0 \dots\dots\dots\dots\dots (3),$$

any one of the solutions obtained will be a solution of (2), since it will make one of its factors to vanish. And if we express the different solutions thus obtained, each with its arbitrary constant annexed, in the forms

$$V_1 - C_1 = 0, \quad V_2 - C_2 = 0 \dots V_n - C_n = 0,$$

any product of two or more of these equations will also be a solution of (2), since it will cause two or more of its factors to vanish.

Ex. Given the differential equation

$$\left(\frac{dy}{dx}\right)^2 - a^2y^2 = 0 \dots\dots\dots\dots\dots\dots\dots\dots\dots (4).$$

Here the component equations are

$$\frac{dy}{dx} - ay = 0,$$

$$\frac{dy}{dx} + ay = 0,$$

and their respective solutions are

$$\log y - ax - c_1 = 0 \dots\dots\dots\dots\dots\dots\dots\dots (5),$$

$$\log y + ax - c_2 = 0 \dots\dots\dots\dots\dots\dots\dots\dots (6).$$

Either of these equations is a solution of the given equation, and so is their product

$$(\log y - ax - c_1)(\log y + ax - c_2) = 0 \dots\dots\dots\dots\dots (7).$$

2. And here two important questions are suggested. First, how is it that two arbitrary constants present themselves in the solution of an equation of the first order? Secondly, is it possible to express with equal generality the solution of the equation by a primitive containing a *single* arbitrary constant in accordance with what has been said of the genesis of differential equations of the first order, Chap. I. Art. 6? These are connected questions, and they will be answered together.

The equation (7) implies that y admits of two values each involving an arbitrary constant, but it does not imply that y admits of a value involving two arbitrary constants. The component factors of the solution separately equated to 0, as in (5) and (6), give respectively

$$y = C_1\epsilon^{ax}, \quad y = C_2\epsilon^{-ax} \ldots\ldots\ldots\ldots\ldots\ldots\ldots(8),$$

each of which involves one arbitrary constant only, and each of which corresponds to a single factor of the given differential equation. The true canon is, not that a general solution of an equation of the first order can involve only one arbitrary constant in its expression, but that *each value of* y which such a solution establishes involves in its expression only a single arbitrary constant.

At the same time it remains true that every differential equation of the first order implies the existence of a primitive involving a *single* arbitrary constant,—it remains true that such primitive constitutes a *general* solution of the differential equation. To reconcile these seeming anomalies we shall shew that if we suppose the arbitrary constants c_1 and c_2 in (7) identical, and accordingly replace each of them by c, we shall have an equation which will be, first the true primitive of (4), in that it will generate that equation by differentiation and the elimination of c, secondly its general solution, in that no particular relation is deducible from the solution (7) involving two arbitrary constants which may not also be deduced from it.

Thus replacing c_1 and c_2 by c, we have

$$(\log y - ax - c)(\log y + ax - c) = 0 \ldots\ldots\ldots(9),$$

whence $\quad (\log y)^2 - a^2x^2 - 2c\log y + c^2 = 0.$

Differentiating, and representing $\frac{dy}{dx}$ by p,

$$2 \log y \frac{p}{y} - 2a^2x - 2c\frac{p}{y} = 0,$$

$$\text{whence } c = -\frac{a^2xy}{p} + \log y.$$

Substituting this value in (9), we have

$$\left(\frac{a^2xy}{p} - ax\right)\left(\frac{a^2xy}{p} + ax\right) = 0,$$

which reduces to

$$a^2x^2(a^2y^2 - p^2) = 0.$$

Or, rejecting the factor a^2x^2 which does not contain p, and replacing p by $\frac{dy}{dx}$,

$$\left(\frac{dy}{dx}\right)^2 - a^2y^2 = 0,$$

the differential equation given. Thus (9) is its complete primitive.

Let it be remarked that if we eliminated the constant c_1 from (5) or c_2 from (6) we should, as is evident from the origin of these equations, obtain not the given differential equation but only one of the component differential equations into which it is resolvable.

Again, the solution (9) is *general*. The two relations between y and x which it furnishes are

$$y = C\epsilon^{ax}, \quad y = C\epsilon^{-ax} \dots\dots\dots\dots\dots\dots (10),$$

and these differ in expression from (8) only in that the arbitrary constant is here supposed to be the same in one as in the other, but as it is arbitrary and admits of any value, there is no single relation implied in (8) which is not also implied in (10). And it is *in this sense* that the generality of the solution has been affirmed.

3. These illustrations will prepare the way for the demonstration of the general theorem which they exemplify.

Theorem. *If the differential equation of the first order and* n^{th} *degree be resolved into its component equations*

$$\frac{dy}{dx}-p_1=0,\quad \frac{dy}{dx}-p_2=0, \ldots \frac{dy}{dx}-p_n=0,$$

and if the complete primitives of these equations are $V_1=c_1$, $V_2=c_2, \ldots V_n=c_n$, *then the complete primitive of the given equation will be*

$$(V_1-c)(V_2-c)\ldots(V_n-c)=0.$$

Let us first examine the case in which the proposed differential equation is of the second degree, and therefore expressible in the form $\left(\frac{dy}{dx}-p_1\right)\left(\frac{dy}{dx}-p_2\right)=0$. Suppose that the integral $V_1=c_1$ is derived from the equation $\frac{dy}{dx}-p_1=0$ by means of an integrating factor μ_1. Then $dV_1=\mu_1\left(\frac{dy}{dx}-p_1\right)dx$ In like manner we shall have $dV_2=\mu_2\left(\frac{dy}{dx}-p_2\right)dx$. Now taking the equation

$$(V_1-c)(V_2-c)=0 \quad\ldots\ldots\ldots\ldots\ldots\ldots (11)$$

as a primitive, we have, on differentiating with respect to x and y,

$$(V_1-c)\,dV_2+(V_2-c)\,dV_1=0 \quad\ldots\ldots\ldots\ldots\ldots (12).$$

Therefore $$c=\frac{V_1dV_2+V_2dV_1}{dV_1+dV_2},$$

whence $$V_1-c=\frac{(V_1-V_2)\,dV_1}{dV_1+dV_2},$$

$$V_2-c=\frac{(V_2-V_1)\,dV_2}{dV_1+dV_2}.$$

Substituting these values in (11), we have

$$(V_1 - V_2)^2 dV_1 dV_2 = 0 \dots\dots\dots\dots\dots\dots (13),$$

or $$(V_1 - V_2)^2 \mu_1\mu_2 \left(\frac{dy}{dx} - p_1\right)\left(\frac{dy}{dx} - p_2\right) = 0 \dots\dots\dots\dots (14).$$

And this, on rejecting the factor $(V_1 - V_2)^2 \mu_1\mu_2$ which does not contain any differential coefficients, becomes identical with the given differential equation. Hence $(V_1 - c)(V_2 - c) = 0$ is the complete primitive of that equation.

To generalize this particular demonstration it would be necessary to eliminate c between the equation

$$(V_1 - c)(V_2 - c) \dots (V_n - c) = 0 \dots\dots\dots\dots (15),$$

and the equation thence derived by differentiation with respect to x and y. The ordinary process of elimination, as exemplified above in the particular case in which $n = 2$, would be complex, but the result may be determined without difficulty by logical considerations. It will suffice for this purpose to consider the case in which $n = 3$.

We have then as the supposed primitive

$$(V_1 - c)(V_2 - c)(V_3 - c) = 0 \dots\dots\dots\dots\dots (16),$$

and as the derived equation

$$(V_2 - c)(V_3 - c)\frac{dV_1}{dc} + (V_3 - c)(V_1 - c)\frac{dV_2}{dc}$$
$$+ (V_1 - c)(V_2 - c)\frac{dV_3}{dc} = 0 \dots\dots (17).$$

Now (16) implies that some one at least of the equations

$$V_1 - c = 0, \quad V_2 - c = 0, \quad V_3 - c = 0,$$

is satisfied.

If the first of these equations is satisfied we have $c = V_1$, and substituting this value in (17) there results

$$(V_2 - V_1)(V_3 - V_1)\,dV_1 = 0 \quad\text{..............(18)}.$$

If the second equation of the above system is satisfied, we have in like manner

$$(V_3 - V_2)(V_1 - V_2)\,dV_2 = 0 \quad\text{..............(19)}.$$

If the third equation of the system is satisfied we have

$$(V_1 - V_3)(V_2 - V_3)\,dV_3 = 0 \quad\text{..............(20)}.$$

Hence the existence of (16) as primitive supposes the existence of some one at least of the equations (18), (19), (20), and therefore of the equation

$$(V_2 - V_3)^2 (V_3 - V_1)^2 (V_1 - V_2)^2\,dV_1\,dV_2\,dV_3 = 0 \quad\text{......(21)},$$

which is found by multiplying those equations together.

Conversely the supposition that the equation (21) is true, involves the supposition that one at least of the equations (18), (19), (20) is true.

The equation (21) is therefore *equivalent* to the result which ordinary elimination applied to (16) and (17) would give. The same process of reasoning applied to the more general equation (15) as supposed primitive, would lead to a result of the form

$$K\,dV_1\,dV_2 \ldots dV_n = 0 \quad\text{.................(22)},$$

K being the product of the squares of the differences of $V_1, V_2 \ldots V_n$.

On comparison with (13) we see that in the particular case of $n = 2$, this is not only equivalent to but identical with the result of ordinary elimination in that case. And this identity of form, though it is not necessary to our present purpose to establish it, might be demonstrated generally.

Now $dV_1 = \mu_1\left(\dfrac{dy}{dx} - p_1\right)$, $dV_2 = \mu_2\left(\dfrac{dy}{dx} - p_2\right)$, &c. Hence

(22) gives

$$K\mu_1\mu_2\ldots\mu_n\left(\frac{dy}{dx}-p_1\right)\left(\frac{dy}{dx}-p_2\right)\ldots\left(\frac{dy}{dx}-p_n\right)=0,$$

or, rejecting the factor $K\mu_1\mu_2\ldots\mu_n$, which does not contain differential coefficients,

$$\left(\frac{dy}{dx}-p_1\right)\left(\frac{dy}{dx}-p_2\right)\ldots\left(\frac{dy}{dx}-p_n\right)=0.$$

Of this equation it has therefore been shewn, as was required, that $(V_1-c)(V_2-c)\ldots(V_n-c)=0$ constitutes the complete primitive.

Ex. Given $$\left(\frac{dy}{dx}\right)^2-\frac{a}{x}=0 \qquad (1).$$

Here the component equations are

$$\frac{dy}{dx}-\left(\frac{a}{x}\right)^{\frac{1}{2}}=0,\quad \frac{dy}{dx}+\left(\frac{a}{x}\right)^{\frac{1}{2}}=0,$$

and their respective integrals are

$$y-c_1-2\sqrt{(ax)}=0 \qquad (2),$$

$$y-c_2+2\sqrt{(ax)}=0 \qquad (3).$$

Replacing both constants by c and multiplying the equations together, we have

$$(y-c)^2-4ax=0 \qquad (4),$$

as the complete primitive.

Now this primitive represents a series of parabolas, the parameters of which are constant and equal to $4a$, and the axes of which are coincident with the axis of x; but the vertices of which are situated at different points of that axis, corresponding to the different values which may be given to the arbitrary constant c. Of these parabolas the equations (2) and (3), which may be written in the more usual forms

$$y-c_1=2\sqrt{(ax)},\quad y-c_2=-2\sqrt{(ax)},$$

represent respectively the positive and the negative branches, while the equation

$$\{y - c_1 - 2\sqrt{(ax)}\}\{y - c_2 + 2\sqrt{(ax)}\} = 0 \text{.........} (5),$$

represents the terms which would be found by taking one positive and one negative branch, *but not necessarily from the same parabola.* Thus there is no portion of the loci represented by the apparently more general solution (5), which is not also represented by the complete primitive (4). Nor is it a defect of generality, if, when every branch of every curve in the series is represented, those branches which belong to the same curve are paired together.

4. There are certain cases in which differential equations of the first order can be solved without the resolution of the first member into its component factors. Of these the most important are the following.

1st. When the given equation contains only one of the variables x and y in addition to $\frac{dy}{dx}$, being either of the form

$$F\left(x, \frac{dy}{dx}\right) = 0,$$

or of the form

$$F\left(y, \frac{dy}{dx}\right) = 0.$$

2ndly. When, involving x and y only in the first degree, it is expressible in the form

$$x\phi(p) + y\psi(p) = \chi(p), \text{ where } p = \frac{dy}{dx}.$$

3rdly. When the equation is homogeneous with respect to x and y.

These cases we shall consider separately.

Equations involving only one of the variables x and y with $\frac{dy}{dx}$.

In this case if, representing $\frac{dy}{dx}$ by p, and regarding p as a new variable, we form a differential equation between p and the variable which does not enter into the original equation, and integrate the equation thus formed, the elimination of p between the resulting integral and the original equation will give the complete primitive required. For it will express a relation between x, y, and the arbitrary constant introduced by integration.

Thus if from the equation $F(x, p) = 0$ we deduce $x = f(p)$, then, since $dy = pdx$, we have

$$dy = pf'(p)\,dp;$$

$$\therefore\ y = \int pf'(p)\,dp + C \ldots\ldots\ldots\ldots\ldots\ldots(1).$$

After the integration here implied y will be expressed as a function of p and c, and between that result and the original equation p must be eliminated.

In like manner, if from $F(y, p) = 0$ we deduce $y = f(p)$, the equation $dy = pdx$ gives $f'(p)\,dp = pdx$, whence

$$dx = \frac{f'(p)}{p}\,dp.$$

whence

$$x = \int \frac{f'(p)\,dp}{p} + c \ldots\ldots\ldots\ldots\ldots\ldots\ldots\ldots\ldots\ldots(2),$$

between which (after the integration has been performed) and the original equation, p must be eliminated.

But these methods, though always permissible, are only advantageous when it is more easy to solve the given equation, with respect to the variable x or y which it involves, than with respect to p.

Ex. 1. Given $x = 1 + p^3$.

Here $dy = pdx = p \times 3p^2dp = 3p^3dp$;

$$\therefore\ y = \frac{3p^4}{4} + c \ldots\ldots\ldots\ldots\ldots\ldots\ldots\ldots\ldots\ldots(3).$$

Now as the original equation gives $p = (x-1)^{\frac{1}{3}}$, the complete primitive found by substitution of this value in (3) will be

$$y = \frac{3}{4}(x-1)^{\frac{4}{3}} + c \ldots\ldots\ldots\ldots (4),$$

and it would be directly obtained in this form by integrating the original equation reduced by algebraic solution to the form

$$\frac{dy}{dx} = (x-1)^{\frac{1}{3}}.$$

This example illustrates the process but not its advantages.

Ex. 2. Given $x = 1 + p + p^3$.

$$\text{Here } dy = pdx = pdp + 3p^3dp;$$

$$\therefore\ y = \frac{p^2}{2} + \frac{3p^4}{4} + c \ldots\ldots\ldots\ldots (5),$$

between which and the original equation p must be eliminated. We may do this so as to obtain the final equation between x and y in a rational form; but, if this object is not deemed important, we may, by the solution of a quadratic, determine p from (5) and substitute its value in the given equation.

Ex. 3. Given $y = p^2 + 2p^3$.

Here since $pdx = dy$ we have

$$dx = \frac{1}{p}dy = 2dp + 6pdp;$$

$$\therefore\ x = 2p + 3p^2 + c.$$

From this equation we find

$$p = \frac{-1 \pm \sqrt{(3x + C)}}{3},$$

C being an arbitrary constant introduced in the place of $1 - 3c$; and y will be found by substituting this value of p in the original equation.

5. It is worth while to notice that processes virtually equivalent to the above, would be suggested by the forms of

the differential coefficients in the application of the more general method of Art. 1.

Thus when the equation $x = f(p)$ does not admit of algebraic solution with respect to p, we can only express the value p by the inverse functional notation,

$$p = f^{-1}(x),$$

$$\text{Whence } y = \int f^{-1}(x)\, dx.$$

To get rid of the inverse functional sign in the second member, let $f^{-1}x = t$, then $x = f(t)$, $dx = f'(t)\, dt$, whence

$$y = \int tf'(t)\, dt.$$

Now this equation only differs from (1) in that t takes the place of p.

Equations in which x and y are involved only in the first degree, the typical form being $x\phi(p) + y\psi(p) = \chi(p)$.

6. Any equation of the above class may be reduced to a linear differential equation between x and p, after the solution of which, p must be eliminated.

The reduced equation is found by differentiating the given equation and then eliminating, if necessary, the variable y. It may happen that such elimination is unnecessary, y disappearing through differentiation.

Ex. Let us apply this method to the equation

$$y = xp + f(p) \ldots\ldots\ldots\ldots\ldots\ldots \text{(1)},$$

usually termed Clairaut's equation.

Differentiating, we have

$$p = p + x\frac{dp}{dx} + f'(p)\frac{dp}{dx},$$

whence $$\{x + f'(p)\}\frac{dp}{dx} = 0.$$

Now this is resolvable into the two equations,

$$x+f'(p)=0 \qquad (2),$$

$$\frac{dp}{dx}=0 \qquad (3).$$

The second of these, which alone contains differentials of the new variables x and p, is the true *differential* equation between x and p.

Integrating it we have $p=c$,

and substituting this value of p in (1),

$$y=cx+f(c) \qquad (4),$$

which is the complete primitive required.

But what relation does the rejected equation (3) bear to the given differential equation (1), and what relation to its complete primitive just obtained?

If we eliminate p between (1) and (3) we obtain a new relation between x and y not included in the complete primitive already found, i.e. not deducible from that primitive by assigning a particular value to its arbitrary constant, and yet satisfying the same differential equation, and, as we shall hereafter see, connected in a remarkable manner with the complete primitive. Such a relation between x and y is called a singular solution. We shall enter more fully into the theory of singular solutions in a distinct chapter, but the following example will throw some light upon their nature, as well as illustrate the process above described.

Ex. Given $y=xp+\dfrac{m}{p}$.

Here differentiating we have

$$0=\left(x-\frac{m}{p^2}\right)\frac{dp}{dx}.$$

From the equation $\dfrac{dp}{dx}=0$, we have $p=c$, whence

$$y=cx+\frac{m}{c} \qquad (5),$$

the complete primitive. From the equation $x-\frac{m}{p^2}=0$, we have

$$p=\sqrt{\left(\frac{m}{x}\right)},$$

and this value substituted in the original equation gives, after freeing the result from radical signs,

$$y^2=4mx \quad\text{.............................(6)},$$

the singular solution.

Here the singular solution (6) is the equation of a parabola whose parameter is $4m$, and the complete primitive (5) is the well-known equation of that tangent to the same parabola which makes with the axis of x an angle whose trigonometrical tangent is c.

Now, for the infinitesimal element in which the curve and its tangent coincide, the values of x, y, and $\frac{dy}{dx}$ are the same in both. And thus it is that the algebraic equations of the curve and of its tangent satisfy the same differential equation of the first order.

On the other hand, if (5) be regarded as the general equation of a system of straight lines, each straight line in that system being determined by giving a special value to c in the equation, the envelop or boundary curve of the system will be determined by (6). Here the singular solution is presented as the equation of the envelop of the system of lines defined by the complete primitive.

7. In the second place let us consider the more general equation

$$y=xf(p)+\phi(p).$$

Differentiating, we have

$$p=f(p)+\{xf'(p)+\phi'(p)\}\frac{dp}{dx},$$

whence
$$\{p-f(p)\}\frac{dx}{dp}-f'(p)x=\phi'(p),$$

or $$\frac{dx}{dp} - \frac{f'(p)}{p-f(p)}\, x = \frac{\phi'(p)}{p-f(p)},$$

a linear equation of the first order by which x may be determined as a function of p. The elimination of p between the resulting equation and the given one will give the complete primitive.

The typical equation

$$x\phi(p) + y\psi(p) = \chi(p)$$

may be reduced to the above form by dividing by $\phi(p)$, but it may also be treated independently by direct differentiation.

Instead however of forming a differential equation between x and p, we may form a differential equation between y and p. Or, with greater generality, representing any proposed function of p by t, we may form a differential equation between either of the primitive variables and t. Such a differential equation will necessarily be linear, and its solution must of course be followed by the elimination of t. And this general procedure, more fully to be exemplified when we come to treat of some of the inverse problems of Geometry and of Optics, is often attended with signal advantage.

Ex. Given $x + yp = ap^2$.

We shall reduce this to a differential equation between x and p.

Differentiating, we have

$$1 + p^2 + y\frac{dp}{dx} = 2ap\frac{dp}{dx},$$

then eliminating y by means of the given equation, we have

$$1 + p^2 + \left(\frac{ap^2 - x}{p}\right)\frac{dp}{dx} = 2ap\frac{dp}{dx},$$

which may be reduced to the linear form

$$\frac{dx}{dp} - \frac{x}{p(1+p^2)} = \frac{ap}{1+p^2},$$

its integral being

$$x = \frac{p}{\sqrt{(1+p^2)}}\,[C + a \log\{p + \sqrt{(1+p^2)}\}].$$

If in this equation we substitute for p its value in terms of x and y furnished by the given equation, i.e. if we make

$$p = \frac{y \pm \sqrt{(y^2 + 4ax)}}{2a}$$

we shall be in possession of the complete primitive.

Had we chosen to form a differential equation between y and p, we should have, on differentiating the given equation while regarding y as independent variable,

$$\frac{dx}{dy} + p + y\frac{dp}{dy} = 2ap\frac{dp}{dy},$$

whence, replacing $\frac{dx}{dy}$ by $\frac{1}{p}$ and reducing,

$$\frac{dy}{dp} + \frac{p}{1+p^2}y = \frac{2ap^2}{1+p^2},$$

therefore on integration

$$y = \frac{1}{\sqrt{(1+p^2)}}\,[C + ap\sqrt{(1+p^2)} - a\log\{p + \sqrt{(1+p^2)}\}],$$

from which, as before, p must be eliminated. The final results are of course identical.

Homogeneous Equations of the first order.

8. Equations which are homogeneous with respect to x and y may be prepared for solution by assuming $y = vx$.

The typical form of such equations is

$$x\,\phi\left(\frac{y}{x}, p\right) = 0 \qquad\qquad (1).$$

Assuming then $\frac{y}{x}=v$, and dividing by x^n, we have

$$\phi(v, p)=0 \ldots\ldots\ldots\ldots\ldots\ldots\ldots(2).$$

If we can solve this equation with respect to p, we have

$$p=f(v).$$

But, since $y=xv$

$$p=x\frac{dv}{dx}+v.$$

Thus the transformed equation becomes

$$x\frac{dv}{dx}+v=f(v),$$

whence
$$\frac{dv}{v-f(v)}+\frac{dx}{x}=0,$$

an equation in which the variables are separated, and in the integral of which it will only remain to substitute for v its value $\frac{y}{x}$.

But if it be more easy to solve (2) with respect to v than with respect to p, and if the result be

$$v=f(p),$$

then restoring to v its value $\frac{y}{x}$, we have

$$y=xf(p),$$

which is a particular case of the equation of the previous section. Hence differentiating, we have

$$p=f(p)+xf'(p)\frac{dp}{dx},$$

from which results

$$\frac{dx}{x}+\frac{f'(p)dp}{f(p)-p}=0,$$

an equation in which the variables x and p are separated. Between the integral of this equation and the given equation p must be eliminated, and the relation between x and y which results will be the complete primitive.

Ex. Given $yp+nx=\sqrt{(y^2+nx^2)}\sqrt{(1+p^2)}$.

Assuming $y=vx$, we have

$$vp+n=\sqrt{(v^2+n)}\sqrt{(1+p^2)},$$

the solution of which with respect to p gives

$$p=v\pm\sqrt{\left(\frac{n-1}{n}\right)}\sqrt{(v^2+n)}.$$

But $$p=x\frac{dv}{dx}+v.$$

Therefore $$x\frac{dv}{dx}=\pm\sqrt{\frac{(n-1)}{n}}\sqrt{(v^2+n)},$$

$$\frac{dv}{\sqrt{(v^2+n)}}=\pm\sqrt{\left(\frac{n-1}{n}\right)}\frac{dx}{x}.$$

Integrating, we have

$$\log\{v+\sqrt{(v^2+n)}\}=\pm\sqrt{\left(\frac{n-1}{n}\right)}\log x+C;$$

$$\therefore\ v+\sqrt{(v^2+n)}=cx^{\pm\sqrt{\left(\frac{n-1}{n}\right)}},$$

or, replacing v by $\dfrac{y}{x}$,

$$y+\sqrt{(y^2+nx^2)}=cx^{1\pm\sqrt{\left(\frac{n-1}{n}\right)}}$$

the complete primitive.

Equations solvable by differentiation.

9. A remarkable class of equations, the theory of which has been fully discussed by Lagrange, deserves attention.

It has been shewn, Chap. I. Art. 9, that if two differential equations of the first order, each involving a distinct arbitrary constant, give rise to the same differential equation of the second order, they are derived from a common primitive involving both the arbitrary constants in question.

Let us suppose these differential equations of the first order to be reduced to the forms

$$\phi\left(x,\ y,\ \frac{dy}{dx}\right)=a\ldots\ldots\ldots\ldots\ldots\ldots\ldots(1),$$

$$\psi\left(x,\ y,\ \frac{dy}{dx}\right)=b\ldots\ldots\ldots\ldots\ldots\ldots\ldots(2),$$

and let the primitive obtained by the elimination of $\frac{dy}{dx}$ be $\Phi(x,\ y,\ a,\ b)=0$. Lagrange has then observed that if we have any differential equation of the first order of the form

$$F\left\{\phi\left(x,\ y,\ \frac{dy}{dx}\right),\ \ \psi\left(x,\ y,\ \frac{dy}{dx}\right)\right\}=0\ldots\ldots(3),$$

its complete primitive will still be $\Phi(x,\ y,\ a,\ b)=0$, but with the condition that a and b are no longer independent constants, but are connected by the relation

$$F(a,\ b)=0.$$

This is an obvious truth. For as, by hypothesis, the supposed primitive $\Phi(x,\ y,\ a,\ b)=0$ gives

$$\phi\left(x,\ y,\ \frac{dy}{dx}\right)=a,\quad \psi\left(x,\ y,\ \frac{dy}{dx}\right)=b,$$

it will convert (3) into $F(a,\ b)=0$, and will therefore *satisfy* that equation if a and b are connected by the relation

$$F(a,\ b)=0.$$

Moreover it contains virtually only one arbitrary constant, for the relation $F(a, b) = 0$ permits us to determine b as a function of a. Hence it will constitute the complete primitive of (3). See also Chap. I. Art. 10.

This result may be expressed in the following theorem.

If any differential equation of the first order be expressible in the form

$$\Phi(\phi, \psi) = 0 \text{ ....} (4),$$

where ϕ and ψ are functions of x, y, $\frac{dy}{dx}$, such that the differential equations

$$\phi = a, \quad \psi = b,$$

are derivable from a single primitive involving a and b as arbitrary constants, the solution of the given differential equation will be found by limiting that primitive by the condition

$$\Phi(a, b) = 0,$$

so as actually or virtually to eliminate one of the arbitrary constants.

Ex. Suppose that the given equation is

$$y\sqrt{\left\{1 + \left(\frac{dy}{dx}\right)^2\right\}} = f\left(x + y\frac{dy}{dx}\right) \text{ ....} (1).$$

Now the differential equations of the first order

$$x + y\frac{dy}{dx} = a \text{ ....} (2),$$

$$y\sqrt{\left\{1 + \left(\frac{dy}{dx}\right)^2\right\}} = b \text{ ....} (3),$$

are derivable from a common primitive. For, on differentiating them, we have respectively

$$1 + \left(\frac{dy}{dx}\right)^2 + y\frac{d^2y}{dx^2} = 0,$$

$$\frac{\frac{dy}{dx}}{\sqrt{\left\{1 + \left(\frac{dy}{dx}\right)^2\right\}}}\left\{1 + \left(\frac{dy}{dx}\right)^2 + y\frac{d^2y}{dx^2}\right\} = 0,$$

and these agree as differential equations of the second order, Chap. I. Art. 9. That common primitive, found by eliminating $\frac{dy}{dx}$ between (2) and (3), is

$$y^2 + (x - a)^2 = b^2.$$

Hence the primitive of the given equation is

$$y^2 + (x - a)^2 = \{f(a)\}^2 \text{.................} (4).$$

We might also proceed as in the solution of Clairaut's equation. Differentiating the given equation, we have

$$\left\{\frac{\frac{dy}{dx}}{\sqrt{\left\{1+\left(\frac{dy}{dx}\right)^2\right\}}} - f'\left(x + y\frac{dy}{dx}\right)\right\} \times \left\{1 + \left(\frac{dy}{dx}\right)^2 + y\frac{d^2y}{dx^2}\right\} = 0.$$

The second factor, which alone involves $\frac{d^2y}{dx^2}$, equated to 0, gives on integration the primitive

$$y^2 + (x - a)^2 = b^2,$$

in which the relation between b and a remains to be determined as before. The first factor equated to 0 constitutes the differential equation of the *singular solution*, which will be obtained by eliminating $\frac{dy}{dx}$ between that equation and the equation given.

Clairaut's equation belongs to the above class. We may express it in the form

$$y - x\frac{dy}{dx} = f\left(\frac{dy}{dx}\right).$$

Now the differential equations

$$y - x\frac{dy}{dx} = a,$$

$$\frac{dy}{dx} = b,$$

generate the same differential equation of the second order

$$\frac{d^2y}{dx^2}=0,$$

and are derivable from the same primitive

$$y=bx+a.$$

Examples of Transformation.

10. Well-chosen transformations facilitate much the solution of differential equations of the first order.

Ex. 1. Given $\dfrac{y-xp}{\sqrt{(1+p^2)}}=f(x^2+y^2)^{\frac{1}{2}}$. *Lacroix*, Tom. II. p. 292.

Assuming $x=r\cos\theta$, $y=r\sin\theta$, we have

$$\frac{-r^2}{\sqrt{\left\{r^2+\left(\frac{dr}{d\theta}\right)^2\right\}}}=f(r),$$

whence

$$\frac{dr}{d\theta}=\frac{r\sqrt{[r^2-\{f(r)\}^2]}}{f(r)}.$$

Consequently

$$\theta=\int\frac{f(r)\,dr}{r\sqrt{[r^2-\{f(r)\}^2]}}+C.$$

As $\dfrac{y-xp}{\sqrt{(1+p^2)}}$ is the expression for the length of the perpendicular let fall from the origin upon the tangent to a curve, the above is the solution of the problem which proposes to determine the equation of a curve in which that perpendicular is a given function of the distance of the point of contact from the origin.

By the same transformation we may solve the equation

$$\frac{y - xp}{\sqrt{(1 + p^2)}} = \sqrt{(x^2 + y^2)}\, f\left\{\frac{x}{\sqrt{(x^2 + y^2)}}\right\}.$$

Ex. 2. Given $\left(\frac{dy}{dx}\right)^k = Ax^\alpha + By^\beta$.

To render the above equation homogeneous if possible, let $y = z^n$; we find

$$\left(nz^{n-1}\frac{dz}{dx}\right)^k = Ax^\alpha + Bz^{n\beta}.$$

This will be homogeneous with respect to z and x, if we have

$$k(n-1) = \alpha = n\beta,$$

equations from which we deduce

$$k = \frac{\alpha\beta}{\alpha - \beta}, \quad n = \frac{\alpha}{\beta},$$

the former of which expresses a condition between the indices of the given equation, the latter the value which must be given to n when that condition is satisfied.

It appears then that the equation

$$\left(\frac{dy}{dx}\right)^{\frac{\alpha\beta}{\alpha-\beta}} = Ax^\alpha + Bx^\beta,$$

can be rendered homogeneous by the assumption $y = z^{\frac{\alpha}{\beta}}$.

If the more general transformation $y = z^n$, $x = t^m$, which seems at first sight to put us in possession of two disposable constants, be employed, the necessity for the fulfilment of the same condition between α, β, and k, will not be evaded, the ratio of the constants m and n, not their absolute values, proving to be alone available.

Ex. 3. The equation of the projection on the plane xy of the lines of curvature of the ellipsoid is

$$Axy\left(\frac{dy}{dx}\right)^2 + (x^2 - Ay^2 - B)\frac{dy}{dx} - xy = 0 \text{.........}(1).$$

Assuming $x^2 = s$, $y^2 = t$, the equation is reduced to one of Clairaut's form, Art. 6. Its solution is

$$y^2 - Cx^2 = -\frac{BC}{AC+1}.$$

The equation may also, without preliminary transformation, be integrated by Lagrange's method, Art. 9. We may express it in the form

$$A\phi\psi + B\phi + \psi = 0 \text{..........................} (2),$$

$$\text{where } \phi = \frac{yp}{x}, \quad \psi = y^2 - ypx.$$

Now $$\frac{y}{x}p = a, \quad y^2 - ypx = b,$$

are derived from a common primitive $y^2 - ax^2 = b$. The solution of (2) will therefore be,

$$y^2 - ax^2 = b$$

with the connecting relation between the constants,

$$Aab + Ba + b = 0.$$

And this will be found to agree with the previous result.

EXERCISES.

The following examples are chiefly in illustration of Arts. 1, 2, and 3.

1. $$\left(\frac{dy}{dx}\right)^2 - 5\left(\frac{dy}{dx}\right) + 6 = 0.$$

2. $$\left(\frac{dy}{dx}\right)^2 - \frac{a^2}{x^2} = 0.$$

3. $$\left(\frac{dy}{dx}\right)^2 = \frac{1-x}{x}.$$

4 $$\left(\frac{dy}{dx}\right)^2 + 2\frac{x}{y}\frac{dy}{dx} - 1 = 0.$$

The following examples are in illustration of Arts. 4 and 5, as well as the preceding Articles.

5. $$y = a\frac{dy}{dx} + b\left(\frac{dy}{dx}\right)^2.$$

6. $$x = a\frac{dy}{dx} + b\left(\frac{dy}{dx}\right)^2.$$

7. $$y = \frac{dy}{dx} + \sqrt{\left\{1 + \left(\frac{dy}{dx}\right)^2\right\}}.$$

8. $$x = \frac{dy}{dx} + \sqrt{\left\{1 + \left(\frac{dy}{dx}\right)^2\right\}}.$$

9. $$\frac{dy}{dx} - \frac{1}{x}\sqrt{\left\{1 + \left(\frac{dy}{dx}\right)^2\right\}} = 0.$$

10. $$x^2\left\{1 + \left(\frac{dy}{dx}\right)^2\right\}^3 - a^2 = 0.$$

11. $$1 + \left(\frac{dy}{dx}\right)^2 = \frac{(x+c)^2}{x^2 + 2cx}.$$

The following examples are intended to illustrate Art. 6. The singular solutions as well as the complete primitives are to be determined.

12. $$y = x\frac{dy}{dx} + \frac{dy}{dx} - \left(\frac{dy}{dx}\right)^2.$$

13. $$y = x\frac{dy}{dx} + \sqrt{\left\{b^2 - a^2\left(\frac{dy}{dx}\right)^2\right\}}.$$

The following examples are in illustration of Arts. 7 and 8.

14. $$y = x\frac{dy}{dx} + x\sqrt{\left\{1 + \left(\frac{dy}{dx}\right)^2\right\}}.$$

15. $$y = x\frac{dy}{dx} + nx\sqrt{\left\{1 + \left(\frac{dy}{dx}\right)^2\right\}}.$$

16. $$x + y\frac{dy}{dx} = m\left(\frac{dy}{dx}\right)^2.$$

17. $$x + y\frac{dy}{dx} = a\sqrt{\left\{1 + \left(\frac{dy}{dx}\right)^2\right\}}.$$

The following examples are in illustration of Art. 9.

18. $$y\frac{dy}{dx} = xf\left\{y^2 - y^2\left(\frac{dy}{dx}\right)^2\right\}.$$

19. $$y - \frac{1}{\sqrt{\left\{1 + \left(\frac{dy}{dx}\right)^2\right\}}} = f\left\{x - \frac{\frac{dy}{dx}}{\sqrt{\left\{1 + \left(\frac{dy}{dx}\right)^2\right\}}}\right\}.$$

20. $$y - 2x\frac{dy}{dx} = f\left\{x\left(\frac{dy}{dx}\right)^2\right\}.$$

21. $$\frac{y - x\frac{dy}{dx}}{y^2 + \frac{dy}{dx}} = f\left(\frac{y - x\frac{dy}{dx}}{1 + x^2\frac{dy}{dx}}\right).$$

CHAPTER VIII.

ON THE SINGULAR SOLUTIONS OF DIFFERENTIAL EQUATIONS OF THE FIRST ORDER.

1. In the largest sense which has been given to the term, a singular solution of a differential equation is a relation between the variables which reduces the two members of the equation to an identity, but which is not included in the complete primitive.

In this sense, the relation obtained by equating to 0 some common algebraic factor of the terms of the equation would be called a singular solution. Thus $x-y=0$ would present itself as a singular solution of the equation

$$(x-y)\,dx+(x-y)\,dy=0.$$

But, in a juster and more restricted sense, a singular solution of a differential equation is a relation between x and y, which satisfies the differential equation *by means of the value which it gives to the differential coefficients* $\frac{dy}{dx}$, $\frac{d^2y}{dx^2}$, &c., but is not included in the complete primitive. In this sense the equation $x^2+y^2=n^2$, is a singular solution of the differential equation of the first order

$$y-x\frac{dy}{dx}=n\sqrt{\left\{1+\left(\frac{dy}{dx}\right)^2\right\}}.$$

It reduces the members of that equation to an identity, but not by causing any algebraic factor of them both to vanish. At the same time it is not included in the complete primitive

$$y-cx=n\sqrt{(1+c^2)}.$$

And this is the juster definition, because that which is essential in the singular solution is thus in a direct manner connected with that which is essential in the differential equation. Def. Chap. I.

When it is said that a singular solution of a differential equation is not included in the complete primitive, it is meant that it is not deducible from that primitive by a particular determination of the arbitrary constant wholly independent of the value of x. The full import of the last clause will hereafter be seen. But although a singular solution is not *included* in the complete primitive, it is still *implied* by it. Upon the possibility of satisfying a differential equation by an infinite number of particular equations, each formed by the particular determination of an arbitrary constant, rests the possibility of satisfying it by another equation, to the formation of which each particular solution has contributed an element. We have seen in Chap. VII. how a singular solution, as representing the envelope of the loci defined by the series of particular solutions, possesses a differential element common with each of them. We shall now see that this property is not accidental—that it is intimately connected with the definition of a singular solution.

It is important that the two marks, positive and negative, by the union of which a singular solution of a differential equation of the first order is characterized, and by the expression of which its definition is formed, should be clearly apprehended. 1st. It must give the same value of $\frac{dy}{dx}$ in terms of x and y, as the differential equation itself does. This is its *positive* mark, a mark which it possesses in common with the complete primitive, and with each included particular primitive. 2ndly. It must not be included in the complete primitive. This is its negative mark. Upon the analytical expression of these characters the entire theory of this class of solutions depends.

Among the different objects to which that theory has reference, the two following are the most important. 1st. The derivation of the singular solution from the complete primitive. 2ndly. The deduction of the singular solution from the differential equation without the previous knowledge of the complete primitive. The theory of the latter process is so dependent upon that of the former that it is necessary to consider them in the order above stated.

Derivation of the singular solution from the complete primitive.

2. Two preliminary propositions are first to be noticed.

PROP. 1. *A differential equation of the first order cannot have two complete primitives,* i.e. *it cannot have two distinct and independent solutions each involving an arbitrary constant.*

If possible let there exist two such distinct primitives represented by

$$\Phi(x, y) = c_1, \quad \Psi(x, y) = c_2 \ldots\ldots\ldots\ldots\ldots\ldots\ldots\ldots (1),$$

supposed to belong to the same differential equation, and therefore giving rise to the same value of $\frac{dy}{dx}$ as a function of x and y. Differentiating these equations, and, for brevity, representing their first members by Φ and Ψ, we have

$$\frac{d\Phi}{dx} + \frac{d\Phi}{dy}\frac{dy}{dx} = 0,$$

$$\frac{d\Psi}{dx} + \frac{d\Psi}{dy}\frac{dy}{dx} = 0,$$

whence, eliminating $\frac{dy}{dx}$ which is by hypothesis common, we must have identically

$$\frac{d\Phi}{dx}\frac{d\Psi}{dy} - \frac{d\Phi}{dy}\frac{d\Psi}{dx} = 0 \ldots\ldots\ldots\ldots\ldots\ldots\ldots\ldots (2),$$

an equation which, by Chap. IV. Art. 3, indicates that Ψ is a function of Φ, and therefore that the supposed primitives are not independent. For if $\Psi = f(\Phi)$, then the primitive $\Phi = c_1$ giving $f(\Phi) = f(c_1)$, gives $\Psi = f(c_1)$, and this, since a function of an arbitrary constant is itself an arbitrary constant, is equivalent to $\Psi = c_2$, so that one of the supposed primitives is seen to be a consequence of the other.

PROP. 2. *A given primitive equation involving x, y, and c, may, by the conversion of the arbitrary constant c into a function of x, be made to represent any proposed relation between x and y.*

Let $\chi(x, y) = 0$ be a proposed equation between x and y, into which it is required to convert the supposed primitive

$$\phi(x, y, c) = 0.$$

Eliminating y between the two equations, we obtain an equation between x and c, expressing the condition under which the given equation $\phi(x, y, c) = 0$ and the proposed equation $\chi(x, y) = 0$ admit of being simultaneously true. From the equation thus found we can determine c as a function of x, and this value of c substituted in the primitive $\phi(x, y, c) = 0$, will reduce it either to the form $\chi(x, y) = 0$, or to a form thereto equivalent.

Ex. Let it be required to convert the equation $y = cx$ into $x^2 + y^2 = 1$, by the conversion of c into a function of x.

Eliminating y from the given and the proposed equation, we have

$$x^2 + c^2x^2 = 1,$$

$$\text{whence } c = \frac{\sqrt{(1 - x^2)}}{x}.$$

This value of c substituted in $y = cx$, converts it into

$$y = \sqrt{(1 - x^2)},$$

which is equivalent to $x^2 + y^2 = 1$.

Cor. The same course of demonstration shews that a primitive containing only y and c may, by the conversion of c into a function of x, be made to represent any proposed relation between x and y. And the *general* proposition evidently is that a primitive containing y *at least*, together with c, may, by the conversion of c into a function of x,—and a primitive containing x at least, together with c, may, by the conversion of c into a function of y,—be made to represent any proposed equation between x and y.

3. Let us now examine the consequences of the above propositions.

From Prop. 1 it appears that if we have obtained a solution under the general form $\phi(x, y, c) = 0$ of a differential equation of the first order, no other solution involving an arbitrary constant exists. Here then the inquiry is suggested whether the differential equation can be satisfied if any other than a constant value be given to c in its complete primitive. By Prop. 2 it is seen that when the primitive involves x, y, and c, or, still more generally, if it involve y at least, together with c, the hypothesis of the variation of c as a function of x is sufficiently wide to embrace every possible supposition as to the nature of the relation between x and y. We propose therefore to inquire whether in the equation

$$\phi(x, y, c) = 0,$$

assumed as primitive, it is possible so to determine c as a function of x, that the resulting expression for $\frac{dy}{dx}$ in the differential equation shall be the same as if c were a constant.

Representing the primitive and the derived equation in the forms

$$y = f(x, c), \quad \frac{dy}{dx} = \frac{df(x, c)}{dx} \quad \text{..................} \quad (1),$$

the differential equation is obtained from these by eliminating c. It will therefore be unaffected by any change in the *nature* of c, provided that the *form* of the relation between x, y, and c in the primitive and between $\frac{dy}{dx}$, x, and c in the derived equation remains unchanged.

Now differentiating the primitive on the hypothesis that c is a function of x, and representing the differential coefficient of c thus considered by $\left(\frac{dc}{dx}\right)$, we have

$$\frac{dy}{dx} = \frac{df(x, c)}{dx} + \frac{df(x, c)}{dc}\left(\frac{dc}{dx}\right) \quad \text{........................} \quad (2).$$

And this will agree in form with the expression for $\frac{dy}{dx}$ in (1) if $\frac{df(x, c)}{dc}\left(\frac{dc}{dx}\right)=0$. But to suppose $\left(\frac{dc}{dx}\right)=0$ would be to suppose c a constant and to return to the ordinary primitive. It remains therefore that for a singular solution we have

$$\frac{df(x, c)}{dc}=0, \quad \text{or } \frac{dy}{dc}=0 \text{..............(3)}.$$

This is the first analytical condition. What it means is that if a fixed value be given to x in the primitive, y must not vary for an infinitesimal variation of c. And by this condition c is to be determined as a function of x.

Now the substitution of any function of x for c in a primitive which contains y at least, cannot lead to a resulting equation not containing y, though it may lead to a resulting equation not containing x. Hence the condition $\frac{dy}{dc}=0$ can only lead to those singular solutions in the expression of which y at least is involved. Had we reduced the primitive to the form $x=f(y, c)$ we should, as is evident from the principle of symmetry, have arrived at the analytical condition

$$\frac{dx}{dc}=0 \text{..............................(4)},$$

a condition by which c would be determined as a function of y. And the substitution of such value or values of c in the primitive would lead to all singular solutions in the expression of which x at least is involved.

It will be remembered that what is essential to a singular solution is that c should not admit of determination as a constant wholly independent of the variables. But whether it be determined as a function of x or as a function of y is indifferent. The one form is usually, but not always, convertible into the other by means of the primitive. Thus, if the primitive be in the form $\phi(x, y, c)=0$, and c be determined in the form $c=f(y)$, the elimination of y between these equations will generally enable us to determine c as a function of x; but it will not do so if, in the elimination of y, c should disappear.

Thus if the primitive were

$$x = (y - c)^2,$$

the value of c determined as a function of y by the condition $\frac{dx}{dc} = 0$ would be $c = y$, and *this* value of c is not expressible by means of x, for on attempting to eliminate y between the above equations c also disappears. Nor is it indeed possible in the above case to satisfy the condition $\frac{dy}{dc} = 0$. Hence it is necessary in establishing a general method to take account of *both* the conditions (3) and (4).

And these conditions are sufficient. No other is implied. The comparison of (1) and (2) from which the condition $\frac{dy}{dc} = 0$ was derived, leads also to the condition $\frac{dx}{dc} = 0$, but not to any other condition. The expressions which they furnish for $\frac{dy}{dx}$ become equivalent in two cases only, viz. 1st if $\frac{df(x, c)}{dc} = 0$, the case first considered; 2ndly, if without supposing $\frac{df(x, c)}{dc} = 0$, we have $\frac{df(x, c)}{dc}\left(\frac{dc}{dx}\right)$ infinitesimal in comparison with $\frac{df(x, c)}{dx}$, and therefore if we have

$$\frac{df(x, c)}{dc} \div \frac{df(x, c)}{dx} = 0 \quad\ldots\ldots\ldots\ldots\ldots\ldots (5),$$

for, c being regarded as a function of x, and therefore variable, the factor $\left(\frac{dc}{dx}\right)$ cannot be continuously infinite. Now differentiating the equation $y = f(x, c)$ we have

$$dy = \frac{df(x, c)}{dx} dx + \frac{df(x, c)}{dc} dc \quad\ldots\ldots\ldots\ldots (6).$$

Hence, if we make $dy = 0$, we have

$$\frac{dx}{dc} = -\frac{df(x, c)}{dx} \div \frac{df(x, c)}{dx} \text{ ....(7)},$$

so that (5) assumes the form $\frac{dx}{dc} = 0$. But, as a *demonstration* of this condition, the above method is less general than the previous one, for it assumes the possibility of expressing as a function of x the value of c determined by the condition $\frac{dx}{dc} = 0$. Now that value is primarily a function of y, and may not be expressible at all by means of x.

It is well to note that the final criteria $\frac{dy}{dc} = 0$, $\frac{dx}{dc} = 0$ are in effect analytical expressions of what logicians term conditional propositions. The former expresses that *if* x is assumed constant, y will not vary for an infinitesimal variation of c; the latter that *if* y be assumed constant, x will not vary for an infinitesimal variation of c.

4. We have shewn that each of these conditions has its special case of failure. It may be proper to shew that except in such cases of failure they are equivalent.

As expressed by means of the primitive $y = f(x, c)$, these conditions assume the forms

$$\frac{df(x, c)}{dc} = 0, \quad \frac{df(x, c)}{dc} \div \frac{df(x, c)}{dx} = 0,$$

or

$$\frac{dy}{dc} = 0, \quad \frac{dy}{dc} \div \frac{dy}{dx} = 0,$$

and these are equivalent unless $\frac{dy}{dx}$ be 0 or infinite.

But $\frac{dy}{dx} = 0$ implies that the singular solution is of the form

$$y = \text{a definite constant},$$

and this is precisely that form of singular solution which the condition $\frac{dx}{dc}=0$ fails to give.

Similarly $\frac{dy}{dx}=\infty$, being equivalent to $\frac{dx}{dy}=0$, implies that the singular solution is of the form

$$x = \text{a definite constant},$$

and this is that form of singular solution which the condition $\frac{dy}{dc}=0$ fails to give.

Thus the conditions $\frac{dy}{dc}=0$, $\frac{dx}{dc}=0$, although not necessarily equivalent, do not lead to *conflicting* results.

When we cannot solve the primitive equation with respect to y and x so as to enable us to form directly the expressions for $\frac{dy}{dc}$ and $\frac{dx}{dc}$, we may proceed thus. Representing the primitive by $\phi=0$, we have on differentiation

$$\frac{d\phi}{dx}dx+\frac{d\phi}{dy}dy+\frac{d\phi}{dc}dc=0.$$

Hence, remembering what is meant by $\frac{dy}{dc}$ and $\frac{dx}{dc}$,

$$\frac{dy}{dc}=-\frac{\frac{d\phi}{dc}}{\frac{d\phi}{dy}},\quad \frac{dx}{dc}=-\frac{\frac{d\phi}{dc}}{\frac{d\phi}{dx}}\quad\text{.......................(8)},$$

and the second members of these equations must be equated to 0.

We see that these second members will usually vanish if $\frac{d\phi}{dc}=0$. And this equation $\frac{d\phi}{dc}=0$ is adopted by most writers as the sole expression of the rule for the derivation of the singular solution from the complete primitive, unrestricted by any accompanying condition. (Lagrange, *Calcul des Fonctions*, p. 207). Thus stated however it can only mislead. The vanishing of $\frac{dy}{dc}$ or $\frac{dx}{dc}$ in (8) may be due not to the vanishing of the numerator $\frac{d\phi}{dc}$, but to the assumption of an infinite value by the denominator $\frac{d\phi}{dy}$ or $\frac{d\phi}{dx}$. The latter is indeed quite as probable a cause as the former when ϕ is not expressed as a rational and integral function of x and y. And even when ϕ is thus expressed the condition $\frac{d\phi}{dc}=0$ may fail through its involving a factor contained in $\frac{d\phi}{dy}$ or $\frac{d\phi}{dx}$. We conclude that while the true tests of a singular solution are $\frac{dy}{dc}=0$ and $\frac{dx}{dc}=0$, any subsidiary conditions such as $\frac{d\phi}{dc}=0$, $\frac{d\phi}{dy}=\infty$, $\frac{d\phi}{dx}=\infty$, are only to be used for purposes of convenience, and never without reference to the more fundamental relations of which they take the place.

The following is a legitimate example of the application of the subsidiary condition $\frac{d\phi}{dc}=0$.

The complete primitive of the differential equation $\frac{dy}{dx}=2y^{\frac{1}{2}}$ is $y=(x-c)^2$. Here $\phi=y-(x-c)^2$, and, this being rational and integral, the condition $\frac{d\phi}{dc}=0$ gives $2(x-c)=0$, whence $c=x$, a value which, substituted in the primitive, gives $y=0$ a singular solution.

The condition $\frac{dy}{dc}=0$ also gives $c=x$, and leads to the same result. But, since the primitive solved with respect to x gives $x=c+y^{\frac{1}{2}}$, the condition $\frac{dx}{dc}=0$ cannot be satisfied. Thus the singular solution is here obtained by means of the condition $\frac{dy}{dc}=0$, and not by the condition $\frac{dx}{dc}=0$.

5. The chief results of the above investigation are combined in the following Theorem.

THEOREM. *Every singular solution of a differential equation of the first order may be deduced from its complete primitive by giving therein to c a variable value determined from that primitive by either or both of the equations*

$$\frac{dy}{dc}=0,\ \frac{dx}{dc}=0 \quad\quad (1).$$

And any solution which is thus obtained, and which cannot be also obtained by giving to c in the primitive a constant value, is a singular solution.

The conditions (1) are equivalent, except when one only of the variables x and y is involved in the singular solution; solutions involving only the variable y resulting only from the condition $\frac{dy}{dc}=0$, *and those involving only the variable x resulting only from the condition* $\frac{dx}{dc}=0$.

When the primitive, represented by $\phi=0$, *is rational and integral we may for convenience employ the single condition* $\frac{d\phi}{dc}=0$; *but never without reference to the fundamental conditions* (1).

In the statement of the above theorem the two following particulars should be noticed.

1st. It supposes c to be determined as a variable quantity. Now if c be obtained as a function of both x and y, as it generally will be if the condition $\frac{d\phi}{dc}=0$ be made use of, it may be necessary by a subsequent elimination to reduce it to a function of *one* of the variables, in order to assure ourselves that it is not constant in virtue of the relation between x and y established in the primitive.

2ndly. The theorem takes account equally of the positive and of the negative characters of a singular solution. The existence of a variable value of c determined by either of the conditions (1) does not assure us that the resulting solution is singular, unless constant values of c are at the same time excluded.

Ex. 1. The equation $y^2 - 2xy\frac{dy}{dx} + (1+x^2)\left(\frac{dy}{dx}\right)^2 = 1$, has for its complete primitive $y = cx + \sqrt{(1-c^2)}$. Its singular solution is required.

Here $\frac{dy}{dc} = x - \frac{c}{\sqrt{(1-c^2)}}$. Hence $\frac{dy}{dc} = 0$ gives for c the *variable* value $c = \frac{x}{\sqrt{(x^2+1)}}$, the substitution of which in the primitive gives

$$y = \sqrt{(x^2+1)} \text{ .... (1).}$$

This value of y satisfies the given differential equation, and it is evident on inspection that it is not included in the complete primitive. Formally to establish this, we find on eliminating y between that equation and (1)

$$cx + \sqrt{(1-c^2)} = \sqrt{(x^2+1)};$$

solving which with respect to c, we have the unique value $c = \frac{x}{\sqrt{(x^2+1)}}$, which, agreeing with the value of c before employed, shews that c admits of no other value, and in particular that it admits of no constant value. The solution is therefore singular.

The condition $\frac{dx}{dc} = 0$ would, in the above example, give $c = \frac{(y^2+1)^{\frac{1}{2}}}{y}$, and lead to the same final result.

We must be careful not to rely upon the condition $\frac{d\phi}{dc}=0$, except under the circumstances specified in the general theorem. This remark will be illustrated in the following example.

Ex. 2. The complete primitive of the differential equation $y=px+\frac{m}{p}$, where p stands for $\frac{dy}{dx}$, is $y-cx-\frac{m}{c}=0$, and, if we represent its first member by ϕ, the elimination of c between the equations $\phi=0$, $\frac{d\phi}{dc}=0$, gives the singular solution $y^2=4mx$.

But if we reduce the primitive by solution to the form

$$\frac{y\pm\sqrt{(y^2-4mx)}}{x}-2c=0,$$

and then represent its first member by ϕ, we shall have

$$\begin{aligned}\frac{dy}{dc}&=-\frac{d\phi}{dc}\div\frac{d\phi}{dy}\\ &=-2\div\left\{\frac{1}{x}\pm\frac{y}{x\sqrt{(y^2-4mx)}}\right\}.\end{aligned}$$

And here the singular solution $y^2-4mx=0$, before obtained, is seen to be dependent, not upon the vanishing of $\frac{d\phi}{dc}$, but upon the assumption of an infinite value by $\frac{d\phi}{dy}$.

The true ground of preference for the conditions $\frac{dy}{dc}=0$, $\frac{dx}{dc}=0$, consists, however, not in the directness of their application to irrational forms of the primitive, but in the plainness of their geometrical interpretation, and still more in their fundamental relation to the problem of the derivation of the singular solution from the differential equation—points hereafter to be discussed.

The following example is intended to illustrate that portion of the theorem which relates to the negative character of a singular solution.

Ex. 3. The complete primitive of the differential equation

$$\left(\frac{dy}{dx}\right)^3 - 4xy\frac{dy}{dx} + 8y^2 = 0,$$

is $y = c(x-c)^2$. The singular solution is required.

Here the condition $\dfrac{dy}{dc} = 0$ gives

$$(x-c)(x-3c) = 0,$$

whence $c = x$, or $\dfrac{x}{3}$. These values of c, both of which are variable, reduce the primitive to the forms

$$y = 0, \quad y = \frac{4x^3}{27},$$

and both these are solutions of the differential equation. But while the latter of the two is not included in the complete primitive, the former is included in it. If between the equations

$$y = c(x-c)^2, \quad y = 0,$$

we eliminate y, the resulting values of c will be

$$c = 0, \quad c = x.$$

We see therefore that the solution to which we were led by the assumption $c = x$ was a particular integral. The only singular solution is $y = \dfrac{4x^3}{27}$.

Geometrical Interpretation.

6. Let $y=f(x, c)$ represent a family of curves the individual members of which are determined by giving different values to c. Then, adopting for a moment the language of infinitesimals, the differentiation of y with respect to c implies the transition from an ordinate y of one curve to an ordinate $y+\frac{dy}{dc}dc$, corresponding to the *same* value of x, but belonging to another curve of the series; viz. the curve obtained by changing c into $c+dc$.

When we impose the condition $\frac{dy}{dc}=0$, we demand that this transition shall not affect the value of the ordinate y corresponding to a value of x determined by the equation $\frac{dy}{dc}=0$.

Hence the singular equation obtained by the elimination of c between the equations $y=f(x, c)$, $\frac{dy}{dc}=0$, represents the *locus* of such points of successive intersection.

In stricter language, the singular solution represents the locus of those points which constitute the *limits* of position of the points of *actual* intersection of the different members of the family of curves represented by the equation $y=f(x, c)$, always excepting the case in which that locus coincides with a particular curve of the system.

And as at these limiting points the value of $\frac{dy}{dx}$ is the same for the locus of the singular solution and the loci of primitives, it follows that the former has *contact* with each of the latter. The locus of the singular solution is seen to be the envelope of the loci of primitives. The envelope of the loci of primitives is the locus of a singular solution, except when it coincides with one of the particular loci, of which it forms the connecting bond.

Similar observations may be made with reference to the condition $\frac{dx}{dc}=0$.

Derivation of the singular solution from the differential equation.

7. We have found that the singular solution of a differential equation considered as derived from its complete primitive possesses the following characters.

A. It satisfies one of the conditions $\frac{dy}{dc}=0$, $\frac{dx}{dc}=0$.

B. It is not possible to deduce it from the complete primitive by giving to c a constant value.

It has also been shewn that the conditions in A are equivalent except when the singular solution involves only one of the variables in its expression.

Now we shall endeavour to translate the above characters from a language whose elements are x, y, and c to a language whose elements are x, y, and $\frac{dy}{dx}$,—from the language of the complete primitive to the language of the differential equation.

If we differentiate with respect to x the complete primitive expressed in the form

$$y=f(x, c) \quad\quad\quad (1),$$

we obtain the *derived* equation

$$p=\frac{df(x, c)}{dx} \quad\quad\quad (2),$$

and substituting in this for c its expression in terms of x and y given by the primitive (1), we have finally the *differential equation* in the form

$$p=\phi(x, y) \quad\quad\quad (3).$$

Thus the differential equation (3) is the same as the derived equation (2), provided that c be considered therein as a function of x and y determined by (1).

Accordingly we have

$$\frac{dp}{dy} \text{ in } (3)=\frac{dp}{dc} \text{ in } (2) \times \frac{dc}{dy} \text{ in } (1),$$

or $$\frac{dp}{dy} \text{ in } (3) = \frac{d^2 f(x, c)}{dx dc} \div \frac{df(x, c)}{dc};$$

since in (1) $$\frac{dc}{dy} = 1 \div \frac{dy}{dc} = 1 \div \frac{df(x, c)}{dc}.$$

Hence $$\frac{dp}{dy} \text{ in } (3) = \frac{d}{dx} \log \frac{df(x, c)}{dc},$$

or finally $$\frac{dp}{dy} = \frac{d}{dx} \log \frac{dy}{dc} \dots\dots\dots\dots\dots\dots (4),$$

provided that the value of the first member be derived from the differential equation, that of the second member from the complete primitive.

In like manner if we suppose the complete primitive expressed in the form

$$x = \phi(y, c),$$

we shall have through symmetry the relation,

$$\frac{d}{dx}\left(\frac{1}{p}\right) = \frac{d}{dy} \log \frac{dx}{dc} \dots\dots\dots\dots\dots (5),$$

the first member referring to the differential equation, the second to the complete primitive.

The equations (4) and (5), which are rigorous and fundamental, establish a connexion between the differential equation and the complete primitive, and it now only remains to introduce the conditions $\frac{dy}{dc} = 0$, $\frac{dx}{dc} = 0$. We begin with the former.

We have seen that when $\frac{dy}{dc} = 0$ leads to a singular solution it does so by enabling us to determine c as a function of x, suppose $c = X$. Before proceeding to more general considerations it will be instructive to make a particular hypothesis as to the *form* of the equation $\frac{dy}{dc} = 0$.

Suppose then this equation to be of the form

$$Q(c-X)^m=0 \dots\dots\dots\dots\dots\dots (6),$$

m being a positive constant and Q a function of x and c, which neither vanishes nor becomes infinite when $c=x$ This hypothesis is at least sufficiently general to include all the cases in which $\frac{dy}{dc}=0$ is algebraic.

By (6) we have then

$$\frac{dp}{dy}=\frac{d}{dx}\log\frac{dy}{dc}=\frac{\frac{dQ}{dx}}{Q}-m\frac{\frac{dX}{dx}}{c-X} \dots\dots\dots (7),$$

and the second term of the right-hand member having $c-X$ for its denominator and not containing c at all in its numerator, is infinite. At the same time, we see that no such infinite term would present itself were c determined as a constant.

For let $\frac{dy}{dc}=Q(c-a)^m$, then $\frac{d}{dx}\log\frac{dy}{dc}=\frac{dQ}{dx}\div Q$, the right-hand member of (7) being now reduced to its first term.

The conclusion to which this points is that $\frac{dp}{dy}$ is infinite for a singular solution, but finite for a particular integral.

Again, suppose the value of c in terms of x and y furnished by algebraic solution of the complete primitive to be $c=\phi(x, y)$, then substituting this value in the equation $c-X=0$, we obtain the singular solution in the form

$$\phi(x, y)-X=0.$$

Now the same substitution gives to the infinite term in the value of $\frac{dp}{dy}$ the form

$$\frac{-m\frac{dX}{dx}}{\phi(x, y)-X} \dots\dots\dots\dots\dots\dots (8).$$

We see then, in the case of a singular solution corresponding to a determination $c = X$, that $\frac{dp}{dy}$ as derived from the differential equation becomes infinite owing to $\phi(x, y) - X$ occurring in a denominator. And, whatever modification of form may be made by clearing of fractions or radicals, we may still infer that, if $u = 0$ be a singular solution derived from an algebraic primitive, the function $\frac{dp}{dy}$ will become infinite, owing to u presenting itself under a negative index.

The analysis does not however warrant the conclusion that *any* relation between x and y which makes $\frac{dp}{dy}$ infinite will be a solution. If m be a negative constant, the second term in the expression of $\frac{dp}{dy}$ is still infinite, but the prior condition $\frac{dy}{dc} = 0$ is no longer satisfied. All we can affirm is that if $\frac{dp}{dy} = \infty$ gives a solution at all it will be a singular solution.

Since $\frac{dx}{dy} = \frac{1}{p}$, it is evident that a singular solution originating in a determination of c in the form $c = Y$ will make $\frac{d}{dx}\left(\frac{1}{p}\right)$ infinite.

A contrast between the conditions $\frac{dy}{dc} = 0$, $\frac{dx}{dc} = 0$, and the conditions $\frac{dp}{dy} = \infty$, $\frac{d}{dx}\left(\frac{1}{p}\right) = \infty$, is also developed. The former lead to solutions, but not necessarily to singular solutions; the latter do not necessarily lead to solutions, but when they do, those solutions are singular.

Ex. 1. Given $p^2 - 2xp + 2y = 0$.

Here
$$p = x \pm \sqrt{(x^2 - 2y)},$$
$$\frac{dp}{dy} = \mp (x^2 - 2y)^{-\frac{1}{2}},$$

which becomes infinite if $y=\frac{x^2}{2}$, and this satisfies the differential equation. It is therefore a singular solution.

It may be objected against the above reasoning, not only that it involves an assumption as to the form of the equation $\frac{dy}{dc}=0$, but also that it takes no account of any possibilities arising from the first term in the expression of $\frac{dp}{dy}$. But it serves well to illustrate what, in the vast majority of instances, is the *actual* mode of transition from the one set of conditions to the other. We proceed to consider the question in a more strict and general manner.

8. When $\frac{dy}{dc}=0$ determines c as a function of x, it reciprocally determines x as a function of c, so that if a definite value be given to c, a corresponding definite value or values will be given to x. Let $\frac{dy}{dc}$ be represented by $\psi(x, c)$, then

$$\frac{dp}{dy}=\frac{d}{dx}\log\frac{dy}{dc}$$

$$=\text{limit of } \frac{\log\psi(x+h, c)-\log\psi(x, c)}{h} \dots\dots(9),$$

h approaching to 0.

Now for a singular solution $\psi(x, c)=0$, and this being, from what precedes, satisfied only by definite values of x, corresponding to our assumed definite value of c, it follows that $\psi(x+h, c)$ will not be equal to 0 for any continuous series of values of h however small; neither then will $\log\psi(x+h, c)$ retain continuously the value of $\log\psi(x, c)$, viz. $-\infty$. Thus the numerator of the fraction in the second member being equal to the difference between a finite and an infinite quantity is infinite, and the limit of the fraction therefore infinite. Hence we conclude that a singular solution considered as derived from the primitive by the conversion of c into a function of x, satisfies relatively to the differential equation the condition

$$\frac{dp}{dy}=\infty.$$

And in the same way it may be shewn that a singular solution derivable from the primitive by the conversion of c into a function of y satisfies the condition $\frac{d}{dx}\left(\frac{1}{p}\right)=\infty$.

Changing the order of the enquiry, let us now examine whether there exist any other forms of solution satisfying the condition $\frac{dp}{dy}=\infty$, $\frac{d}{dx}\left(\frac{1}{p}\right)=\infty$. If there be, it will be made evident that more is involved in the definition of a singular solution than we have yet recognized in our processes of deduction, or else that the definition must be enlarged.

Expressing the condition $\frac{dp}{dy}=\infty$, in the form

$$\frac{d}{dx}\log\frac{dy}{dc}=\infty \ldots\ldots\ldots\ldots\ldots (10),$$

we observe that it can be satisfied only in one of two ways, viz. either independently of c, or by some determination of c, and if the latter again only in one of two ways, viz. either by the determination of c as a function of x, or by the determination of c as a constant.

We may pass over the case in which the above equation is satisfied independently of c, because the relation obtained would involve x only, while it is a condition accompanying the use of $\frac{dp}{dy}=\infty$ that it leads to solutions involving y at least. We may also pass over the case in which it is satisfied by the assumption $c=X$, because such a value of c, if it lead to a solution at all, can only do so by satisfying the condition $\frac{dy}{dc}=0$, and thus lead to the form of singular solution already investigated. There remains only the case in which the equation (10) is satisfied by a constant value of c.

Let then the equation (10) be satisfied by $c=a$. The most general assumption we can make respecting the form of its first member is the following, viz.

$$\frac{d}{dx}\log\frac{dy}{dc}=\phi(c)\,\psi(x, c),$$

where $\phi(c)$ is a function of c which becomes infinite when c assumes the constant value in question, and $\psi(x, c)$ becomes neither 0 nor infinite for such value. Hence the most general form of $\log \frac{dy}{dc}$ is

$$\log \frac{dy}{dc} = \int \phi(c)\, \psi(x, c)\, dx = \phi(c) \int \psi(x, c)\, dx.$$

To give to this expression the utmost generality, we must, on effecting the integration with respect to x, add an arbitrary function of c. Thus we shall have

$$\log \frac{dy}{dc} = \phi(c) \{\int \psi(x, c)\, dx + \chi(c)\}.$$

Therefore $$\frac{dy}{dc} = \epsilon^{\phi(c) \{\int \psi(x, c)\, dx + \chi(c)\}},$$

or, representing the function $\int \psi(x, c)\, dx + \chi(c)$ by $\Phi(x, c)$,

$$\frac{dy}{dc} = \epsilon^{\phi(c)\, \Phi(x, c)} \quad \text{.........................} (11).$$

This is the most general form of $\frac{dy}{dc}$, as determined from the primitive, which is consistent with the hypothesis that $\frac{d}{dx} \log \frac{dy}{dc}$ becomes infinite for a constant value of c. Accordingly if, supposing the primitive to be given, we sought o determine the singular solution by the condition $\frac{dy}{dc} = 0$, we should be led to an equation of the form

$$\epsilon^{\phi(c)\, \Phi(x, c)} = 0,$$

or $$\phi(c)\, \Phi(x, c) = -\infty \quad \text{.................} (12).$$

Now this equation is not satisfied by any value of c which makes $\phi(c)$ infinite, unless it give to $\Phi(x, c)$ an opposite sign to that of $\phi(c)$. But this indicates in general the existence of a *relation* between x and c. Thus suppose

$$\phi(c) = c, \quad \Phi(x, c) = x.$$

Then (12) becomes

$$cx = -\infty,$$

which demands that c should receive the value $-\infty$ or $+\infty$ according as x is positive or negative. In either case c is constant, but it is a *dependent* constant—dependent for its sign upon the sign of x. Thus the condition $\frac{dp}{dy} = \infty$ may indicate the existence of a species of singular solution derived from the complete primitive by regarding c, not as a continuous function of x, but as a discontinuous constant, the law of its discontinuity being however such as to connect it with the variations of x.

Ex. 2. Given $p = \frac{y \log y}{x}$.

Here we find

$$\frac{dp}{dy} = \frac{1}{x}(1 + \log y) \text{ .... } (13),$$

which is infinite if $y = 0$. And this proves on trial to be a solution of the differential equation, the true value of the indeterminate function in the second member when $y = 0$ being 0 (Todhunter's *Diff. Cal.* p. 123). Now the complete primitive is $y = \epsilon^{cx}$. Hence we see that $y = 0$ is not a particular integral in the strict sense of that term. The value to be assigned to c is not *wholly* independent of x. We may therefore regard $y = 0$ as a singular solution satisfying the condition $\frac{dp}{dy} = \infty$.

9. We have said that, in general, the equation (12) indicates the existence of a relation between x and c. A case of exception however exists. Representing $\phi(c)$ by C, suppose $\Phi(x, c)$, expressed in terms of x and C, to be capable of development in descending powers of C: suppose, too, that

the first term of the development is of the form AC^r, where A is constant and $r > -1$. Then as C approaches infinity, (12) tends to assume the form

$$AC^{r+1} = -\infty,$$

indicating that C, and therefore c, possesses more than one value, real or imaginary. Here, then, the condition $\frac{dp}{dy} = \infty$ would accompany a solution possessing this singularity, viz. that it corresponds to a *multiple value* of c, the arbitrary constant in the complete primitive. It is in fact a species of *multiple particular integral.*

Ex. 3. Given $p^2 - pxy + y^2 \log y = 0$.

Here $$p = \frac{xy \pm y\sqrt{(x^2 - 4\log y)}}{2};$$

$$\therefore \frac{dp}{dy} = \frac{x \pm \sqrt{(x^2 - 4\log y)}}{2} \mp \frac{1}{\sqrt{(x^2 - 4\log y)}} \quad \text{........ (14)},$$

and this is satisfied by $y = 0$ and by $x^2 - 4\log y = 0$, that is by

$$y = 0, \quad y = \epsilon^{\frac{x^2}{4}}.$$

Both these satisfy the differential equation, and the second is obviously a singular solution. To determine the nature of the first let it be observed that the complete primitive is

$$y = \epsilon^{cx - c^2},$$

and that this reduces to $y = 0$, irrespectively of the value of x, by the assumptions $c = +\infty$ and $c = -\infty$. Now this is the *only* case in which two particular integrals agree. We might in *any* case, by changing in the complete primitive of an equation c into c^2, get two values of c for a particular integral, but then it would be for *every* particular integral. It is only when the property is *singular*, that the condition $\frac{dp}{dy} = \infty$ is satisfied.

It is obvious that one negative feature marks all the cases in which a solution involving y satisfies the condition $\frac{dp}{dy}=\infty$. It is, that the solution, while expressed by a single equation, is not connected with the complete primitive by a single and absolutely constant value of c. In the first, or as it might be termed *envelope* species of singular solutions, c receives an infinite number of different values connected with the values of x by a law. In the second it receives a finite number of values also connected with the values of x by a law. In the third species it receives a finite number of values, determinate, but not connected with the values of x.

If we observe that all the above cases, while agreeing in the point which has been noted, possess true singularity, we shall be led to the following definition.

DEFINITION. A singular solution of a differential equation of the first order is a solution, the connexion of which with the complete primitive does not consist in the giving to c of a single constant value absolutely independent of the value of x.

Criterion of species.

10. It is a question of some interest to determine whether a given singular solution, $u=0$, of a differential equation, is of the envelope species or not.

On the particular hypotheses assumed in Art. 7, it is shewn that singular solutions of the envelope species possess the following character, viz. if $u=0$ be such a solution, then $\frac{dp}{dy}$ becomes infinite through containing a term in which u is presented under a negative index.

Now inquiries which are scarcely of a sufficiently elementary character to find a place in this work, indicate (with very high probability) that this character is universal and independent of any particular hypothesis, and that it constitutes a *criterion* for distinguishing solutions of the envelope species from others.

As an example of an hypothesis different from that of Art. 7, let us suppose

$$\frac{dy}{dc} = \frac{Q}{\log (c - X)},$$

which vanishes when $c = x$

We find

$$\frac{d}{dx} \log \frac{dy}{dc} = \frac{\frac{dQ}{dx}}{Q} + \frac{\frac{dX}{dx}}{(c-X)\log (c-X)}.$$

The second term in the right-hand member becomes indeterminate when $c = X$, but its true value is ∞, and it assumes this value in consequence of $c - X$ presenting itself with a negative index. We remark that the fraction $\frac{1}{\log (c-X)}$ is one which vanishes with $c - X$ in *whatever manner* $c - X$ approaches to 0,—a consideration which is quite of essential importance.

Applying the above criterion to some of the previous examples, we see from the form of $\frac{dp}{dy}$ in Ex. 1, Art. 7, that the singular solution belongs to the envelope species; in (13) Art. 8, it is implied that the solution is not of that species; in (14) Art. 9 two species are indicated, the solution $y = 0$ resulting from $\log y = -\infty$ being not of the envelope species, while the other solution is of that species.

11. The collected results of the above analysis are contained in the following theorem.

THEOREM. *The singular solutions of a differential equation of the first order* (Def. Art. 9) *consist of all relations which belong to one or both of the following classes, viz.*

1st. *Relations involving y, with or without x, which make $\frac{dp}{dy}$ infinite and only infinite, and satisfy the differential equation.*

2nd. *Relations involving x, with or without y, which make $\frac{d}{dx}\left(\frac{1}{p}\right)$ infinite and only infinite, and satisfy the differential equation.*

When a solution as above defined is actually obtained by equating to 0 a factor which appears under a negative index in the expression of $\frac{dp}{dy}$ or $\frac{d}{dx}\left(\frac{1}{p}\right)$ it may be considered to belong to the envelope species of singular solutions. In other cases it is deducible from the complete primitive by regarding c as a constant of multiple value,—its particular values being either 1st dependent in some way on the value of x, or 2ndly independent of x, but still such as to render the property a singular one.

We may add that there exist cases in which the characters of different species of solutions seem to be blended together. Thus $\frac{dp}{dy}$ may admit of both a finite and an infinite value, indicating a duplex genesis of the solution from the complete primitive. It may also happen that the assumption of an infinite value by $\frac{dp}{dy}$ may be attributed, indifferently, either to a negative index or to a logarithm. And then it should be inquired whether or not the solution is of the envelope species, but marked with some peculiarity arising from a breach of continuity in the mode of its derivation from the complete primitive.

The following examples are intended to elucidate particular points either of theory or of method.

Ex. 1. Given $(1+x^2)\left(\frac{dy}{dx}\right)^2 - 2xy\frac{dy}{dx} + y^2 - 1 = 0.$

This equation, first discussed in Brooke Taylor's *Methodus Incrementorum*, is remarkable as having afforded the earliest instance of the actual deduction of a singular solution from a differential equation (Lagrange, *Calcul des Fonctions*, p. 276). We shall first explain Taylor's procedure, and afterwards apply the above general Theorem.

Taylor differentiates the equation, and finding

$$\left\{2(1+x^2)\frac{dy}{dx}-2xy\right\}\frac{d^2y}{dx^2}=0,$$

resolves this into the two equations

$$(1+x^2)\frac{dy}{dx}-xy=0,\quad \frac{d^2y}{dx^2}=0\ldots\ldots\ldots\ldots (1).$$

The second of these gives $y=ax+b$, which satisfies the differential equation provided that $b=\sqrt{(1-a^2)}$. Thus the complete primitive is

$$y=ax+\sqrt{(1-a^2)}.$$

The first equation of (1) gives, on eliminating $\frac{dy}{dx}$ by means of the differential equation,

$$y^2=x^2+1,$$

and this he terms the singular solution, (*singularis quædam solutio problematis*).

To apply to this example the general method, we find

$$p=\frac{xy\pm\sqrt{(x^2-y^2+1)}}{x^2+1}.$$

Hence, $$\frac{dp}{dy}=\frac{1}{x^2+1}\left\{x\mp\frac{y}{\sqrt{(x^2-y^2+1)}}\right\}.$$

Introducing the condition $\frac{dp}{dy}=\infty$, we should *apparently* have the equations

$$x^2-y^2+1=0,$$
$$x^2+1=0,$$

but of the second of these, as it does not involve y in its expression, no account is to be taken. The first making $\frac{dp}{dy}$ infinite whether the upper or the lower sign be taken, and satisfying the differential equation, is a singular solution. Again, as also it is derived from the vanishing of a function under a negative index, it belongs to the envelope species.

We may add that it might be found but less readily from the condition $\frac{d}{dx}\left(\frac{1}{p}\right) = \infty$.

The following example is intended to illustrate the use of the latter condition.

Ex. 2. Given $\frac{dy}{dx} = x^{-n}$.

Hence, since $p = x^{-n}$, the condition $\frac{dp}{dy} = \infty$ cannot be satisfied.

The condition $\frac{d}{dx}\left(\frac{1}{p}\right) = \infty$ gives

$$nx^{n-1} = \infty,$$

and this is satisfied by $x = 0$ if n be less than 1, but is not satisfied by $x = 0$ if n be equal to or greater than 1.

Now the differential equation is satisfied by $x = 0$, whatever positive value we give to n, as may be seen by expressing it in the form $\frac{dx}{dy} = x^n$. We conclude therefore that $x = 0$, is a singular solution of the proposed equation if n be positive and less than 1, but a particular integral if n be equal to or greater than 1. We infer too that the solution, when singular, belongs to the envelope species.

In verification, it may be observed that, if n be not equal to 1, the complete primitive is

$$y = \frac{x^{1-n}}{1-n} + c,$$

or

$$x = \{(1-n)(y-c)\}^{\frac{1}{1-n}}.$$

Now if n is less than 1, the index in the second member is positive, and we cannot have $x = 0$ unless the quantity under the index be made equal to 0. But this would give $c=y$. Hence, $x=0$ is a singular solution.

If n be greater than 1, the index in the second member being negative we cannot have $x=0$ unless the quantity under the index becomes infinite. But this it does if c is infinite. Here then $x=0$ is a particular integral.

If n be equal to 1, the complete primitive is

$$x=ce^{y},$$

and this is reduced to $x=0$ by the assumption $c=0$. Here then also $x=0$ is a particular integral.

The following example is intended to illustrate a class of problems in which $\frac{dp}{dy}$ admits of both a finite and an infinite value.

Ex. 3. Given $p^2-2xy^{\frac{1}{2}}p+4y^{\frac{3}{2}}=0$.

Here we find

$$p=xy^{\frac{1}{2}}\pm\sqrt{(x^2y-4y^{\frac{3}{2}})} \text{ .... } (1).$$

Therefore

$$\frac{dp}{dy}=\frac{1}{2y^{\frac{1}{2}}}\left\{x\pm\frac{x^2-6y^{\frac{1}{2}}}{\sqrt{(x^2-4y^{\frac{1}{2}})}}\right\} \text{ .... } (2),$$

and this *apparently* becomes infinite when $y=0$, and when $x^2-4y^{\frac{1}{2}}=0$, i.e. for

$$y=0,\ \ y=\frac{x^4}{16}.$$

Let us inquire what are the true values of $\frac{dp}{dy}$.

1st. If $y=\frac{x^4}{16}$, we find, on substitution and reduction,

$$\frac{dp}{dy}=\frac{2}{x^2}\left(x\pm\frac{\;\frac{-x^2}{2}\;}{0}\right),$$

which becomes infinite whichsoever sign be taken. Hence, $y=\frac{x^4}{16}$ is a singular solution; and, from the mode of its origin, it is of the envelope species.

2ndly. If $y=0$, the value of $\frac{dp}{dy}$ in (2) becomes infinite if the upper sign be taken, but assumes the ambiguous form $\frac{0}{0}$ if the lower sign be taken. To determine its true value, we may expand the fraction $\frac{x^2-6y^{\frac{1}{2}}}{\sqrt{(x^2-4y^{\frac{1}{2}})}}$ in ascending powers of $y^{\frac{1}{2}}$. We thus find

$$\frac{dp}{dy}=\frac{1}{2y^{\frac{1}{2}}}\left\{x\pm\left(x-\frac{4y^{\frac{1}{2}}}{x}+\&c.\right)\right\},$$

which, as before, gives $\frac{dp}{dy}=\infty$ when, taking the upper sign, we make $y=0$, but on taking the lower sign gives

$$\frac{dp}{dy}=\frac{1}{2y^{\frac{1}{2}}}\left(\frac{4y^{\frac{1}{2}}}{x}+\&c.\right)$$

$$=\frac{2}{x}+\text{terms containing positive powers of } y.$$

And this expression, on making $y=0$, assumes the value $\frac{2}{x}$.

These results lead us to infer that the solution $y=0$, originates in two distinct ways from the primitive, which is in this case $y=c^2(x-c)^2$. It is evident that this is reduced to $y=0$, by either of the assumptions $c=0$ and $c=x$. Hence the solution $y=0$ is a particular integral.

At the same time it is to be noted that this solution possesses all the geometrical properties of a singular solution. The complete primitive represents an infinite system of parabolas whose axes are parallel to the axis of y,—whose vertices all touch the axis, of x which thus constitutes a branch of their complete envelope, — and of whose parameters each is inversely as the square of the distance of the corresponding vertex from the origin of coordinates. The nearer any particular vertex is to the origin, the more does the curve to which it belongs approach to a straight line; and the curve, if we may continue thus to speak, whose vertex is at the origin coincides with the axis of x which is the envelope of the series. It might in a certain real sense be said that the particular and the general are here united.

The following example shews, though by no means in the most extreme case, how slight may be the difference between a singular solution and a particular integral.

Ex. 4. Given $x\dfrac{dy}{dx} = y(\log x + \log y - 1)$.

Representing $\dfrac{dy}{dx}$ by p, we have

$$p = \frac{y(\log x + \log y - 1)}{x};$$

$$\therefore \frac{dp}{dy} = \frac{\log x + \log y}{x},$$

and this becomes infinite, 1st, if $y = 0$, 2ndly, if $y = \infty$, 3rdly, if $x = 0$.

The first only of these satisfies the differential equation, the assumption $y = 0$ reducing the indeterminate function $y \log y$ in the second member to 0 (Todhunter's *Differential Calculus*, p. 115). We conclude, that $y = 0$ is a singular solution, but from the nature of its origin not of the envelope species.

Now the complete primitive is $y = \dfrac{\epsilon^{cx}}{x}$, and, judging from this, it might at first sight seem as if $y = 0$ were a particular integral corresponding to $c = -\infty$. We remark however that the primitive is not reduced to $y = 0$, by the assumption $c = -\infty$, *unless x be positive.* If x is negative we must make $c = +\infty$ to effect that reduction. In fact, the value of c which reduces the complete primitive to the form $y = 0$, though independent of x in all other respects, is dependent upon x for its sign, which must always be opposite to the sign of x. And this connexion, slight as it is, determines the character of the solution.

The following example illustrates a mode of procedure which may be adopted when $\frac{dp}{dy}$ presents itself in the ambiguous form $\frac{0}{0}$, while the differential equation cannot readily be solved with respect to p.

Ex. 5. Given $p^3 - 4xyp + 8y^2 = 0$.

Differentiating with respect to y and p, we find

$$\frac{dp}{dy} = \frac{4xp - 16y}{3p^2 - 4xy} \quad \text{(1)}.$$

Equating to 0 the denominator, we have $p = \frac{2x^{\frac12}y^{\frac12}}{\sqrt{3}}$, and, substituting this value in the differential equation, we obtain a result resolvable into the following equations, viz.

$$y = \frac{4}{27}x^3, \quad y = 0 \quad \text{(2)},$$

either of which satisfies the differential equation. On substitution in (1), the former of these values of y makes $\frac{dp}{dy}$ infinite, and is evidently a singular solution. The latter value of y reduces $\frac{dp}{dy}$ to the form $\frac{0}{0}$.

To determine the real value or values of $\frac{dp}{dy}$ when $y = 0$, we must obtain from the differential equation, regarded as a cubic with respect to p, the three expressions for that quantity in ascending powers of y, substitute them in the second member of (1), and then after reduction make $y = 0$.

It will somewhat simplify the process if we transform the expressions by assuming $p = 2ty^{\frac12}$. We shall have

$$\frac{dp}{dy} = \frac{2xt - 4y^{\frac12}}{3t^2y^{\frac12} - 4xy^{\frac12}} \quad \text{(3)},$$

while the differential equation will become

$$t^3 - xt + y^{\frac{1}{2}} = 0 \text{ .... } (4),$$

which, expressed in the form

$$t = \frac{y^{\frac{1}{2}}}{x} + \frac{t^3}{x},$$

gives, by Lagrange's theorem,

$$t = \frac{y^{\frac{1}{2}}}{x} + \frac{y^{\frac{3}{2}}}{x^4} + \&c.$$

Substituting in (3), and retaining those terms only which contain the lowest power of y, we have

$$\frac{dp}{dy} = \frac{-2y^{\frac{1}{2}}}{-4xy^{\frac{1}{2}}} = \frac{2}{x}.$$

Such is the value of $\frac{dp}{dy}$ corresponding to the value of t which is given by Lagrange's theorem.

That value of t vanishes with y. Its other values do not vanish with y, but approach the limits $\pm x^{\frac{1}{2}}$ as y approaches to 0; for if in (4) we make $y = 0$, we find 0 and $\pm x^{\frac{1}{2}}$ for the corresponding values of t. Now if in (3) we make $y = 0$, $t = \pm\sqrt{x}$, we have

$$\frac{dp}{dy} = \infty.$$

From these results combined we infer that $y = 0$ is a particular integral, possessing the geometrical characters of a singular solution. It originates in fact from the complete primitive $y = c(x-c)^2$, either by making $c = 0$ or $c = x$. And that primitive, like the primitive of Ex. 3, represents a system of parabolas enveloped by one of their own number, only with a different relation of magnitude and position.

Setting out from the primitive we find

$$\frac{d}{dx}\log\frac{dy}{dc} = \frac{1}{x-c} + \frac{1}{x-3c}.$$

This expression becomes infinite when $c = \frac{x}{3}$ corresponding to the singular solution $y = \frac{4}{27}x^3$. It becomes infinite when $c = x$, and assumes the value $\frac{2}{x}$ when $c = 0$,—these cases belonging to the particular integral $y = 0$. All these determinations agree with those of $\frac{dp}{dy}$ obtained from the differential equation.

The following is an example of a special geometrical problem generalized.

Ex. 6. Determine a curve such, that the area intercepted between its tangent and the rectangular coordinate axes shall be constant and equal to $\frac{a^2}{2}$.

The supposed area is a right-angled triangle whose base and perpendicular, being the intercepts cut off by the tangents from the coordinate axes, are expressed by $x - \frac{y}{p}$, and $y - xp$ respectively. We have therefore

$$(y - xp)\left(x - \frac{y}{p}\right) = a^2.$$

Proceeding in the usual way the singular solution will be found to be

$$xy = \frac{a^2}{4},$$

representing an hyperbola, while the complete primitive represents the series of tangents by whose successive intersection the curve is generated.

To generalize the above problem we might suppose a *functional* relation given between the intercepts. The differential equation would assume the form

$$y - xp = f\left(x - \frac{y}{p}\right).$$

Its complete primitive would always be determinable by the method of Art. 9, Chap. VII. Or, since $x - \frac{y}{p} = -\frac{y - xp}{p}$, it is easily seen that the equation is reducible to Clairaut's form

$$y - xp = \phi(p).$$

The singular solution may then be found either as in Chap. VII., or by the direct application of the condition $\frac{dp}{dy} = \infty$.

Geometrical problems which are of a truly symmetrical character frequently admit of this kind of generalization.

Remarks on the foregoing theory.

12. As the theory of the tests of singular solutions which has been developed in this chapter differs in many material respects from any that have been given before, it is proper to shew in what its peculiarity consists. To this end it will be necessary briefly to sketch the history of this portion of analysis.

Leibnitz in 1694, Taylor in 1715, (see Ex. 1, Art. 11), and Clairaut in 1734, had in special problems, and Euler in 1756 had in a distinct memoir entitled *Exposition de quelques Paradoxes du Calcul Intégral*, examined, more or less deeply, various questions connected with the singular solutions of differential equations. Taylor in particular had first recognised the distinctive character of such solutions as set forth in their definition. The problem of the deduction of the singular solution from the differential equation seems however to have been first considered in its general form by Laplace. The same problem was subsequently investigated in a different manner by Lagrange, and again in a still different way by Cauchy. The state of the theory up to the present time will be adequately represented by a summary of the results to which these several investigations have led.

1st. Laplace (*Mémoires de l'Académie des Sciences*, 1772), employing the method of expansions, arrived at results which agree, so far as they go, with those of this chapter. They apply only to the envelope species of solutions, and the demonstrations of them rest essentially on the hypothesis expressed in (6), Art. 7.

Lagrange, with whom originated a more fundamental idea of the method of the inquiry, was led to the less exact criteria

$$\frac{dp}{dy}=\infty, \quad \frac{dp}{dx}=\infty.$$

(*Calcul des Fonctions*, Leçons XIV—XVII.)

Cauchy, whose method was founded on the study of the cases of failure of certain processes for obtaining the complete primitive in the form of a series, was led to the conclusion that a singular solution must satisfy one of the two following conditions, viz.

$$\frac{dp}{dy}=\frac{0}{0}, \quad \frac{dp}{dy}=\infty,$$

together with a certain further condition, the application of which depends upon a process of integration (Moigno, *Calcul*, Vol. II. p. 435).

Upon these results the following observations may be made,

1st. Although Laplace recognised the necessity of employing in certain cases the condition $\frac{d}{dx}\left(\frac{1}{p}\right)=\infty$, for $\frac{dp}{dy}=\infty$, subsequent writers who have employed his method seem to have invariably omitted this qualification.

2ndly. The supposed criterion $\frac{dp}{dx}=\infty$, introduced by Lagrange, and since very generally adopted, as the proper accompaniment of $\frac{dp}{dy}=\infty$, is erroneous. If we should apply it to Ex. 2, Art. 11, viz. $p=x^{-n}$, we should be led to the conclusion that $x=0$ is a singular solution whenever n is positive. We have seen however, both from the application of the true test, and by verification from the complete primitive, that $x=0$ is a singular solution only when n is less than 1.

The principle of Lagrange's method was the same as that adopted in the present chapter, and consisted in expressing $\frac{dp}{dx}$ and $\frac{dp}{dy}$ as derived from the differential equation, by means of differential coefficients derived from the complete primitive before the elimination of c. The fallacy which vitiated his results consisted in assuming that these expressions become infinite in consequence of the appearance of a vanishing factor in their denominators (*Calcul des Fonctions*, pp. 229, 232). Moigno, the expositor of Cauchy's views, also quotes Lagrange's method and results as presented by Caraffa, but without involving any essential variation (*Calcul*, Tom. II. p. 719). Professor de Morgan, in perhaps the latest publication on the subject, adopts Lagrange's results, expressing, however, only a qualified confidence in his method (*Cambridge Philosophical Transactions*, Vol. IX. Pt. II. 'On some points of the Integral Calculus"). And he illustrates these results by geometrical considerations which are sufficient to shew that they contain at least a considerable element of truth. Nor should this be thought surprising. For it is plain that Lagrange's condition $\frac{dp}{dx} = \infty$, and the true condition $\frac{d}{dx}\left(\frac{1}{p}\right) = \infty$, are equivalent, except when the singular solution makes p assume one of the forms 0 and ∞. And such cases do exist. Perhaps the peculiar difficulty of this subject has consisted in the faint and shadowy character of the line by which truth and error are separated.

13. Of Cauchy's tests the first, viz. $\frac{dp}{dy} = \frac{0}{0}$, may certainly be set aside. Whenever $\frac{dp}{dy}$ assumes an ambiguous form its true value or values must be determined. This is illustrated in some of the foregoing examples. Professor De Morgan's observations on this subject in the memoir above referred to, are deserving of attention. The final criterion, which is peculiar to Cauchy's theory, seems to be founded upon what we cannot but regard as an unauthorized position as to the meaning of

a singular solution. Thus $y=0$, the solution deduced by the criterion $\frac{dp}{dy}=\infty$ from the differential equation $p=y\log y$, is regarded by Cauchy as a particular integral. Now although when x is real the complete primitive $\log y = c\epsilon^x$ reduces to $y=0$ by the assumption $c=-\infty$, it does not necessarily do so when x is imaginary. Thus, if $x=\pi\sqrt{(-1)}$, we must make $c=\infty$, in order to give $y=0$. Cauchy's rule seems indeed to have been designed, contrary to the general spirit of his own writings, to exclude the consideration of imaginary values.

Properties of Singular Solutions.

14. Various properties of singular solutions of the envelope species have been demonstrated. Of these we shall notice the most important.

1st. *An exact differential equation does not admit of a singular solution.*

Let the supposed equation be

$$\frac{d\phi(x, y)}{dx}+\frac{d\phi(x, y)}{dy}\frac{dy}{dx}=0 \ldots\ldots\ldots\ldots\ldots (1),$$

and let $y=f(x)$ be a relation actually satisfying it and assumed to be singular. On this assumption the primitive $\phi(x, y)=c$ must, on substituting for y its value $f(x)$, determine c as a function of x and not a constant. Let $F(x)$ be the value of c thus determined, then $\phi(x, y)=F(x)$ whence

$$\frac{d\phi(x, y)}{dx}+\frac{d\phi(x, y)}{dy}\frac{dy}{dx}=\frac{dF(x)}{dx} \ldots\ldots\ldots\ldots (2),$$

which contradicts (1), since $\frac{dF(x)}{dx}$ cannot be permanently equal to 0, unless $F(x)$ is constant.

2ndly. *It follows directly from the above that a singular solution of a differential equation of the first order and degree, makes its integrating factor's infinite.*

For let the proposed equation be

$$Mdx+Ndy=0 \ldots\ldots\ldots\ldots\ldots\ldots\ldots\ldots (3),$$

and let μ be an integrating factor. Then

$$\mu\,(Mdx+Ndy)=0\ldots\ldots\ldots\ldots\ldots\ldots\ldots(4),$$

will be an exact differential equation. Hence, a singular solution of (3), while it makes the first member of that equation to vanish, will not make the first member of (4) to vanish. Now comparing these members, this can only be through its making μ infinite.

Ex. The equation $x+y\dfrac{dy}{dx}=\dfrac{dy}{dx}\sqrt{(x^2+y^2-a^2)}$ has for its singular solution $x^2+y^2=a^2$. An integrating factor is

$$(x^2+y^2-a^2)^{-\frac{1}{2}},$$

and this the singular solution evidently makes infinite. Multiplying the equation by its integrating factor and transposing we have the *exact* differential equation

$$\frac{x+y\dfrac{dy}{dx}}{\sqrt{(x^2+y^2-a^2)}}-\frac{dy}{dx}=0;$$

or

$$\frac{d}{dx}\sqrt{(x^2+y^2-a^2)}-\frac{dy}{dx}=0,$$

and this is not satisfied by $x^2+y^2=a^2$, the singular solution of the unrestricted differential equation.

3rdly. *Even when we are unable to discover its integrating factor, a differential equation may be so prepared as to cease to admit of a given singular solution of the envelope species.*

This proposition is due to Poisson, and the following demonstration, which is purposely given in order to illustrate the nature of the assumption usually employed in the theory of singular solutions, does not essentially differ from his.

Let us represent the singular solution by $u=0$, and transform the differential equation by assuming u and x as variables in place of y and x. Suppose the new equation reduced to the form

$$p=f(x,u)\ldots\ldots\ldots\ldots\ldots\ldots\ldots\ldots(5),$$

where p stands for $\dfrac{du}{dx}$.

This equation is either satisfied or not satisfied by $u = 0$.

If it is not satisfied, the preparation in question has already been effected.

If it is satisfied, the second member $f(x, u)$ contains some positive power of u as a factor. Assuming that it can be developed in ascending *positive* powers of u it becomes

$$p = Au^{\alpha} + Bu^{\beta} + \ldots + \&c.$$

when A, B, C, &c. are functions of x.

Now, for a singular solution $\dfrac{dp}{du} = \infty$. Hence $u = 0$ must render

$$A\alpha u^{\alpha-1} + B\beta u^{\beta-1} + \&c. = \infty.$$

But this demands that there should exist at least one negative power of u in the above development; therefore $\alpha - 1$, which is the lowest index, must be negative; therefore α being already positive must fall between 0 and 1.

Hence we are permitted to express the differential equation in the form

$$p = Qu^{\alpha},$$

where α is a positive fraction, and Q does not involve u either as a factor or as a divisor.

Dividing by u^{α}, we have

$$u^{-\alpha}\frac{du}{dx} = Q,$$

$$\text{or } \frac{1}{1-\alpha}\frac{d}{dx}u^{1-\alpha} = Q.$$

Now $u = 0$ makes $u^{1-\alpha} = 0$, since $1 - \alpha$ is positive. Hence the first member of the above equation vanishes, while the second, not containing u as a factor, does not vanish. In its present form then the equation is no longer satisfied by $u = 0$.

We see also that the property of being satisfied by $u = 0$ has been lost in consequence of a transformation which, exhibiting the singular solution in the form of a distinct algebraic factor of the equation, permitted its rejection. See Art. 1. It has been shewn in the remarks on Clairaut's equation how, in the process of ascending by differentiation to an equation of a higher order, a somewhat analogous effect is produced, the singular solution seeming to drop aside under changed conditions.

4thly. *Lagrange has noticed that a singular solution will generally make the value of* $\frac{d^2y}{dx^2}$, *as deduced from the differential equation, assume the ambiguous form* $\frac{0}{0}$. His demonstration, in the statement of which we shall endeavour to exhibit distinctly the assumptions which it really involves, is substantially as follows. Let the differential equation expressed in a rational and integral form be

$$F(x, y, p) = 0 \quad\text{.....................(1)},$$

then differentiating

$$\frac{dF}{dx}dx + \frac{dF}{dy}dy + \frac{dF}{dp}dp = 0 \quad\text{...............(2)}.$$

Hence
$$\frac{dp}{dy} = -\frac{dF}{dy} \div \frac{dF}{dp} = \infty \quad\text{.................. (3)}.$$

Now F, being rational and integral, $\frac{dF}{dy}$ and $\frac{dF}{dp}$ are so also, and therefore the above can only become infinite for finite values of x, y, and p, by supposing $\frac{dF}{dp} = 0$. This reduces (2) to the form

$$\frac{dF}{dx}dx + \frac{dF}{dy}dy = 0 \quad\text{...................... (4)}.$$

Now, as obtained from the differential equation,

$$\frac{d^2y}{dx^2} = \frac{dp}{dx} + \frac{dp}{dy}\frac{dy}{dx}$$

$$= -\frac{\frac{dF}{dx} + \frac{dF}{dy}\frac{dy}{dx}}{\frac{dF}{dp}},$$

an expression which the previous results reduce to the form $\frac{0}{0}$.

We may remark that the condition $\frac{dp}{dy} = \infty$ does not involve as a consequence $dp = \infty$ in (2), so as to affect the legitimacy of the deduction of (4). For $\frac{dp}{dy} = \infty$ expresses a *conditional* proposition, whose antecedent is: If x be constant. Now in the deduction of (4) x is not supposed to be constant.

Lagrange's demonstration is certainly only applicable to the envelope species of singular solutions. Of such solutions it expresses however an interesting property. For the differential equation being geometrically common both to the locus of the singular solution and to the locus of each particular primitive, the ambiguity of value of $\frac{d^2y}{dx^2}$ at the point of contact shews that that contact is not generally of the second order.

In like manner, $F(x, y, p)$ still being supposed rational and integral, the equation

$$\frac{dF(x, y, p)}{dp} = 0 \dots\dots\dots\dots\dots\dots(5),$$

shews by the theory of equations that the existence of a singular solution implies in general the existence of a series of points for which two values of $\frac{dy}{dx}$, usually different, come to agree, viz. the values of $\frac{dy}{dx}$ in any particular primitive, and in the singular solution.

15. Mr De Morgan has made the very interesting remark, that when the condition $\frac{dp}{dy} = \infty$, or $\frac{dp}{dx}\left(\text{in strictness } \frac{d}{dx}\frac{1}{p}\right) = \infty$, does not lead to a *solution* of the differential equation, what it does lead to is the equation of a curve which constitutes the locus of points of infinite curvature (most commonly cusps)

in the system of curves represented by the complete primitive (*Transactions of the Cambridge Philosophical Society*, Vol. IX. Part. II.). Geometrical illustrations will be found in the memoir referred to.

EXERCISES.

1. The complete primitive of a differential equation is $y+c=\sqrt{(x^2+y^2-a^2)}$, where c is the arbitrary constant. Shew that the singular solution is $x^2+y^2=a^2$, and that it may be connected with the primitive by either of the equivalent relations $c=-y$ and $c=\sqrt{(a^2-x^2)}$.

2. Why is the above singular solution deducible, by the application of *either* of the conditions $\frac{dx}{dc}=0$, $\frac{dy}{dc}=0$?

3. Expressing the primitive in Ex. 1 in a rational and integral form $\phi(x, y, c)=0$, deduce the singular solution by the application of the condition $\frac{d\phi}{dc}=0$.

4. The complete primitive of a differential equation being $x-a=(y-c)^2$, shew that the singular solution is deducible by the application of the condition $\frac{dx}{dc}=0$ but not by that of the condition $\frac{dy}{dc}=0$, and explain the circumstance.

5. The differential equation, whose complete primitive is given in Ex. 1, may be exhibited in the form

$$(x^2-a^2)p^2-2xyp-x^2=0.$$

Hence also deduce its singular solution and thereby verify the previous result.

6. Form the differential equation whose complete primitive is given in Ex. 4, and shew that the singular solution is deducible by the application of the condition $\frac{d}{dx}\frac{1}{p}=\infty$ but not

by that of the condition $\frac{dp}{dy} = \infty$, and explain this circumstance.

7. Shew that the singular solutions in the last two examples are of the envelope species.

8. The differential equation $y = px + \frac{m}{p}$ (Ex. 2, Art. 5) has $y = cx + \frac{m}{c}$ for its complete primitive, and $y^2 = 4mx$ for its singular solution. Verify in this example the fundamental relation $\frac{dp}{dy} = \frac{d}{dx} \log \frac{dy}{dc}$.

9. Deduce both the singular solution and the complete primitive of the differential equation $y = px + \sqrt{(b^2 + a^2p^2)}$, and interpret each, as well as the connexion of the two, geometrically.

10. The following differential equations admit of singular solutions of the envelope species. Deduce them.

$$x^2p^2 - 2(xy - 2)p + y^2 = 0,$$

$$(y - xp)(mp - n) = mnp$$

$$y = (x - 1)p - p^2,$$

11. The equation $(1 - x^2)p + xy - a = 0$ is satisfied by the equation $y = ax$. Is this a singular solution or a particular integral?

12. The equation $y = \frac{xp}{2}$ is satisfied by $y = 0$, which also makes $\frac{d}{dx}\left(\frac{1}{p}\right) = \infty$. Nevertheless $y = 0$ is a particular integral. Shew that this conclusion is in accordance with the general Theorem, (Art. 11).

13. The equation $p(x^2 - 1) = 2xy \log y$ has a singular solution which is not of the envelope species. Determine it.

14. Determine also the complete primitive in the last example, and shew how the singular solution arises. In particular shew that $y=0$ *accompanied by the condition that* x *is real* will be a particular integral.

15. The equation

$$(p-y)^2 - 2x^3y(p-y) = 4x^3y^2 - 4x^2y^2 \log y$$

is satisfied by $y=0$. Shew that this is a singular solution but not of the envelope species.

16. Find singular solutions of each of the following equations, and determine whether or not they are of the envelope species.

1. $p^2 + 2px^3 = 4x^2y$.
2. $xp^2 - 2yp + 4x = 0$.
3. $xp = n\{x^n + (y-x^n)\log(y-x^n)\}$.

Geometrical Applications.

In solving the following problems, the differential equation being formed, its complete primitive as well as its singular solution is to be found and interpreted.

17. Determine a curve such that the sum of the intercepts made by the tangent on the axes of co-ordinates shall be constant and equal to a.

18. Determine a curve such that the portion of its tangent intercepted between the axes of x and y shall be constant and equal to a.

19. Find a curve always touched by the same diameter of a circle rolling along a straight line.

20. Find a curve such that the product of the perpendiculars from two fixed points upon a tangent shall be constant. (Euler. See Lagrange, *Calc. des Fonctions*, p. 282.)

(Representing the product by k^2, and the distance between the given points by $2m$, making the axis of x coincide with the straight line joining them and taking for the origin of co-ordinates the middle point, the differential equation is

$$\frac{\{y-(x+m)\,p\}\,\{y-(x-m)\,p\}}{1+p^2}=k^2.$$

Its singular solution is

$$y^2+\frac{k^2x^2}{k^2+m^2}=k^2).$$

21. Deduce also the complete primitive of the above differential equation.

22. If the primitive of a differential equation be expressed in the form $\phi\,(x, y, a)=0$, the condition $\frac{dy}{da}=0$ may be expressed in the form $\frac{d\phi\,(x, y, a)}{da}\div\frac{d\phi\,(x, y, a)}{dy}=0$. Art. 5. Hence it has sometimes been laid down that $\frac{d\phi\,(x, y, a)}{dy}=\infty$ will lead to a singular solution. Raabe, in *Crelle's Journal* (*Ueber singuläre integrale*, Tom. 48), points out that this rule may fail if at the same time $\frac{d\phi\,(x, y, a)}{da}$ should become infinite. Can it fail in any other case?

23. Exemplify Raabe's observation in the equation

$$x+c-\sqrt{(6cy-3c^2)}=0,$$

which is the complete primitive of $3xp^2-6yp+x+2y=0$. At the same time shew that the singular solutions are

$$y-x=0 \text{ and } 3y+x=0. \quad (\textit{Crelle, Ib.})$$

24. The complete primitive of a differential equation is

$$(c-x+y)^3-3\,(x+y)\,(c-x+y)^2+1=0.$$

Representing its first member, which is rational and integral, by ϕ, the condition $\frac{d\phi}{dc}=0$ assumes the form

$$3(c-x+y)(c-3x-y)=0.$$

Shew that $c-x+y=0$ will not lead to a solution of the differential equation at all, while $c-3x-y=0$ will, and explain this circumstance by a reference to Art. 4.

NOTE. The reader is reminded that in all references to the general conditions $\frac{dp}{dy}=\infty$ and $\frac{d}{dx}\left(\frac{1}{p}\right)=\infty$, the ∞ means simply "infinity" irrespectively of sign. See General Theorem, Art. 11.

CHAPTER IX.

ON DIFFERENTIAL EQUATIONS OF AN ORDER HIGHER THAN THE FIRST.

1. THE typical form of a differential equation of the nth order is given in Chap. I. Art. 2. We may, by solving it algebraically with respect to its highest differential coefficient, present it in the form

$$\frac{d^n y}{dx^n} = f\left(x, y, \frac{dy}{dx}, \frac{d^2 y}{dx^2} \cdots \frac{d^{n-1} y}{dx^{n-1}}\right) \dots\dots\dots\dots (1).$$

Its genesis from a complete primitive involving n arbitrary constants has been explained, Chap. I. Art. 8.

Conversely, the existence of a differential equation of the above type implies the existence of a primitive involving n arbitrary constants and no more; and a primitive possessing this character is termed complete.

The converse proposition above stated, is one to which various and distinct modes of consideration point, but concerning the rigid proof of which opinion has differed. The view which appears the most fundamental is the following. If, as in Chap. II. Art. 2, we represent by $\Delta\phi(x)$ the increment which the function $\phi(x)$ receives when x receives the fixed increment Δx, and if we go on to represent by $\Delta^2\phi(x)$ the increment which the function $\Delta\phi(x)$ receives when x again receives the same fixed increment Δx, and so on, then it is evident that the values of $\Delta\phi(x)$, $\Delta^2\phi(x)$, &c., are fully determinable if the successive values of the function $\phi(x)$ in its successive states of increase are known. Thus since

$$\Delta\phi(x) = \phi(x + \Delta x) - \phi(x),$$

we have by definition

$$\begin{aligned}\Delta^2\phi(x) &= \Delta\{\phi(x + \Delta x) - \phi(x)\} \\ &= \{\phi(x + 2\Delta x) - \phi(x + \Delta x)\} - \{\phi(x + \Delta x) - \phi(x)\} \\ &= \phi(x + 2\Delta x) - 2\phi(x + \Delta x) + \phi(x),\end{aligned}$$

and so on. Conversely if

$$\phi(x),\ \Delta\phi(x),\ \Delta^2\phi(x),\ \&c.$$

are given, the successive values of the function $\phi(x)$, viz. the values $\phi(x+\Delta x)$, $\phi(x+2\Delta x)$, &c., are thereby made determinate. Geometrically we may represent $\phi(x)$ by y, the ordinate of a curve, or of a series of points in the plane x, y, and therefore functionally connected with the abscissa x.

Now the view to which reference has been made is that which, 1st, presents the differential equation (1) as the limiting form of the relation expressed by the equation

$$\frac{\Delta^n y}{\Delta x^n}=f\left(x,\ y,\ \frac{\Delta y}{\Delta x},\ \frac{\Delta^2 y}{\Delta x^2}\cdots\frac{\Delta^{n-1}y}{\Delta x^{n-1}}\right) \text{............} (2),$$

Δx approaching to 0; 2ndly, constructs the latter equation in geometry (the arithmetical or purely quantitative construction being therein implied) by a series of points on a plane, of which the n first, viz. those which answer to the co-ordinates x, $x+\Delta x, \ldots x+(n-1)\Delta x$, have the corresponding values of y arbitrary, while for all the rest the values of y are determined; 3dly, represents the solution of the differential equation as the curve which the above series of points in their limiting state tend to form. According to this view, the n arbitrary points in the constructed solution of the equation of differences (2) give rise to one arbitrary point in the limiting curve, accompanied by $n-1$ arbitrary values for the first $n-1$ differential coefficients of its ordinate. And this mode of consideration appears the most fundamental, because it assumes no more than the definition itself demands of us when we attempt to realize the geometrical meaning of a differential coefficient as a limit. We may however add that when by the consideration of the limit, the mere existence of a primitive has been established, other considerations would suffice to shew that in its complete form it will involve n arbitrary constants and no more. The fact that each integration introduces a single constant is a direct indication of the fact. An indirect proof of a more formal character will be found in a memoir by Professor de Morgan (*Transactions of the Cambridge Philosophical Society*, Vol. IX. Pt. II.)

The above theory may be illustrated by the form in which Taylor's Theorem enables us to present the solution of a differential equation of the nth order, as will be seen in the following Article.

Solution by development in a series.

2. Reducing the proposed equation to the form

$$\frac{d^n y}{dx^n} = f\left(x, y, \frac{dy}{dx} \cdots \frac{d^{n-1}y}{dx^{n-1}}\right) \dots\dots\dots\dots\dots (3),$$

and differentiating with respect to x, the first member becomes $\frac{d^{n+1}y}{dx^{n+1}}$, while the second member will in general involve all the differential coefficients of y up to $\frac{d^n y}{dx^n}$. If for the last we substitute its value given in (3), the equation will assume the form

$$\frac{d^{n+1}y}{dx^{n+1}} = f_1\left(x, y, \frac{dy}{dx} \cdots \frac{d^{n-1}y}{dx^{n-1}}\right) \dots\dots\dots\dots\dots (4).$$

Thus $\frac{d^{n+1}y}{dx^{n+1}}$ is expressible in the same manner as $\frac{d^n y}{dx^n}$, viz. in terms of x, y, and the first $n-1$ differential coefficients of y.

Differentiating (4) and again reducing the second member by means of (3) we have a result of the form

$$\frac{d^{n+2}y}{dx^{n+2}} = f_2\left(x, y, \frac{dy}{dx} \cdots \frac{d^{n-1}y}{dx^{n-1}}\right) \dots\dots\dots\dots\dots (5),$$

and in this form and by the same method all succeeding differential coefficients may be expressed.

Hence reasoning as in Chap. II. Art. 12, we see that supposing y to be developed in a series of ascending powers of x, or more generally of $x - x_0$, where x_0 is an assumed arbitrary value of x, the coefficients of the higher powers of $x - x_0$ beginning with $(x - x_0)^n$ will have a *determinate* connexion, established by means of the differential equation, with the coefficients of the inferior powers of $x - x_0$. The latter coefficients, n in number, beginning with the constant term which corresponds to the index 0, and ending with $\frac{1}{1.2.3\ldots n-1}\frac{d^{n-1}y}{dx^{n-1}}$, which is the coefficient of $(x-x_0)^{n-1}$, will be perfectly arbitrary in value.

To exhibit the actual form of the development let $y_0, y_1, \ldots y_{n-1}$ be the arbitrary values assigned to $y, \frac{dy}{dx}, \ldots \frac{d^{n-1}y}{dx^{n-1}}$ when $x = x_0$. Also let f, f_1, f_2, &c. represent the values which the second members of the series of equations (3), (4), (5) assume when we make them $x = x_0$; then

$$y = y_0 + y_1(x-x_0) + \frac{y_2}{1.2}(x-x_0)^2 \ldots + \frac{y_{n-1}}{1.2\ldots n-1}(x-x_0)^{n-1}$$

$$+\frac{f_n}{1.2\ldots n}(x-x_0)^n + \frac{f_{n+1}}{1.2\ldots(n+1)}(x-x_0)^{n+1}\ldots\&\text{c. } ad\ inf. \ldots (6).$$

In this expression the arbitrary values of y and its $n-1$ first differential coefficients corresponding to an assumed and *definite* value of x, viz. $y_0, y_1, \ldots y_{n-1}$ are the n arbitrary constants of the solution, the values of f_n, f_{n+1}, &c., being determinate functions of these, and therefore not involving any arbitrary element.

Any function of arbitrary constants is itself an arbitrary constant, and thus it may be that an equation has *effectively* a smaller number of arbitrary constants than it appears to have from the mere enumeration of its symbols. As a general principle we may affirm, that the number of effective arbitrary constants in the solution of a differential equation while on the one hand equal to the index of the order of the equation, is on the other hand to be measured by the number of conditions which they enable us to satisfy. Systems of conditions to be thus satisfied will indeed vary in form, but there is one system which we may consider as normal and to which all other systems are in fact reducible. It is that which is described above, and which demands that to a given value of x a given set of simultaneous values of y and of its differential coefficients up to an order less by 1 than the order of the equation shall correspond. Conversely, the arbitrary constants of a solution may be said to be *normal*, when they actually represent a simultaneous system of values of y and its successive differential coefficients up to the number required.

Ex. Given $\frac{d^2y}{dx^2} = \frac{dy}{dx} + y^2$. Required an expression for y in the form of a series such that when $x=0$, y and $\frac{dy}{dx}$ shall assume the respective values c and c'.

Differentiating, we have

$$\frac{d^3y}{dx^3} = \frac{d^2y}{dx^2} + 2y\frac{dy}{dx}$$

$$= \frac{dy}{dx} + y^2 + 2y\frac{dy}{dx}, \text{ by the given equation,}$$

$$= y^2 + (1+2y)\frac{dy}{dx},$$

$$\frac{d^4y}{dx^4} = 2y\frac{dy}{dx} + 2\left(\frac{dy}{dx}\right)^2 + (1+2y)\frac{d^2y}{dx^2}$$

$$= y^2 + 2y^3 + (1+4y)\frac{dy}{dx} + 2\left(\frac{dy}{dx}\right)^2,$$

by similar reduction, and so on. Hence, corresponding to $x=0$, we have the series of values,

$$y = c, \quad \frac{dy}{dx} = c', \quad \frac{d^2x}{dx^2} = c' + c^2,$$

$$\frac{d^3y}{dx^3} = c^2 + (1+2c)\,c',$$

$$\frac{d^4y}{dx^4} = c^2 + 2c^3 + (1+4c)\,c' + 2c'^2,$$

and so on. Hence,

$$y = c + c'x + \frac{c^2+c'}{2}x^2$$

$$+ \frac{c^2 + (1+2c)\,c'}{2\,.\,3}x^3 + \frac{c^2 + 2c^3 + (1+4c)\,c' + 2c'^2}{2\,.\,3\,.\,4}x^4 + \&c.$$

Finitely Integrable Forms.

3. As the difficulty of the finite integration of differential equations increases as their order is more elevated, it becomes important to classify the chief cases in which that difficulty has been overcome.

It will be found that for the most part these cases are characterized by some one or more of the following marks, viz. 1st, Linearity, the coefficients being at the same time either constant or subject to some restriction as to form; 2ndly, Absence of one or more of the variables or their differential coefficients; 3rdly, Homogeneity; 4thly, Expressibility in the form of an exact differential or in a form easily reducible thereto by means of a multiplier.

The subject of linear equations being of primary importance, we shall devote the remainder of this chapter to its discussion. But as it will be resumed in another part of this work, and in connexion with a higher method, we propose to notice here only the more important general properties of linear equations, and to illustrate them in the solution of equations with constant coefficients.

Linear Equations.

4. The type of a linear differential equation of the nth order, (Chap. I. Art. 4), is

$$\frac{d^n y}{dx^n} + X_1 \frac{d^{n-1} y}{dx^{n-1}} + X_2 \frac{d^{n-2} y}{dx^{n-2}} \dots + X_n y = X \;\dots\dots\; (7),$$

in which the coefficients X_1, $X_2 \dots X_n$ and the second member X are either constant quantities or functions of the independent variable x.

Considering, first, the case in which the second member is 0, the following important proposition may be established.

Prop. If y_1, $y_2 \dots y_n$ represent n distinct values of y, each containing an arbitrary constant, which individually satisfy the linear equation,

$$\frac{d^n y}{dx^n} + X_1 \frac{d^{n-1} y}{dx^{n-1}} + X_2 \frac{d^{n-2} y}{dx^{n-2}} \dots + X_n y = 0 \;\dots\dots\; (8),$$

then will the complete value of y be

$$y = y_1 + y_2 \dots + y_n.$$

In other words the *complete* value of y is the sum of n distinct particular values of y, each containing an arbitrary constant.

For on substitution of the assumed general value of y in (8), we have a result which may be arranged in the following form, viz.

$$\left.\begin{array}{l} \dfrac{d^n y_1}{dx^n} + X_1 \dfrac{d^{n-1} y_1}{dx^{n-1}} + X_2 \dfrac{d^{n-2} y_1}{dx^{n-2}} \dots + X_n y_1 \\ + \dfrac{d^n y_2}{dx^n} + X_1 \dfrac{d^{n-1} y_2}{dx^{n-1}} + X_2 \dfrac{d^{n-2} y_2}{dx^{n-2}} \dots + X_n y_2 \\ \dots\dots\dots\dots\dots\dots\dots\dots\dots\dots \\ + \dfrac{d^n y_n}{dx^n} + X_1 \dfrac{d^{n-1} y_n}{dx^{n-1}} + X_2 \dfrac{d^{n-2} y_n}{dx^{n-2}} \dots + X_n y_n \end{array}\right\} = 0 \dots (9).$$

Now each line in the left-hand member of the above equation, being merely the result of substituting some one of the particular values of y in the left-hand member of (8), is by hypothesis equal to 0. Hence the equation (9) reduces to an identity, and the theorem is established.

The problem of the complete solution of a linear equation of the nth order whose second member is equal to 0 is, therefore, reduced to that of finding n distinct particular solutions, each involving an arbitrary constant. More than that number do not exist.

5. Prop. To solve the linear equation with constant coefficients when the second member is 0.

Were the proposed equation of the first order and of the form

$$\frac{dy}{dx} - my = 0,$$

its solution would be

$$y = c\epsilon^{mx}.$$

From this result, and from the known constancy of form of the differential coefficients of exponentials, we are led to examine the effect of such a substitution in the equation

$$\frac{d^n y}{dx^n} + a_1 \frac{d^{n-1} y}{dx^{n-1}} + a_2 \frac{d^{n-2} y}{dx^{n-2}} \ldots + a_n y = 0 \ldots\ldots\ldots (10).$$

Assuming then $y = C\epsilon^{mx}$, and observing that

$$\frac{d^n (C\epsilon^{mx})}{dx^n} = m^n C\epsilon^{mx},$$

we have, on rejection of the common factor $C\epsilon^{mx}$, the equation,

$$m^n + a_1 m^{n-1} + a_2 m^{n-2} \ldots + a_n = 0 \ldots\ldots\ldots\ldots (11),$$

the different roots of which determine the different values of m which make $y = C\epsilon^{mx}$ a solution of the equation given.

When those roots are real and unequal, we have, therefore, on representing them by $m_1, m_2, \ldots m_n$, the system of n particular solutions,

$$y = C_1 \epsilon^{m_1 x},\ y = C_2 \epsilon^{m_2 x}, \ldots y = C_n \epsilon^{m_n x} \ldots\ldots\ldots (12),$$

from which by the foregoing theorem we may construct the general solution,

$$y = C_1 \epsilon^{m_1 x} + C_2 \epsilon^{m_2 x} \ldots + C_n \epsilon^{m_n x} \ldots\ldots\ldots\ldots (13).$$

The equation (11) by which the values of m are determined is usually called the auxiliary equation.

Ex. Given $\dfrac{d^2 y}{dx^2} - 3\dfrac{dy}{dx} + 2y = 0.$

Here, assuming $y = C\epsilon^{mx}$, we obtain as the auxiliary equation

$$m^2 - 3m + 2 = 0.$$

Whence the values of m are 1 and 2. The corresponding particular integrals are $y = C_1 \epsilon^x$, and $y = C_2 \epsilon^{2x}$, and the complete primitive is

$$y = C_1 \epsilon^x + C_2 \epsilon^{2x}.$$

6. If among the roots, still supposed unequal, imaginary pairs present themselves, the above solution, though formally correct, needs transformation. Let $a \pm b\sqrt{-1}$ represent one of these pairs, then will the second member of (13) contain a corresponding pair of terms of the form

$$C\epsilon^{(a+b\sqrt{-1})x} + C'\epsilon^{(a-b\sqrt{-1})x},$$

which we may reduce as follows,

$$C\epsilon^{ax+bx\sqrt{-1}} + C'\epsilon^{ax-bx\sqrt{-1}}$$

$$= C\epsilon^{ax}(\cos bx + \sqrt{-1}\sin bx) + C'\epsilon^{ax}(\cos bx - \sqrt{-1}\sin bx)$$

$$= (C + C')\,\epsilon^{ax}\cos bx + (C - C')\sqrt{(-1)}\,\epsilon^{ax}\sin bx,$$

or, replacing $C + C'$ and $(C - C')\sqrt{(-1)}$ by new arbitrary constants A and B,

$$A\epsilon^{ax}\cos bx + B\epsilon^{ax}\sin bx \ldots\ldots\ldots\ldots\ldots (14).$$

Ex. Given $\dfrac{d^2y}{dx^2} - 4\dfrac{dy}{dx} + 13y = 0$.

Assuming $y = C\epsilon^{mx}$, the auxiliary equation is

$$m^2 - 4m + 13 = 0,$$

whence $m = 2 \pm 3\sqrt{(-1)}$. The complete solution therefore is

$$y = A\epsilon^{2x}\cos 3x + B\epsilon^{2x}\sin 3x.$$

7. Lastly, let the auxiliary equation have equal roots whether real or imaginary, e.g. suppose $m_2 = m_1$. Then in the general solution (13) the terms $C_1\epsilon^{m_1x} + C_2\epsilon^{m_2x}$ reduce to a single term $(C_1 + C_2)\,\epsilon^{m_1x}$, and the number of arbitrary constants is effectively diminished, since $C_1 + C_2$ is only equivalent to a single one. Here then the form (13) ceases to be general.

To deduce the general solution when $m_2 = m_1$ let us begin by supposing m_2 to differ from m_1 by a finite quantity h, and examine the limit to which the terms of the solution, then really general, approach as h approaches to 0. Now

$$C_1\epsilon^{m_1x} + C_2\epsilon^{(m_1+h)x} = \epsilon^{m_1x}(C_1 + C_2\epsilon^{hx})$$

$$= \epsilon^{m_1x}\left(C_1 + C_2 + C_2hx + C_2\frac{h^2x^2}{1.2}\dots\right)$$

$$= \epsilon^{m_1x}\left(A + Bx + Bh\frac{x^2}{1.2} + \&c.\right);$$

on replacing $C_1 + C_2$ and C_2h by A and B, new arbitrary constants. This change it is permitted to make, however small h may be, provided that it is not equal to 0. The limit to which the last member of the above equation approaches as h approaches to 0 is

$$\epsilon^{m_1x}(A + Bx).$$

And in precisely the same way, were there r roots equal to m_1, we should have for the corresponding part of the *complete* value of y, the expressions

$$\epsilon^{m_1x}(A_1 + A_2x + A_3x^2 \dots + A_rx^{r-1}) \dots\dots\dots\dots (15).$$

Thus the difference which the repetition of a particular root m_1 produces is that the coefficient of the exponential ϵ^{m_1x} is no longer an arbitrary constant, but a polynomial of the form $A_1 + A_2x + \&c.$, the number of arbitrary constants involved being equal to the number of times that the supposed root presents itself.

Ex. Given $\dfrac{d^3y}{dx^3} - \dfrac{d^2y}{dx^2} - \dfrac{dy}{dx} + y = 0$.

Here, assuming $y = C\epsilon^{mx}$, the auxiliary equation is

$$m^3 - m^2 - m + 1 = 0,$$

the roots of which are -1, 1, 1. Thus, corresponding to the root -1, we have in y the term $C\epsilon^{-x}$, while to the two roots 1, we have the term $(A + Bx)\epsilon^x$. The complete primitive therefore is

$$y = C\epsilon^{-x} + (A + Bx)\epsilon^x.$$

8. It follows from (15), that if a pair of imaginary roots $a \pm b\sqrt{-1}$ present itself n times, the corresponding portion of the complete value of y will be

$$(C_1 + C_2 x \ldots + C_r x^{r-1})\,\epsilon^{ax+bx\sqrt{-1}} + (C_1' + C_2' x \ldots + C_r' x^{r-1})\,\epsilon^{ax-bx\sqrt{-1}},$$

which, substituting for $\epsilon^{bx\sqrt{-1}}$ and $\epsilon^{-bx\sqrt{-1}}$ their trigonometrical values and finally making

$$C_1 + C_1' = A_1, \quad (C_1 - C_1')\sqrt{-1} = B_1, \text{ \&c.},$$

assumes the form

$$(A_1 + A_2 x \ldots + A_r x^{r-1})\,\epsilon^{ax}\cos bx$$
$$+ (B_1 + B_2 x \ldots + B_r x^{r-1})\,\epsilon^{ax}\sin bx \ \ldots\ldots\ (16).$$

Hence, therefore, the repetition of a pair of imaginary roots $a \pm b\sqrt{-1}$ changes also the two arbitrary constants of the ordinary real solution into polynomials, each of which involves a number of constants equal to the number of times that the imaginary pair presents itself.

Ex. Given $\dfrac{d^4 y}{dx^4} + 2n^2 \dfrac{d^2 y}{dx^2} + n^4 y = 0.$

Assuming $y = C\epsilon^{mx}$, the auxiliary equation is

$$m^4 + 2n^2 m^2 + n^4 = 0,$$

whence m has two pairs of roots of the form $\pm n\sqrt{(-1)}$.

For one such pair the form of solution would be

$$y = A\cos x + B\sin x.$$

For the actual case it therefore is

$$y = (A_1 + A_2 x)\cos x + (B_1 + B_2 x)\sin x.$$

9. The above is the ordinary method of investigating the form of the complete solution when the auxiliary equation involves equal roots, and we have therefore thought it proper to give it a place here. We must, however, remember that it involves the assumption that a law of continuity connects the form of solution when roots are equal with the form of solution when the roots are unequal. Now, though it is perfectly true that such a law does exist, its assumption with-

out proof of that existence must be regarded as opposed to the requirements of a strict logic. In all legitimate applications of the Differential Calculus it is with a limit that we are directly concerned. Here it is with something which exists, and which admits of being determined independently of the notion of a limit. Such determination shews, however, that the continuity assumed to exist exists in reality.

Thus if we take as an example $\frac{d^2y}{dx^2} - 2\frac{dy}{dx} + y = 0$, in which the auxiliary equation $m^2 - 2m + 1 = 0$ shews that the values of m are each equal to 1, we are entitled to assume as a particular solution

$$y = C\epsilon^x.$$

Let us now substitute this value of y in the given equation regarding C as variable, and inquire whether it admits of any more general determination than it has received above. On substitution we find simply

$$\frac{d^2C}{dx^2} = 0,$$

whence $C = A + Bx$. Thus while the correctness of the solution furnished by the assumption of continuity is established, it is made manifest that this assumption is not indispensable. We shall endeavour to establish upon other grounds the theory of these cases of failure, in a future chapter.

10. The results of the previous investigation may be summed up in the following rule.

RULE. *The coefficients being constant and the second member* 0, *form an auxiliary equation by assuming* $y = C\epsilon^{mx}$, *and determine the values of* m. *Then the complete value of* y *will be expressed by a series of terms characterized as follows, viz. For each real distinct value of* m *there will exist a term* $C\epsilon^{mx}$; *for each pair of imaginary values* $a \pm b\sqrt{(-1)}$, *a term*

$$A\epsilon^{ax} \cos bx + B\epsilon^{ax} \sin bx;$$

each of the coefficients A, B, C *being an arbitrary constant if the corresponding root occur only once, but a polynomial of the*

$r-1^{\text{th}}$ degree with arbitrary constant coefficients, if the root occur r times.

Ex. Given $\dfrac{d^5y}{dx^5} - \dfrac{d^3y}{dx^3} - 2\dfrac{d^2y}{dx^2} + 2\dfrac{dy}{dx} = 0.$

Here the auxiliary equation is

$$m^5 - m^3 - 2m^2 + 2m = 0,$$

whence it will be found that the values of m are

$$0, 1, 1, -1 \pm \sqrt{(-1)}.$$

The complete primitive therefore is

$$y = C + (C_1 + C_2x)\epsilon^x + C_3\epsilon^{-x}\cos x + C_4\epsilon^{-x}\sin x.$$

11. To solve the linear equation with constant coefficients when its second member is not equal to 0.

The usual mode of solution is 1st to determine the complete value of y on the hypothesis that the second member is 0; 2ndly, to substitute its expression in the given equation regarding the arbitrary constants as variable parameters; 3rdly, to determine those parameters so as to satisfy the equation given.

Supposing the given equation to be of the nth degree, n parameters will be employed. These may evidently be subjected to any $n-1$ arbitrary conditions. Now that system of conditions which renders the discovery of the remaining relation (involved in the condition that the given differential equation shall be satisfied) the most easy, is that which demands that the formal expression of the $n-1$ differential coefficients

$$\frac{dy}{dx}, \ \frac{d^2y}{dx^2}, \ \frac{d^{n-1}y}{dx^{n-1}},$$

shall, like the formal expression of y, be the same in the system in which $c_1, c_2, \ldots c_n$ represent variable parameters, as in the system in which they represent arbitrary constants.

The above method is commonly called the method of the variation of parameters. It is, as we shall hereafter see, far from being the easiest mode of solving the class of equations under consideration; but it is interesting as being probably the first general method discovered, and still more so from its containing an application of a principle successfully employed in higher problems.

Ex. Given $\frac{d^2y}{dx^2}+n^2y=\cos ax$.

Were the second member 0, the solution would be

$$y=c_1\cos nx+c_2\sin nx\ldots\ldots\ldots\ldots\ldots\ldots(a).$$

Assume this then to be the form of the solution of the equation given, c_1, c_2 being variable parameters, but such that $\frac{dy}{dx}$ shall also retain the same form as if they were constant, viz.

$$\frac{dy}{dx}=-c_1n\sin nx+c_2n\cos nx\ldots\ldots\ldots\ldots\ldots(b).$$

Now the unconditional value of $\frac{dy}{dx}$ derived from (a) is

$$\frac{dy}{dx}=-c_1n\sin nx+c_2n\cos nx+\cos nx\frac{dc_1}{dx}+\sin nx\frac{dc_2}{dx},$$

which reduces to the foregoing form if we assume

$$\cos nx\frac{dc_1}{dx}+\sin nx\frac{dc_2}{dx}=0\ldots\ldots\ldots\ldots\ldots\ldots(c).$$

This then is the condition which must accompany (a).

Now differentiating b and regarding c_1, c_2 as variable, we have

$$\frac{d^2y}{dx^2}=-c_1n^2\cos nx-c_2n^2\sin nx-n\sin nx\frac{dc_1}{dx}+n\cos nx\frac{dc_2}{dx}.$$

Substituting the above values of y, $\frac{dy}{dx}$ and $\frac{d^2y}{dx^2}$ in the given equation, we have

$$-n\sin nx\frac{dc_1}{dx}+n\cos nx\frac{dc_2}{dx}=\cos ax\ldots\ldots(d),$$

and this equation, in combination with (c), gives

$$\frac{dc_1}{dx}=-\frac{1}{n}\cos ax\sin nx,\ \frac{dc_2}{dx}=\frac{1}{n}\cos ax\cos nx,$$

whence $$c_1 = \frac{1}{2n}\left\{\frac{\cos(n+a)x}{n+a} + \frac{\cos(n-a)x}{n-a}\right\} + C_1.$$

$$c_2 = \frac{1}{2n}\left\{\frac{\sin(n+a)x}{n+a} + \frac{\sin(n-a)x}{n-a}\right\} + C_2.$$

Lastly, substituting these values in (a) and reducing, we have

$$y = \frac{\cos ax}{n^2 - a^2} + C_1 \cos nx + C_2 \sin nx \dots\dots\dots (e).$$

This solution fails if $n = a$. But giving to (e) the form

$$y = \frac{\cos ax - \cos nx}{n^2 - a^2} + C_1' \cos nx + C_2 \sin nx,$$

and regarding the first term as a vanishing fraction when $n = a$, we find

$$y = \frac{x \sin nx}{2n} + C_1' \cos nx + C_2 \sin nx.$$

Or we might proceed thus. Differentiating twice the equation

$$\frac{d^2y}{dx^2} + n^2 y = \cos nx,$$

we get $$\frac{d^4y}{dx^4} + n^2 \frac{d^2y}{dx^2} = -n^2 \cos nx.$$

Hence eliminating $\cos nx$

$$\frac{d^4y}{dx^4} + 2n^2 \frac{d^2y}{dx^2} + n^4 y = 0,$$

an equation whose complete solution is

$$y = (A + Bx)\cos nx + (C + Dx)\sin nx.$$

Substituting this in the given equation we find $B = 0$, $D = \frac{1}{2n}$, whence

$$y = A \cos nx + \left(C + \frac{x}{2n}\right)\sin nx,$$

which agrees with the previous solution.

The latter method, which is general, consists in forming a new equation of a higher order, but with its second member free from that term which is the cause of failure. As by the elevation of the order of the equation superfluous constants are introduced, the relations which connect them must be found by substitution of the result in the given equation.

12. To the class of linear equations with constant coefficients all equations of the form

$$(a+bx)^n\frac{d^ny}{dx^n}+A(a+bx)^{n-1}\frac{d^{n-1}y}{dx^{n-1}}+B\,(a+bx)^{n-2}\frac{d^{n-2}y}{dx^{n-2}}\ldots+Ly=X,$$

$A, B, \ldots L$ being constant and X a function of x, may be reduced. It suffices to change the independent variable by assuming $a+bx=\epsilon^t$.

Ex. Given $(a+bx)^2\dfrac{d^2y}{dx^2}+b\,(a+bx)\dfrac{dy}{dx}+n^2y=0.$

Assuming $a+bx=\epsilon^t$, we find

$$\frac{dy}{dx}=b\epsilon^{-t}.\frac{dy}{dt},$$

$$\frac{d^2y}{dx^2}=b^2\epsilon^{-2t}\left(\frac{d^2y}{dt^2}-\frac{dy}{dt}\right).$$

Hence, by substitution in the given equation, we have

$$b^2\frac{d^2y}{dt^2}+n^2y=0,$$

the solution of which is

$$y=C\cos\frac{nt}{b}+C'\sin\frac{nt}{b},$$

in which it only remains to substitute for t its value $\log(a+bx)$.

13. Beside the properties upon which the above methods are founded, linear equations possess many others, of which we shall notice the most important. We suppose, as before, y to be the dependent, x the independent variable.

1st. The complete value of y when the linear equation has a second member X will be found by adding to any particular value of y that complementary function which would express its complete value were the second member 0.

Representing the linear equation in the form (7), let y_1 be the particular value of y which satisfies it, Y the complete value which would satisfy it were the second member 0; and assume $y = y_1 + Y$. The equation then becomes

$$\left.\begin{array}{l} \dfrac{d^n y_1}{dx^n} + X_1 \dfrac{d^{n-1} y_1}{dx^{n-1}} \ldots + X_n y_1 \\ + \dfrac{d^n Y}{dx^n} + X_1 \dfrac{d^{n-1} Y}{dx^{n-1}} \ldots + X_n Y \end{array}\right\} = X \ldots\ldots\ldots (17),$$

and this becomes an identity, the first line of its left-hand member being by hypothesis equal to X, and the second line equal to 0.

Ex. Thus a particular integral of the equation

$$\frac{d^2y}{dx^2} - a^2 y = x + 1$$

being $y = -\dfrac{x+1}{a^2}$, its complete integral is

$$y = C\epsilon^{ax} + C'\epsilon^{-ax} - \frac{x+1}{a^2}.$$

The above property, which relates to the *generalizing* of a particular solution, is important, because, as we shall hereafter see, a particular solution of a linear equation may often be obtained by a symbolical process which does not involve even the labour of an integration.

2ndly. The order of a linear differential equation may always be depressed by unity if we know a particular value of y which would satisfy the equation were its second number equal to 0.

It will suffice to demonstrate this property for the equation of the second order

$$\frac{d^2y}{dx^2} + X_1 \frac{dy}{dx} + X_2 y = X \ldots\ldots\ldots\ldots\ldots\ldots (18).$$

Let y_1 be a particular value of y when $X=0$, and assume $y=y_1v$. Substituting, we have

$$\left(\frac{d^2y_1}{dx^2}+X_1\frac{dy_1}{dx}+X_2y_1\right)v$$

$$+\left(2\frac{dy_1}{dx}+X_1y_1\right)\frac{dv}{dx}+y_1\frac{d^2v}{dx^2}=X,$$

the first line of which is by hypothesis 0. In the reduced equation let $\frac{dv}{dx}=u$, then we have

$$y_1\frac{du}{dx}+(2\frac{dy_1}{dx}+X_1y_1)\,u=X\,..............(19),$$

a linear equation of the first order for determining u. And this being found, we have

$$v=\int udx+c.$$

In the particular case in which $X=0$, we find from (19)

$$u=\frac{C\epsilon^{-\int X_1dx}}{y_1^2},$$

whence
$$y=y_1\left(C\int\frac{\epsilon^{-\int X_1dx}}{y_1^2}dx+C_1\right)...............(20).$$

3rdly. Linear equations are connected by remarkable analogies with ordinary algebraic equations.

This subject has been investigated chiefly by Libri and Liouville, who have shewn that most of the characteristic properties of algebraic equations have their analogies in linear differential equations.

An algebraic equation can be deprived of its 2nd, 3rd ... r^{th} term by the solution of an algebraic equation of the 1st, 2nd, ... $r-1^{\text{th}}$ degree. A linear differential equation can be deprived of its 2nd, 3rd, ... r^{th} term by the solution of another linear equation of the 1st, 2nd, ... $r-1^{\text{th}}$ order.

This may be proved by assuming $y=vy_1$, and properly determining v so as to make in the resulting equation y_1 assume the required form.

Given two algebraic equations involving the same unknown quantity and of the respective degrees m and n, we can deduce by their combination a new equation of the degree $m \sim n$.

Given two linear equations involving the same variables and of the respective orders m and n, we can deduce by their combination a new linear equation involving the same variables and of the order $m \sim n$.

Many other properties exemplifying the general analogy might be noted. Fuller demonstration of those above noted will be found in Moigno, *Calcul Intégral*, Tom. II. p. 579.

EXERCISES.

1. $\dfrac{d^2y}{dx^2} - 7\dfrac{dy}{dx} + 12y = 0.$

2. $\dfrac{d^2y}{dx^2} - 7\dfrac{dy}{dx} + 12y = x.$

3. Integrate $\dfrac{d^4y}{dx^4} - 4\dfrac{d^3y}{dx^3} + 6\dfrac{d^2y}{dx^2} - 4\dfrac{dy}{dx} + y = 0.$

4. $\dfrac{d^4y}{dx^4} + 2\dfrac{d^2y}{dx^2} + y = 0.$

5. $\dfrac{d^3y}{dx^3} - 3\dfrac{d^2y}{dx^2} + 4y = 0$, it being given that one of the roots of the auxiliary equation, $m^3 - 3m^2 + 4 = 0$, is-1.

6. $\dfrac{d^4y}{dx^4} - 2\dfrac{d^3y}{dx^3} + 2\dfrac{d^2y}{dx^2} - 2\dfrac{dy}{dx} + y = 1.$

7. $\dfrac{d^2y}{dx^2} - 2k\dfrac{dy}{dx} + k^2y = \epsilon^x.$

8. What form does the solution of the above equation assume when $k = 1$?

9. $x^2\dfrac{d^2y}{dx^2} - x\dfrac{dy}{dx} = 3y.$

10. $(x+a)^2 \dfrac{d^2y}{dx^2} - 4(x+a)\dfrac{dy}{dx} + 6y = 0.$

11. Integrate $\dfrac{d^2y}{dx^2} - 2bx\dfrac{dy}{dx} + b^2x^2y = 0.$

12. A particular integral of $(1-x^2)\dfrac{d^2y}{dx^2} - x\dfrac{dy}{dx} - a^2y = 0$ is $y = C\epsilon^{1 \sin^{-1} x}$, find the complete integral by the method of Art. 13.

13. The form of the general integral might in the above case be inferred from that of the particular one without employing the method of Art. 13. Prove this.

14. It being given that

$$y = A\left(\sin x + \frac{\cos x}{x}\right) + B\left(\cos x - \frac{\sin x}{x}\right)$$

is the complete integral of the equation $\dfrac{d^2y}{dx^2} + \left(1 - \dfrac{2}{x^2}\right)y = 0$, find the general integral of $\dfrac{d^2y}{dx^2} + \left(1 - \dfrac{2}{x^2}\right)y = x^2$.

15. Explain on what grounds it is asserted that the complete integral of a differential equation of the nth order contains n arbitrary constants and no more.

16. Mention any circumstances under which it may be advantageous to form, from a proposed differential equation, one of a higher order. In deducing from the solution of the latter that of the former, what kind of limitation must be introduced?

CHAPTER X.

EQUATIONS OF AN ORDER HIGHER THAN THE FIRST, CONTINUED.

1. WE have next to consider certain forms of non-linear equations.

Of the following principle frequent use will be made, viz. *When either of the primitive variables is wanting, the order of the equation may be depressed by assuming as a dependent variable the lowest differential coefficient which presents itself in the equation.*

Thus if the equation be of the form

$$F\left(x, \frac{dy}{dx}, \frac{d^2y}{dx^2}\right) = 0 \ldots\ldots\ldots\ldots\ldots (1),$$

and we assume

$$\frac{dy}{dx} = z \ldots\ldots\ldots\ldots\ldots\ldots\ldots\ldots (2),$$

we have, on substitution, the differential equation of the first order,

$$F\left(x, z, \frac{dz}{dx}\right) = 0 \ldots\ldots\ldots\ldots\ldots\ldots (3).$$

If, by the integration of this equation, z can be determined as a function of x involving an arbitrary constant c, $\{$suppose $z = \phi(x, c)\}$ we have from (2)

$$\frac{dy}{dx} = \phi(x, c),$$

whence integrating

$$y = \int \phi(x, c)\, dx + c'.$$

If the lowest differential coëfficient of y which presents itself be of the second order, the order of the equation can be depressed by 2, and so on.

A similar reduction may be effected when x is wanting. Thus, if in the equation of the second order

$$F\left(y, \frac{dy}{dx}, \frac{d^2y}{dx^2}\right)=0 \ldots\ldots\ldots\ldots (4),$$

we assume $\frac{dy}{dx}=p$, we have

$$\frac{d^2y}{dx^2}=\frac{dp}{dx}=\frac{dp}{dy}\frac{dy}{dx}=p\frac{dp}{dy},$$

by means of which (4) becomes

$$F\left(y, p, p\frac{dp}{dy}\right)=0 \ldots\ldots\ldots\ldots\ldots (5).$$

Should we succeed by the integration of this equation of the first order in determining p as a function of y and c, suppose $p=\phi(y, c)$, the equation $\frac{dy}{dx}=p$, will give

$$dx=\frac{dy}{\phi(y, c)},$$

whence

$$x=\int\frac{dy}{\phi(y, c)}+c' \ldots\ldots\ldots\ldots\ldots (6).$$

2. In close connexion with the above proposition, stand the three following important cases.

CASE I. When but one differential coefficient as well as but one of the primitive variables presents itself in the given equation.

1st. Let the equation be of the form $\frac{d^ny}{dx^n}=X$, we have by successive integrations

$$\frac{d^{n-1}y}{dx^{n-1}}=\int Xdx+c,$$

$$\frac{d^{n-2}y}{dx^{n-2}}=\iint Xdx^2+cx+c',$$

and finally

$$y = \iint \ldots X dx^n + c_1 x^{n-1} + c_2 x^{n-2} \ldots + c_n \ldots\ldots (7).$$

We shall hereafter shew that the first term in the second member may be replaced by a series of n single integrals.

2dly, If the equation be of the form $\dfrac{d^n y}{dx^n} = Y$, it is not generally integrable, but it is so in the case of $n = 2$. Thus there being given

$$\frac{d^2 y}{dx^2} = Y,$$

we have

$$2 \frac{dy}{dx} \frac{d^2 y}{dx^2} = 2 Y \frac{dy}{dx},$$

and integrating

$$\left(\frac{dy}{dx}\right)^2 = 2 \int Y dy + C.$$

Hence

$$\frac{dy}{dx} = (2 \int Y dy + C)^{\frac{1}{2}},$$

$$dx = \frac{dy}{(2 \int Y dy + C)^{\frac{1}{2}}},$$

$$x = \int \frac{dy}{(2 \int Y dy + C)^{\frac{1}{2}}} + C' \ldots\ldots\ldots (8).$$

As a particular example, let $\dfrac{d^2 y}{dx^2} = a^2 y$.

Here

$$x = \int \frac{dy}{(2 \int a^2 y \, dy + C)^{\frac{1}{2}}} + C'$$

$$= \int \frac{dy}{(a^2 y^2 + C)^{\frac{1}{2}}} + C'$$

$$= \frac{1}{a} \log \{ay + \sqrt{(a^2 y^2 + C)}\} + C'.$$

CASE II. When the given equation merely expresses a relation between two consecutive differential coefficients.

Suppose the equation reduced to the form

$$\frac{d^n y}{dx^n} = f\left(\frac{d^{n-1}y}{dx^{n-1}}\right) \quad \ldots\ldots\ldots\ldots\ldots\ldots (9),$$

then, assuming $\frac{d^{n-1}y}{dx^{n-1}} = z$, we have

$$\frac{dz}{dx} = f(z),$$

whence

$$dx = \frac{dz}{f(z)},$$

$$x = \int \frac{dz}{f(z)} + c \quad \ldots\ldots\ldots\ldots\ldots (10).$$

If, after effecting the integration, we can express z in terms of x and c, suppose $z = \phi(x, c)$ we have finally to integrate

$$\frac{d^{n-1}y}{dx^{n-1}} = \phi(x, c) \quad \ldots\ldots\ldots\ldots\ldots (11),$$

which belongs to Case I.

But if, after effecting the integration in (10), we cannot algebraically express z in terms of x and c, we may proceed thus.

From $\frac{d^{n-1}y}{dx^{n-1}} = z$, we have

$$\frac{d^{n-2}y}{dx^{n-2}} = \int z dx$$

$$= \int \frac{z dz}{f(z)}.$$

$$\frac{d^{n-3}y}{dx^{n-3}} = \int dx \int \frac{z dz}{f(z)}$$

$$= \int \frac{dz}{f(z)} \int \frac{z dz}{f(z)},$$

and finally,

$$y = \int \frac{dz}{f(z)} \int \frac{dz}{f(z)} \cdots \int \frac{z dz}{f(z)} \quad \ldots\ldots\ldots\ldots\ldots\ldots (12),$$

the right-hand member indicating the performance of $n-1$ successive integrations, each of which introduces an arbitrary constant. If between this equation and (10) we, after integration, eliminate z, we shall obtain a final relation between y, x, and n arbitrary constants, which will be the integral sought.

Ex. Given $a\dfrac{d^2y}{dx^2}\dfrac{d^3y}{dx^3}=\sqrt{\left\{1+\left(\dfrac{d^2y}{dx^2}\right)^2\right\}}$.

Making $\dfrac{d^2y}{dx^2}=z$, we have $az\dfrac{dz}{dx}=\sqrt{(1+z^2)}$, whence

$$x=c+a\sqrt{(1+z^2)} \text{ .... } (a).$$

According to the first of the above methods, we should now solve this with respect to z, and thus obtaining

$$z=\frac{d^2y}{dx^2}=\sqrt{\left\{\left(\frac{x-c}{a}\right)^2-1\right\}},$$

find hence

$$y=\iint\sqrt{\left\{\left(\frac{x-c}{a}\right)^2-1\right\}}\,dx^2+c_1x+c_2 \text{ .... } (b),$$

to which it only remains to effect the integrations. According to the second method, we should proceed thus. Since $dx=\dfrac{az\,dz}{\sqrt{(1+z^2)}}$, we have

$$\frac{dy}{dx}=\int z\,dx=\int\frac{az^2\,dz}{\sqrt{(1+z^2)}}$$

$$=\frac{az\sqrt{(1+z^2)}}{2}-\frac{a}{2}\log\{z+\sqrt{(1+z^2)}\}+c',$$

whence multiplying the second member by $\dfrac{az\,dz}{\sqrt{(1+z^2)}}$ for dx, and again integrating,

$$y=\frac{a^2z^3}{6}-\frac{a^2}{2}\sqrt{(1+z^2)}\log\{z+\sqrt{(1+z^2)}\}+\frac{a^2}{2}z$$
$$+ac'\sqrt{(1+z^2)}+c'' \text{ ...... } (c).$$

The complete primitive now results from the elimination of z between (a) and (c).

CASE III. When the given equation merely connects two differential coefficients whose orders differ by 2.

Reducing the equation to the form

$$\frac{d^n y}{dx^n} = f\left(\frac{d^{n-2}y}{dx^{n-2}}\right) \quad \text{.......................} \quad (13).$$

Let $\dfrac{d^{n-2}y}{dx^{n-2}} = z$, then

$$\frac{d^2 z}{dx^2} = f(z).$$

This form has been considered under Case I.

It gives

$$x = \int \frac{dz}{\{2\int f(z)\,dz + C\}^{\frac{1}{2}}} + C'.$$

If from this equation z can be determined as a function of x, C, and C',—suppose $z = \phi(x, C, C')$,—then

$$\frac{d^{n-2}y}{dx^{n-2}} = \phi(x, C, C'),$$

the integration of which by Case I. will lead to the required integral. If z cannot be thus determined, we must proceed as under the same circumstances in Case II.

Ex. Given $a^2 \dfrac{d^4 y}{dx^4} = \dfrac{d^2 y}{dx^2}$.

Proceeding as above, the final integral will be found to be

$$y = c_1 \epsilon^{\frac{x}{a}} + c_2 \epsilon^{\frac{-x}{a}} + c_3 x + c_4.$$

Homogeneous Equations.

3. There exist certain classes of homogeneous equations which admit of having their order depressed by unity.

Class I. Equations which, on supposing x and y to be each of the degree 1, $\dfrac{dy}{dx}$ of the degree 0, $\dfrac{d^2y}{dx^2}$ of the degree -1, &c., become homogeneous in the ordinary sense.

Adopting the notion and the language of infinitesimals, the earlier analysts described the above class of equations somewhat more simply as homogeneous with respect to the primitive variables and their differentials, i.e. with respect to x, y, dx, dy, d^2y, &c.

All equations of the above class admit of having their order depressed by unity.

For if we assume $x = \epsilon^\theta$, $y = \epsilon^\theta z$, we shall find by the usual method for the change of variables,

$$\frac{dy}{dx} = \frac{dz}{d\theta} + z \quad \ldots\ldots\ldots\ldots\ldots\ldots\ldots\ldots \quad (14),$$

$$\frac{d^2y}{dx^2} = \epsilon^{-\theta}\left(\frac{d^2z}{d\theta^2} + \frac{dz}{d\theta}\right) \quad \ldots\ldots\ldots\ldots\ldots\ldots \quad (15),$$

and so on. Here y is presented as of the first degree with respect to ϵ^θ which takes the place of x, while $\frac{dy}{dx}$ is of the degree 0, and $\frac{d^2y}{dx^2}$ of the degree -1, with respect to ϵ^θ. And the law of continuation is obvious. Hence, from the supposed constitution of the given equation, it follows that on substitution of these values the resulting equation will be homogeneous with respect to ϵ^θ, which will therefore divide out and leave an equation involving only z, $\frac{dz}{d\theta}$, $\frac{d^2z}{d\theta^2}$, &c. That equation will therefore have its order depressed by unity on assuming

$$\frac{dz}{d\theta} = p \ \ldots. \quad \text{(Art. 1.)}$$

Let us examine the general form of the result for equations of the second order.

Representing the given equations under the form

$$F\left(x,\, y,\, \frac{dy}{dx},\, \frac{d^2y}{dx^2}\right) = 0 \quad \ldots\ldots\ldots\ldots\ldots\ldots \quad (16),$$

we have, on substitution,

$$F\left\{\epsilon^\theta,\, \epsilon^\theta z,\, \frac{dz}{d\theta} + z,\, \epsilon^{-\theta}\left(\frac{d^2z}{d\theta^2} + \frac{dz}{d\theta}\right)\right\} = 0 \ \ldots\ldots \ (17),$$

and from this equation, from what has been above said, ϵ^θ will disappear on division by some power of that quantity, e. g. $\epsilon^{n\theta}$. But the effect of simply removing a factor is the same as that of simply replacing such factor by unity. Now to replace $\epsilon^{n\theta}$ by unity is the same as to replace ϵ^θ by unity, and if we do this simply, i. e. without changing $\frac{dz}{d\theta}$ and $\frac{d^2z}{d\theta^2}$, (17) will become

$$F\left(1,\ z,\ \frac{dz}{d\theta}+z,\ \frac{d^2z}{d\theta^2}+\frac{dz}{d\theta}\right)=0 \qquad (18).$$

Assuming then $\frac{dz}{d\theta}=u$, whence $\frac{d^2z}{d\theta^2}=\frac{du}{d\theta}=u\frac{du}{dz}$, we have

$$F\left(1,\ z,\ u+z,\ u\frac{du}{dz}+u\right)=0 \qquad (19),$$

an equation of the first order, which by integration gives

$$u=\phi(z,c) \qquad (20).$$

Then since $u=\frac{dz}{d\theta}$, we have

$$d\theta=\frac{dz}{\phi(z,c)},$$

$$\theta=\int\frac{dz}{\phi(z,c)}+c' \qquad (21),$$

in which, after effecting the integration, it is only necessary to write

$$\theta=\log x,\quad z=\frac{y}{x} \qquad (22),$$

The solution of the proposed equation is therefore involved in (20), (21), (22).

Ex. Given $nx^3\frac{d^2y}{dx^2}=\left(y-x\frac{dy}{dx}\right)^2$.

Substituting as above $x=\epsilon^\theta$, $y=\epsilon^\theta z$, we find, as the transformed equation,

$$n\left(\frac{d^2z}{d\theta^2}+\frac{dz}{d\theta}\right)=\left(\frac{dz}{d\theta}\right)^2,$$

whence, making $\frac{dz}{d\theta} = u$, we have

$$n\left(u\frac{du}{dz} + u\right) = u^2 \text{ .... } (a),$$

which resolves itself into the two equations,

$$n\left(\frac{du}{dz} + 1\right) = u, \quad u = 0.$$

The former gives on integration

$$u = n + C\epsilon^{\frac{z}{n}}.$$

Now $u = \frac{dz}{d\theta}$, whence

$$d\theta = \frac{dz}{n + C\epsilon^{\frac{z}{n}}};$$

$$\therefore \ \theta = -\log\left(n\epsilon^{\frac{-z}{n}} + C\right) + C',$$

and now replacing θ by x, and z by $\frac{y}{x}$, we have on reduction,

$$y = nx \log \frac{x}{A + Bx} \text{ .... } (b),$$

A and B being arbitrary constants. This is the complete primitive.

The remaining equation $u = 0$, or $\frac{dz}{d\theta} = 0$, gives $z = c$, or $y = cx$, and this is the singular solution.

The equation (a) might have been directly deduced from the given equation by the general theorem (19), which indicates that for such deduction it is only necessary to change x to 1, y to z, $\frac{dy}{dx}$ to $u + z$, and $\frac{d^2y}{dx^2}$ to $u\frac{du}{dz} + u$.

Class II. Equations which on regarding x as of the first degree, y as of the nth degree, $\frac{dy}{dx}$ of the $n - 1^{\text{th}}$ degree, $\frac{d^2y}{dx^2}$ of the $n - 2^{\text{th}}$ degree, &c., are homogeneous.

To effect the proposed reduction assume $x=\epsilon^{\theta}$, $y=\epsilon^{n\theta}z$. The transformed equation will be free from θ, and, on assuming $\frac{dz}{d\theta}=u$, will degenerate into an equation of a degree lower by unity between u and z.

It is easy to establish that, if the given differential equation be

$$F\left(x,\ y,\ \frac{dy}{dx},\ \frac{d^2y}{dx^2}\right)=0 \text{..................} (23),$$

the reduced equation for determining u will be

$$F\left\{1,\ z,\ u+nz,\ u\frac{du}{dz}+(2n-1)\,u+n\,(n-1)\,z\right\}=0\text{...}(24).$$

Suppose that by the solution of this we find

$$u=\phi\,(z,c) \text{........................} (25),$$

then since $\qquad u=\frac{dz}{d\theta}$, we have

$$d\theta=\frac{dz}{\phi\,(z,c)},$$

$$\theta=\int\frac{dz}{\phi\,(z,c)}+c' \text{.....................} (26),$$

in which it only remains to substitute $\log x$ for θ, and $\frac{y}{x^n}$ for z.

Ex. Given $x^4\frac{d^2y}{dx^2}=(x^3+2xy)\frac{dy}{dx}-4y^2$.

This equation proves homogeneous on assuming x to be of the degree 1, y of the degree 2, $\frac{dy}{dx}$ of the degree 1, and $\frac{d^2y}{dx^2}$ of the degree 0.

Changing then, according to the formula (24), x into 1, y into z, $\frac{dy}{dx}$ into $u+2z$, and $\frac{d^2y}{dx^2}$ into $u\frac{du}{dz}+3u+2z$, we have

$$u\frac{du}{dz}+3u+2z=(1+2z)\,(u+2z)-4z^2 \text{........} (a),$$

which is reducible to

$$u\left(\frac{du}{dz}+2-2z\right)=0.$$

This is resolvable into two equations, viz.

$$\frac{du}{dz}+2-2z=0,\ u=0\ldots\ldots\ldots\ldots\ldots\ldots(b).$$

The first gives on integration

$$u=(z-1)^2\pm c^2.$$

Hence, since $\dfrac{dz}{d\theta}=u$, we have

$$d\theta=\frac{dz}{(z-1)^2\pm c^2},$$

$$\theta=\frac{1}{c}\tan^{-1}\frac{z-1}{c}+c',\ \text{or}\ \frac{1}{2c}\log\left(\frac{z-1-c}{z-1+c}\right)+c'.$$

Hence, replacing θ by $\log x$, and z by $\dfrac{y}{x^2}$, we have

$$\log x=\frac{1}{c}\tan^{-1}\frac{y-x^2}{cx^2}+c',\ \text{or}\ \frac{1}{2c}\log\frac{y-(1+c)x^2}{y-(1-c)x^2}+c',$$

the rational forms of the integral required.

The factor $u=0$ in (b) giving $\dfrac{dz}{d\theta}=0$, or $z=c$, leads to the singular solution $y=cx^2$.

Class III. Equations which are homogeneous with respect to y, $\dfrac{dy}{dx}$, $\dfrac{d^2y}{dx^2}$, &c.

Properly speaking, this class constitutes a limit to the class just considered. For when n becomès large, the quantities n, $n-1$, $n-2$, the supposed measures of the degrees of y, $\dfrac{dy}{dx}$, $\dfrac{d^2y}{dx^2}$ approach a ratio of equality.

If we assume $y = \epsilon^z$, we have

$$\frac{dy}{dx} = \epsilon^z \frac{dz}{dx} \quad \text{.................................} (27),$$

$$\frac{d^2y}{dx^2} = \epsilon^z \left\{\frac{d^2z}{dx^2} + \left(\frac{dz}{dx}\right)^2\right\} = 0 \text{..............} (28).$$

All these being of the first degree with respect to ϵ^z, it follows that after substitution in the proposed equation, that function will disappear on division. Thus, if the given equation be

$$F\left(x, y, \frac{dy}{dx}, \frac{d^2y}{dx^2}\right) = 0 \text{ ....} (29),$$

the transformed equation will be

$$F\left\{x, 1, \frac{dz}{dx}, \frac{d^2z}{dx^2} + \left(\frac{dz}{dx}\right)^2\right\} = 0 \text{............} (30),$$

or, on assuming $\dfrac{dz}{dx} = u$,

$$F\left(x, 1, u, \frac{du}{dx} + u^2\right) = 0 \text{.............} (31).$$

Integrating this equation of the first degree, we have

$$u = \phi(x, c);$$

$$\therefore z = \int\phi(x, c)\, dx + c' \text{..................} (32),$$

in which it only remains to substitute for z its value $\log y$.

Or we may assume at once $y = \epsilon^{\int u dx}$. The transformed equation between u and x will be of an order lower by unity than the equation given.

Ex. Given $ay\dfrac{d^2y}{dx^2} + b\left(\dfrac{dy}{dx}\right)^2 = \dfrac{y\dfrac{dy}{dx}}{\sqrt{(e^2 + x^2)}}$.

Assuming $y = \epsilon^{\int u dx}$, we find

$$a\left(\frac{du}{dx} + u^2\right) = \frac{u}{\sqrt{(e^2 + x^2)}} - bu^2,$$

as would directly result from (29) and (30). Expressed in the form

$$\frac{du}{dx}-\frac{u}{a\sqrt{(e^2+x^2)}}=-\left(1+\frac{b}{a}\right)u^2,$$

this equation is seen to belong to the class discussed in Chap. II. Art. 11.

On comparing the above classes of homogeneous equations we see that Class II. is the most general. It includes Class I. as a subordinate species, and Class III. as a limit.

It is proper to observe that Classes I. and II. are usually treated by a different method from that above employed. Thus, in Class I., it is customary to make the assumptions

$$y=xt,\ \frac{dy}{dx}=u,\ \frac{d^2y}{dx^2}=\frac{v}{x},\ \frac{d^3y}{dx^3}=\frac{w}{x^2},\ \&\text{c.}$$

On substitution x divides out, and there remains an equation involving y and the new variables t, u, v, w, &c., which may be reduced by successive eliminations to a differential equation between two variables, and of an order lower by unity than the equation given. But this method is far more complicated than the one which we have preferred to employ.

Exact Differential Equations.

4. A differential equation of the form

$$\phi\left(x,\ y,\ \frac{dy}{dx},\ \frac{d^2y}{dx^2}\cdots\frac{d^ny}{dx^n}\right)=0\ldots\ldots\ldots\ldots(33),$$

is said to be exact if, representing its first member by V, the expression Vdx is the exact differential of a function U, which is therefore necessarily of the form $\psi\left(x,\ y,\ \frac{dy}{dx}\cdots\frac{d^{n-1}y}{dx^{n-1}}\right)$.

Thus $\frac{dy}{dx}\frac{d^2y}{dx^2}-yx^2\frac{dy}{dx}-xy^2=0$, is an exact differential equation, its first member multiplied by dx being the differential of the function $\frac{1}{2}\left\{\left(\frac{dy}{dx}\right)^2-x^2y^2\right\}$, and the first member itself the differential coefficient of that function.

Hence then a first integral of the above equation will be

$$\left(\frac{dy}{dx}\right)^2 - x^2y^2 = c.$$

The method of integrating an exact differential equation which we shall illustrate, and which contains an implicit solution of the question whether a proposed equation is exact or not, appears to be primarily due to M. Sarrus (*Liouville*, Tom. XIV. p. 131, note).

Ex. Given $y + 3x\frac{dy}{dx} + 2y\left(\frac{dy}{dx}\right)^3 + \left(x^2 + 2y^2\frac{dy}{dx}\right)\frac{d^2y}{dx^2} = 0.$

Supposing the above an exact differential, we are by definition permitted to write

$$dU = \left\{y + 3x\frac{dy}{dx} + 2y\left(\frac{dy}{dx}\right)^3 + \left(x^2 + 2y^2\frac{dy}{dx}\right)\frac{d^2y}{dx^2}\right\}dx \ldots (34).$$

Now a first and obvious condition is that the highest differential coefficient in an exact differential equation, being the one introduced by differentiation, can only present itself in the first degree. This condition is seen to be satisfied.

Representing the highest differential coefficient but one by p, we can express (34) in the form

$$dU = (y + 3xp + 2yp^3)\,dx + (x^2 + 2y^2p)\,dp.$$

Now let U_1 represent what the integral of the term containing dp would be were p the only variable. Then

$$U_1 = x^2p + y^2p^2.$$

Assume, then, removing all restriction,

$$U_1 = x^2\frac{dy}{dx} + y^2\left(\frac{dy}{dx}\right)^2,$$

whence $dU_1 = \left\{2x\frac{dy}{dx} + 2y\left(\frac{dy}{dx}\right)^3 + \left(x^2 + 2y^2\frac{dy}{dx}\right)\frac{d^2y}{dx^2}\right\}dx.$

Subtracting this from (34)

$$dU - dU_1 = \left(y + x\frac{dy}{dx}\right)dx \ldots\ldots\ldots\ldots (35).$$

We remark that the highest differential coefficient $\frac{d^2y}{dx^2}$ has now disappeared. We observe too that the next, viz. $\frac{dy}{dx}$ is involved only in the first degree. This is a consequence of the fact that the proposed differential equation was really exact. For the first member of (35) being the difference of two exact differentials, and therefore itself exact, the second member is so, and its highest differential coefficient is therefore of the first degree. The integration of an exact differential involving $\frac{d^2y}{dx^2}$ has, in fact, been reduced to that of an exact differential involving only $\frac{dy}{dx}$ as its highest differential coefficient. And a similar reduction may be effected whatever may be the order of the highest differential coefficient.

The integration of (35) gives

$$U - U_1 = xy,$$

whence

$$U = U_1 + xy = x^2\frac{dy}{dx} + y^2\left(\frac{dy}{dx}\right)^2 + xy.$$

A first integral of the given equation is, therefore,

$$x^2\frac{dy}{dx} + y^2\left(\frac{dy}{dx}\right)^2 + xy = c\text{.............}(36).$$

The general rule for the integration of an exact differential dU involving x, y, $\frac{dy}{dx}$, ... $\frac{d^ny}{dx^n}$, is then as follows. *Integrate the term which involves $\frac{d^ny}{dx^n}$ in the first degree, as if $\frac{d^{n-1}y}{dx^{n-1}}$ were the only variable, and $\frac{d^ny}{dx^n}dx$ its differential. Representing the result by U_1, and removing the restriction, $dU - dU_1$ will be an exact differential involving only x, y, $\frac{dy}{dx}$, ... $\frac{d^{n-1}y}{dx^{n-1}}$. Repeat the process as often as necessary. Then U will be expressed by the sum of its successively determined portions U_1, U_2, U_3, &c.*

For the solution of an exact differential equation, it is therefore only needful to equate to c the integral of the corresponding exact differential as found by the above process.

The failure of that process, through the occurrence of a form in which the highest differential coefficient is not of the first degree, indicates that the proposed function or equation is not 'exact.'

5. There is another mode of proceeding of which it is proper that a brief account should be given.

Representing $\frac{dy}{dx}, \frac{d^2y}{dx^2}, \ldots \frac{d^ny}{dx^n}$, by $y_1, y_2, \ldots y_n$, it is easily shewn by the Calculus of Variations, that if Vdx be an exact differential, V being a function of $x, y, y_1, \ldots y_n$, then identically

$$\frac{dV}{dy} - \left(\frac{d}{dx}\right)\frac{dV}{dy_1} + \left(\frac{d}{dx}\right)^2\frac{dV}{dy_2} \ldots \mp \left(\frac{d}{dx}\right)^n \frac{dV}{dy_n} = 0 \ldots\ldots (37),$$

where $\left(\frac{d}{dx}\right)$ indicates that we differentiate with respect to x regarding $y, y_1, \ldots y_n$ as functions of x. This condition was discovered by Euler.

The researches of Sarrus and De Morgan, not based upon the employment of the Calculus of Variations, have shewn, 1st, that the above condition is not only necessary but sufficient. 2ndly, that it constitutes the last of a series of theorems which enable us, when the above condition is satisfied, to reduce Vdx to an exact differential in *form*, i.e. to express it in the form

$$\frac{dU}{dx}dx + \frac{dU}{dy}dy + \frac{dU}{dy_1}dy_1 \ldots + \frac{dU}{dy_{n-1}}dy_{n-1} \ldots\ldots\ldots (38),$$

where $x, y, y_1, \ldots y_{n-1}$ are regarded as independent. The integration of $Vdx = 0$ in the form $U = c$ is thus reduced to the integration of an exact differential of a function of $n+1$ independent variables,—a subject to be discussed in Chapter XII. (*Cambridge Transactions*, Vol. IX.)

The condition (37) is singly equivalent to the system of conditions implied in the process of Sarrus. The proof of this equivalence *a posteriori* would, as Bertrand has observed, be complicated. (Liouville, Tom. XIV.)

The solution of the differential equations of orders higher than the first is sometimes effected by means of an integrating factor μ, to discover which we might substitute μV for V in (37), and endeavour to solve the resulting partial differential equation. Even here, however, the process of Sarrus would be preferable.

Miscellaneous Methods and Examples.

6. Many forms of equations, besides those above noted, can be integrated by special methods, e. g. by transformations, variation of parameters, reduction to exact differentials, &c. Equations of the classes already considered can also sometimes be integrated by processes more convenient than those above explained.

Ex. 1. Given $\dfrac{d^2y}{dx^2} = ax + by$.

Let $ax + by = t$. We find as the result, $\dfrac{d^2t}{dx^2} = bt$ a linear equation with constant coefficients.

Ex. 2. Given $(1 - x^2)\dfrac{d^2y}{dx^2} - x\dfrac{dy}{dx} + q^2y = 0$.

Changing the independent variable by assuming $\sin^{-1} x = t$, we find $\dfrac{d^2y}{dt^2} + q^2y = 0$, whence the final solution is

$$y = c_1 \cos(q \sin^{-1}x) + c_2 \sin(q \sin^{-1}x) \ldots\ldots\ldots (39).$$

So too the equation $(1 + ax^2)\dfrac{d^2y}{dx^2} + ax\dfrac{dy}{dx} \pm q^2y = 0$, is reducible to the form $\dfrac{d^2y}{dt^2} \pm q^2y = 0$, by the assumption

$$\int \frac{dx}{\sqrt{(1 + ax^2)}} = t.$$

Equations involving the arc s, whether explicitly or implicitly, may be freed from it by differentiation or by change of independent variable.

Ex. 3. Given $s = ax + by$.

Differentiating, we have

$$\sqrt{\left\{1+\left(\frac{dy}{dx}\right)^2\right\}} = a + b\frac{dy}{dx};$$

$$\therefore \frac{dy}{dx} = \frac{ab \pm \sqrt{(a^2+b^2-1)}}{1-b^2},\quad y = \frac{ab \pm \sqrt{(a^2+b^2-1)}}{1-b^2}x + c.$$

Ex. 4. Given $\frac{d^2x}{ds^2} = a$.

Assuming x as independent variable, we have

$$\frac{d^2x}{ds^2} = \frac{dx}{ds}\frac{d}{dx}\frac{dx}{ds} = \left(\frac{ds}{dx}\right)^{-1}\frac{d}{dx}\left(\frac{ds}{dx}\right)^{-1}$$

$$= -\left(\frac{ds}{dx}\right)^{-3}\frac{d^2s}{dx^2} = a.$$

We might here put for $\frac{ds}{dx}$ its value $\sqrt{(1+p^2)}$, and so form a differential equation for determining p. Direct integration, however, gives

$$\left(\frac{ds}{dx}\right)^{-2} = 2ax + c.$$

Whence we find

$$\frac{dy}{dx} = \left(\frac{1}{2ax+c} - 1\right)^{\frac{1}{2}};$$

$$y = \int\left(\frac{1}{2ax+c} - 1\right)^{\frac{1}{2}} dx + c';$$

which indicates a cycloid.

7. M. Liouville has shewn how to integrate the general equation $\frac{d^2y}{dx^2} + f(x)\frac{dy}{dx} + F(y)\left(\frac{dy}{dx}\right)^2 = 0$, (*Journal de Mathématiques*, 1st Series, Tom. VII. p. 134).

Suppressing the last term, the resulting equation

$$\frac{d^2y}{dx^2} + f(x)\frac{dy}{dx} = 0,$$

has for a first integral $\frac{dy}{dx} = C\epsilon^{-\int f(x)\,dx}$. Now assume this to be a first integral of the given equation regarding C as an unknown function of y, then

$$\frac{d^2y}{dx^2} = \epsilon^{-\int f(x)\,dx}\left\{\frac{dC}{dy}\frac{dy}{dx} - Cf(x)\right\}$$

$$= \frac{1}{C}\frac{dy}{dx}\left\{\frac{dC}{dy}\frac{dy}{dx} - Cf(x)\right\}$$

$$= \frac{1}{C}\frac{dC}{dy}\left(\frac{dy}{dx}\right)^2 - f(x)\frac{dy}{dx}.$$

Thus, the given equation becomes

$$\frac{1}{C}\frac{dC}{dy} + F(y) = 0 \;\ldots\ldots\ldots\ldots\ldots\ldots\ldots (40),$$

whence $$C = A\epsilon^{-\int F(y)\,dy}.$$

Therefore $$\frac{dy}{dx} = A\epsilon^{-\int F(y)\,dy} \times \epsilon^{-\int f(x)\,dx};$$

$$\therefore \int \epsilon^{\int F(y)\,dy}\,dy = A\int \epsilon^{-\int f(x)\,dx}\,dx + B \ldots\ldots\ldots\ldots (41),$$

the complete primitive sought.

8. Jacobi has established that when one of the first integrals of a differential equation of the form $\frac{d^2y}{dx^2} = f(x, y)$ is known, the complete primitive may be found. The following demonstration of this proposition is due to Liouville, (*Journal de Mathématiques*, Tom. XIV. p. 225, 2nd Series).

Let the given first integral be $\frac{dy}{dx} = \phi(x, y, c)$. Differentiating, we have

$$\frac{d^2y}{dx^2} = \frac{d\phi}{dx} + \frac{d\phi}{dy}\frac{dy}{dx} = \frac{d\phi}{dx} + \phi\frac{d\phi}{dy},$$

ϕ standing for $\phi(x, y, c)$. Hence, comparing with the given equation,

$$\frac{d\phi}{dx} + \phi\frac{d\phi}{dy} = f(x, y),$$

and differentiating with respect to c,

$$\frac{d^2\phi}{dx\,dc}+\frac{d\phi}{dc}\frac{d\phi}{dy}+\phi\frac{d^2\phi}{dy\,dc}=0.$$

Now this is precisely the condition which must be satisfied in order that the expression $\frac{d\phi}{dc}(dy-\phi dx)$ may be an exact differential. Hence, the first integral expressed in the form $dy-\phi dx=0$, is made an exact differential by means of the factor $\frac{d\phi}{dc}$. The complete primitive therefore is

$$\int\frac{d\phi}{dc}(dy-\phi\,dx)=c' \quad\text{..................}\quad (42).$$

Some equations of great difficulty connected with the theory of the elliptic functions are reduced to the above case in the memoir referred to.

Singular Integrals.

9. Equations of the higher orders, like those of the first order, sometimes admit of singular integrals, i. e. of integrals not derivable from the ordinary ones without making one or more of their constants variable.

We shall term such integrals singular solutions when they connect only the primitive variables, but singular integrals when they present themselves in the form of differential equations inferior in order to the equation given.

And as the entire theory is involved in the theory of singular first integrals, we shall speak chiefly of these, but with less detail than in the corresponding inquiries of Chap. VIII.

PROP. Given a first integral with arbitrary constant of a differential equation of the nth order, required the corresponding singular integral.

Let the given equation be

$$F(x, y, y_1, y_2 \ldots y_n)=0 \quad\text{..................}\quad (43),$$

where y_1 stands for $\frac{dy}{dx}$, y_2 for $\frac{d^2y}{dx^2}$, &c. Suppose the integral given to be expressed in the form

$$y_{n-1} = f(x, y, y_1 \dots y_{n-2}, c) \dots\dots\dots\dots\dots\dots (44),$$

c being an arbitrary constant. Differentiating as if c were an unknown function of x,

$$y_n = \frac{df}{dx} + \frac{df}{dy} y_1 + \frac{df}{dy_1} y_2 \dots + \frac{df}{dy_{n-2}} y_{n-1} + \frac{df}{dc} \frac{dc}{dx}.$$

Now this reduces to the same form, i.e. gives the same expression for y_n in terms of $x, y_1 \dots y_{n-1}, c$, as it would do if c were constant, provided that we have $\frac{df}{dc} = 0$; and therefore, this condition satisfied, the elimination of c will still lead to the given differential equation (43).

An integral of the given equation will therefore be found by attributing to c in the complete first integral (44), such value as will satisfy the condition $\frac{df}{dc} = 0$, or, as we may express it,

$$\frac{dy_{n-1}}{dc} = 0 \dots\dots\dots\dots\dots\dots\dots\dots (45).$$

And unless the value of c thus found is constant, the integral will be singular. The above process amounts to eliminating c between (44) and (45), so that we have the following rule.

Given a first integral of a differential equation of the nth *order, to deduce the corresponding singular integral, we must eliminate* c *between the first integral in question and the equation* $\frac{dy_{n-1}}{dc} = 0$, *where* y_{n-1} *is the value of* $\frac{d^{n-1}y}{dx_{n-1}}$ *expressed in terms of* $x, y \dots \frac{d^{n-2}y}{dx^{n-2}}$, &c. *by means of the given first integral.*

If the proposed first integral is rational and integral in form,

15—2

then representing it by $\phi=0$, it suffices to eliminate c between the equations,

$$\phi=0,\ \frac{d\phi}{dc}=0 \ \ldots\ldots\ldots\ldots\ldots\ldots\ldots\ (46).$$

It is unnecessary to dwell on the particular cases of exception after what has been said on this subject in Chap. VIII.

Ex. 1. The differential equation

$$y-xy_1+\frac{x^2}{2}y_2-(y_1-xy_2)^2-y_2{}^2=0,$$

has for a first integral

$$y-\frac{x}{2}(b+y_1)-\frac{(b-y_1)^2}{x^2}-b^2=0;$$

required the corresponding singular integral.

Differentiating the first integral with respect to b, we find

$$\frac{x}{2}+\frac{2(b-y_1)}{x^2}+2b=0,$$

whence $b=\dfrac{4y_1-x^3}{4(1+x^2)}$, and this value substituted in the given integral, leads to

$$y-\frac{xy_1}{2}-\frac{y_1{}^2}{x^2}+\frac{(4y_1-x^3)^2}{16x^2(1+x^2)}=0,$$

or, on reduction, $16(1+x^2)y-8x^3y_1-16xy_1+x^4-16y_1{}^2=0$.

In connexion with this subject, Lagrange has established the following propositions:

1st. Either of the first two integrals of a differential equation of the second order leads to the same singular integral of that equation.

2nd. The complete primitive of a singular integral of a differential equation of the second order will itself be a singular solution of that equation, but a singular solution of a singular integral will in general not be a solution at all of that equation.

The proof of these propositions will afford an exercise for the student.

10. We proceed to inquire how singular integrals may be determined from the differential equation.

Expressing as before the first integral involving an arbitrary constant in the form

$$y_{n-1} = f(x, y, y_1 \dots y_{n-2}, c) \text{ .... } (47),$$

we have as the derived equation

$$y_n = \left\{\frac{d f(x, y, y_1 \dots y_{n-2}, c)}{dx}\right\} \text{ .... } (48),$$

the brackets in the second member indicating that in effecting the differentiation $y, y_1, \dots y_{n-2}$, are to be regarded as functions of x. The differential equation of the second order is found from (47) by substituting therein, after the differentiation, for c its value in terms of $x, y, y_1, \dots y_{n-1}$, given by (46). The result assumes the form

$$y_n = \phi(x, y, y_1 \dots y_{n-1}) \text{ .... } (49).$$

Hence, we have

$$\frac{dy_n}{dy_{n-1}} \text{ in } (48) = \frac{dy_n}{dc} \text{ in } (47) \times \frac{dc}{dy_{n-1}} \text{ in } (46),$$

or, representing $f(x, y, y_1 \dots y_{n-2}, c)$, by f,

$$\frac{dy_n}{dy_{n-1}} \text{ in } (48) = \left(\frac{d^2 f}{dx\, dc}\right) \div \frac{df}{dc}.$$

Hence,

$$\frac{dy_n}{dy_{n-1}} = \left(\frac{d}{dx} \log \frac{dy_{n-1}}{dc}\right) \text{ .... } (50),$$

provided that the first member be obtained from the differential equation, and the second member from one of its first integrals involving c as arbitrary constant. It is to be borne in mind that in effecting the differentiation with respect to x in the second member, we must regard $y, y_1, \dots y_{n-2}$ as functions of x.

Now reasoning as in Chap. VIII. since a singular solution makes $\frac{dy_{n-1}}{dc}=0$, it makes its logarithm, and in general the differential of its logarithm, infinite. Thus we arrive at the following conclusion.

A singular integral of a differential equation of the nth order will in general satisfy the condition $\frac{dy_n}{dy_{n-1}}=\infty$, and a relation which satisfies both this condition and the differential equation will be a singular integral.

Ex. 2. Applying this method to the equation,

$$y-xy_1+\frac{x^2}{2}y_2-(y_1-xy_2)^2-y_2^{\,2}=0,$$

we find, on differentiating with respect to y_1 and y_2 only,

$$\{x-2(y_1-xy_2)\}\,dy_1+\left\{\frac{x^2}{2}+2x(y_1-xy_2)-2y_2\right\}dy_2=0,$$

whence

$$\frac{dy_2}{dy_1}=\frac{2(y_1-xy_2)-x}{\frac{x^2}{2}+2x(y_1-xy_2)-2y_2}.$$

Equating the denominator of this expression to 0, we find

$$y_2=\frac{x^2+4xy_1}{4(x^2+1)},$$

and substituting this value in the given differential equation, clearing of fractions, and dividing by x^2+1, which will present itself as a common factor,

$$16x^2y+16y-8x^3y_1-16xy_1-16y_1^{\,2}+x^4=0,$$

a singular integral. The equation given and the result agree with those of Ex. 1.

EXERCISES.

1. $\frac{d^2y}{dx^2} = x + \sin x.$

2. $\frac{d^2y}{dx^2} = \frac{-a^2}{(2ax - x^2)^{\frac{3}{2}}}.$

3. $\frac{d^2y}{dx^2} = \frac{1}{\sqrt{(ay)}}.$

4. $\frac{d^2y}{dx^2} + \frac{1}{x}\frac{dy}{dx} = 0.$

5. $\frac{d^2y}{dx^2} = \frac{2y}{x^2}.$

6. $(1 + x^2)\frac{d^2y}{dx^2} + 1 + \left(\frac{dy}{dx}\right)^2 = 0.$

7. $y\frac{d^2y}{dx^2} + \left(\frac{dy}{dx}\right)^2 = 1.$

The two following are reducible to Clairaut's form.

8. $\frac{dy}{dx} - x\frac{d^2y}{dx^2} = f\left(\frac{d^2y}{dx^2}\right).$

9. $\left(\frac{dy}{dx}\right)^2 - y\frac{d^2y}{dx^2} = \frac{dy}{dx}f\left\{\left(\frac{dy}{dx}\right)^{-1}\frac{d^2y}{dx^2}\right\}.$

10. Describe the different kinds of homogeneity in differential equations, and explain their connexion.

The two following homogeneous equations are intended to be solved by the method developed in Art. 3. Solutions obtained by the more usual process will be found in Gregory's *Examples*, p. 334.

11. $x^3\frac{d^2y}{dx^2} = \left(y - x\frac{dy}{dx}\right)^2.$

12. $x^4\frac{d^2y}{dx^2} = (x^2 + 2xy)\frac{dy}{dx} - 4y^2.$

13. Shew that the linear equation $\frac{d^2y}{dx^2} + P\frac{dy}{dx} + Qy = 0$, belongs to one of the homogeneous classes, and is reducible to an equation of the first order by assuming $y = \epsilon^{\int u dx}$.

14. Solve the linear equation $\frac{d^2y}{dx^2} + P\frac{dy}{dx} + \frac{dP}{dx}y = 0$.

15. Mainardi has remarked (Tortolini, Vol. I. p. 76), that Liouville's equation Art. 7, becomes integrable if multiplied by the factor $\left(\frac{dy}{dx}\right)^{-1}$. Applying this method, deduce the complete primitive.

16. Liouville's equation may also be solved by suppressing the *second* term and regarding the arbitrary constant in the first integral of the result as an unknown function of x.

17. Shew that the equation $\frac{d^2y}{dx^2} + P\frac{dy}{dx} + Q\left(\frac{dy}{dx}\right)^2$ is integrable in the following cases, viz. 1st, when P and Q are both functions of x, 2ndly, when they are both functions of y, 3rdly, when P is a function of x, and Q a function of y.

18. Given $\left(\frac{dy}{ds}\right)^2 = a\frac{dx}{ds}\frac{d^2x}{ds^2}$.

19. Given $s = \sqrt{(x^2 + y^2)}$. (Transform to polar coordinates.)

20. Given $s = a\frac{dy}{dx}$. Determine the relation between y and x, so that when $x = 0$, we may have $y = 0$, and $\frac{dy}{dx} = 0$.

21. Equations homogeneous with respect to x, y, and s can be integrated by the assumption $x = \epsilon^\theta$, $y = \epsilon^\theta u$.

22. Given $\frac{ds}{dy} + 3y\frac{d^2s}{dy^2} = 0$, required the complete primitive relation between x and y.

23. $s = \sqrt{(x^2 + 2cx)}$.

24. $s = \sqrt{(y^2 + mx^2)}$.

25. Examine the solution of Ex. 24, when $m = 1$ and when $m = 0$.

26. $\dfrac{d^2x}{dt^2} = -\dfrac{\mu}{x^3} + \dfrac{k}{x^3}\left(\dfrac{dx}{dt}\right)^2$.

27. Shew that $x\dfrac{d^2y}{dx^2}$ is an exact differential coefficient.

28. Shew that $y^2 + (2xy - 1)\dfrac{dy}{dx} + x\dfrac{d^2y}{dx^2} + x^2\dfrac{d^3y}{dx^3} = 0$ is an exact differential equation, and deduce a first integral.

29. The equation $\dfrac{d^2y}{dx^2} + \dfrac{a^2y}{(y^2+x^2)^2}$ becomes integrable by means of the factor $2x^2\dfrac{dy}{dx} - 2xy$. (Moigno, Tom. II. p. 672.) Deduce hence a first integral.

30. Deduce also the complete primitive.

31. Find a singular integral of the equation

$$\left(\frac{d^2y}{dx^2}\right)^2 - \frac{2}{x}\frac{dy}{dx}\frac{d^2y}{dx^2} + 1 = 0.$$

32. Hence deduce a singular solution of the given differential equation.

33. The complete primitive of the differential equation of the second order in Ex. 31 is required.

34. A first integral of the differential equation of the second order $y - xy_1 + \dfrac{x^2}{2}y_2 - (y_1 - xy_2)^2 - y_2^{\,2} = 0$ is

$y + \left(\dfrac{a}{2} - a^2\right)x^2 - (1 - 2a)\,xy_1 - a^2 - y_1^{\,2} = 0$, where y_1 stands for $\dfrac{dy}{dx}$. Hence deduce the singular integral. Shew that it agrees, and ought to agree, with the result obtained in Art. 10.

35. Shew that the complete primitive of the above differential equation is $y = \dfrac{a}{2}x^2 + bx + a^2 + b^2$.

36. The singular integral of the differential equation of the second order, above referred to, has been found to be

$$16(1+x^2)y-8x^3y_1-16xy_1+x^4-16y_1^2=0. \quad \text{Ex. 2, Art. 10.}$$

Shew that this singular integral has for its complete primitive

$$(16y+4x^2+x^4)^{\frac{1}{2}}=x(1+x^2)^{\frac{1}{2}}-\log\{(1+x^2)^{\frac{1}{2}}-x\}+h,$$

h being an arbitrary constant—and that this is a singular solution of the proposed differential equation of the second order.

37. The same singular integral has for its singular solution $16y+4x^2-x^4=0$. Prove this. Have we a right to expect that this will satisfy the differential equation of the second order?

38. By reasoning similar to that of Chap. VIII. Art. 15, shew that a singular integral of a differential equation of the form $y_n+f(x, y, y_1 \cdots y_{n-1})=0$, will render the integrating factor of that equation infinite.

39. Differential equations of the form $\frac{d^2y}{dx^2}=f\left(\frac{dy}{dx}\right)$ can be integrated by obtaining two first integrals of the respective forms $x=f(p, c)$, $y=f_1(p, c)$, and equating the values of p.

40. Prove the assertion in Art. 9, that a singular solution of a singular integral of a differential equation of the second order is in general no solution at all of the equation given.

CHAPTER XI.

GEOMETRICAL APPLICATIONS.

1. In what manner differential equations afford the appropriate expressions of those properties of curves which involve the ideas of direction, tangency or curvature, has been explained in Chap. I. Art. 11.

Of the suggested problem in which from the expression of a property involving some one or more of the above elements it is required to determine, by the solution of a differential equation, the family of curves to which that property belongs, some illustrations have also been given in the foregoing chapters.

Here we propose to consider that problem somewhat more generally.

The following expressions furnished by the Differential Calculus are convenient for reference.

For a plane curve referred to rectangular coordinates x and y, representing also $\frac{dy}{dx}$ by p, $\frac{d^2y}{dx^2}$ by q,

Tangent $=\frac{y(1+p^2)^{\frac{1}{2}}}{p}$. Subtan $=\frac{y}{p}$.

Normal $=y(1+p^2)^{\frac{1}{2}}$. Subnormal $=yp$.

Intercept on axis $x=x-\frac{y}{p}$.

Intercept on axis $y=y-xp$.

Dist. from origin to foot of normal $=x+yp$.

Perpendicular from (a, b) on tangent $=\frac{y-b-(x-a)p}{(1+p^2)^{\frac{1}{2}}}$.

Perpendicular from (a, b) on normal $=\frac{a-x+(b-y)p}{(1+p^2)^{\frac{1}{2}}}$.

Radius of curvature $= \mp \dfrac{(1+p^2)^{\frac{3}{2}}}{q}$.

Coordinates (α, β) of centre of curvature

$$\alpha = x - \frac{p(1+p^2)}{q}, \quad \beta = y + \frac{1+p^2}{q}.$$

To these may be added the well-known formulæ for the differentials of arcs, areas, &c.

It is evident from the above forms that problems which relate only to direction or tangency, give rise to differential equations of the first order—problems which involve the conception of curvature to equations of the second order.

When the conditions of a geometrical problem have been expressed by a differential equation, and that equation has been solved, it will still be necessary to determine the species of the solution—general, particular, or singular, as also its geometrical significance.

2. The class of problems which first presents itself, is that in which it is required to determine a family of curves by the condition that some one of the elements whose expressions are given above shall be constant.

Ex. 1. Required to determine the curves whose subnormal is constant.

Here $y\dfrac{dy}{dx} = a$, and integrating,

$$\frac{y^2}{2} = ax + c,$$

$$y = (2ax + c)^{\frac{1}{2}}.$$

The property is seen to belong to the parabola whose parameter is double of the constant distance in question, and whose axis coincides with the axis of x, while the position of the vertex on that axis is arbitrary.

Ex. 2. Required a curve in which the perpendicular from the origin upon the tangent is constant and equal to a.

Here we have

$$y - xp = a(1+p^2)^{\frac{1}{2}},$$

an equation of Clairaut's form, of which the complete primitive is

$$y = cx + (1 + c^2)^{\frac{1}{2}} a,$$

and the singular solution

$$x^2 + y^2 = a^2.$$

The former denotes a family of straight lines whose distance from the origin is equal to a, the latter a circle whose centre is at the origin, and whose radius is equal to a. And here, as was noted generally by Lagrange, the singular solution seems to be, in relation to geometry, the more important of the two.

3. A more general class of problems is that in which it is required to determine the curves in which some one of the foregoing elements, Art. 1, is equal to a given function of the abscissa x.

Ex. 1. Required the class of curves in which the subtangent is equal to $f(x)$.

Here we have

$$y = f(x) \frac{dy}{dx};$$

whence

$$\frac{dy}{y} = \frac{dx}{f(x)},$$

$$y = C\epsilon^{\int \frac{dx}{f(x)}}.$$

Thus if the proposed function were x^2, we should have

$$y = C\epsilon^{-\frac{1}{x}},$$

as the equation required.

Ex. 2. Required the family of curves in which the radius of curvature is equal to $f(x)$.

Here we have

$$\frac{\frac{d^2y}{dx^2}}{\left\{1 + \left(\frac{dy}{dx}\right)^2\right\}^{\frac{3}{2}}} = \mp \frac{1}{f(x)},$$

whence, multiplying by dx and integrating,

$$\frac{\frac{dy}{dx}}{\left\{1+\left(\frac{dy}{dx}\right)^2\right\}^{\frac{1}{2}}} = \mp \int \frac{dx}{f(x)}$$

$$= \mp (X + C),$$

X representing the integral of $\frac{dx}{f(x)}$. Hence we find by algebraic solution

$$\frac{dy}{dx} = \frac{X + C}{\{1-(X+C)^2\}^{\frac{1}{2}}},$$

$$y = \int \frac{(X+C)\,dx}{\{1-(X+C)^2\}^{\frac{1}{2}}} + C_1,$$

in which it only remains to substitute for X its value, and effect the remaining integration.

If $f(x)$ is constant and equal to a, we find

$$X = \frac{x}{a} + C = \frac{x + aC}{a},$$

$$y = \int \frac{(x+aC)\,dx}{\sqrt{\{a^2-(x+aC)^2\}}} + C_1,$$

$$= -\{a^2-(x+aC)^2\}^{\frac{1}{2}} + C_1,$$

whence $$(y - C_1)^2 + (x + aC)^2 = a^2,$$

and this represents a circle whose centre is arbitrary in position, and whose radius is a.

A yet more general class of problems is that in which it is required that one of the elements expressed in Art. 1 should be expressed by a given function of x and y.

An example of this class is given in Chap. VII. Art. 10.

4. We proceed in the next place to consider certain problems in which more than one of the elements expressed in Art. 1, are involved.

Ex. 1. To determine the curves in which the radius of curvature is equal to the normal.

If the radius of curvature have the same direction as the normal we shall have

$$\frac{-\left\{1+\left(\frac{dy}{dx}\right)^2\right\}^{\frac{3}{2}}}{\frac{d^2y}{dx^2}}=y\left\{1+\left(\frac{dy}{dx}\right)^2\right\}^{\frac{1}{2}}\ldots\ldots\ldots\ldots(1),$$

whence

$$y\frac{d^2y}{dx^2}+\left(\frac{dy}{dx}\right)^2+1=0\ldots\ldots\ldots\ldots\ldots\ldots\ldots(2).$$

The first side multiplied by dx is an exact differential and gives

$$y\frac{dy}{dx}+x=c,$$

whence again integrating

$$y^2+x^2=2cx+c'\ldots\ldots\ldots\ldots\ldots\ldots\ldots(3),$$

the equation of a circle whose centre is on the axis of x.

If the direction of the radius of curvature be opposite to that of the normal, it will be necessary to change the sign of the first member of (1). Instead of (2) we shall have

$$y\frac{d^2y}{dx^2}-\left(\frac{dy}{dx}\right)^2-1=0\ldots\ldots\ldots\ldots\ldots\ldots\ldots(4),$$

and this equation not containing x, we may depress it to the first order by assuming $\frac{dy}{dx}=p$. The transformed equation is

$$yp\frac{dp}{dy}-p^2-1=0,$$

whence,

$$\frac{pdp}{p^2+1}=\frac{dy}{y}$$

$$y=c(1+p^2)^{\frac{1}{2}}.$$

Substituting for p its value $\frac{dy}{dx}$, we find on algebraic solution

$$dx = \frac{c\,dy}{\sqrt{(y^2 - c^2)}},$$

whence, $$x = c' + c \log \{y + (y^2 - c^2)^{\frac{1}{2}}\} \text{..............} (5).$$

This equation, reduced to the exponential form

$$y = \frac{a}{2}\left(\epsilon^{\frac{x-b}{a}} + \epsilon^{-\frac{x-b}{a}}\right), \text{..................} (6),$$

is seen to represent a catenary.

The solution therefore indicates a circle when the direction of the radius of curvature and of the normal are the same, but a catenary when they are opposed. The latter curve has, however, many properties analogous to those of the circle. (Lacroix, Tom. II. p. 459.)

Ex. 2. To find a curve in which the area, as expressed by the formula $\int y\,dx$, is in a constant ratio to the corresponding arc.

We have $$y = C\,(1 + p^2)^{\frac{1}{2}},$$

which, agreeing in form with the last differential equation of the preceding problem, shews that (5) represents the curve required, and connects together the properties noticed in the last two examples.

Ex. 3. Required the class of curves in which the length of the normal is a given function of the distance of its foot from the origin.

The differential equation is

[illegible]

and it belongs to the remarkable class discussed in Chap. VII. Art. 9, where the complete primitive is given, viz.

$$y^2 + (x - a)^2 = \{f(a)\}^2 \text{..................} (2).$$

This represents a circle whose centre is situated on the axis of x at a distance a from the origin, and whose radius is equal to

$f(a)$. It is evident that this circle satisfies the geometrical conditions of the problem.

But there is also a singular solution, found by eliminating the constant a between (2) and the equation thence derived by differentiation with respect to a, viz.

$$x - a + f(a) f'(a) = 0 \text{..................} (3).$$

For instance, if $f(a) = n^{\frac{1}{2}} a^{\frac{1}{2}}$ we have to eliminate a between the equations

$$y^2 + (x-a)^2 = na,$$

$$2(x-a) + n = 0,$$

from which we find

$$y^2 = nx + \frac{n^2}{4},$$

the equation of a parabola. While in this example the complete primitive represents circles only, the singular solution represents an infinite variety of distinct curves, each originating in a distinct form of the function $f(a)$. Other illustrations of this remark will be met with.

The above problem was first discussed by Leibnitz, who did not, however, regard its solution as dependent upon that of a differential equation, but, establishing by independent considerations the equation (2), which constitutes in the above mode of treatment the complete primitive of a differential equation, arrived at a result equivalent to its singular solution by that kind of reasoning which is employed in the geometrical theory of envelopes. Indeed it was in the discussion of this problem that the foundations of that theory were laid (Lagrange, *Calcul des Fonctions*, p. 268).

5. A certain historic interest belongs also to the two following problems, famous in the earlier days of the Calculus, viz. the problem of 'Trajectories' and the problem of 'Curves of pursuit.' These we shall consider next. They will serve to illustrate in some degree the modes of consideration by which the differential equations of a problem are formed when a mere table of analytical expressions suffices no longer.

Trajectories.

Supposing a system of curves to be described, the different members differing only through the differing values given to an arbitrary constant in their common equation—a curve which intersects them all at a constant angle is called a trajectory, and when the angle is right, an orthogonal trajectory.

To determine the orthogonal trajectory of a system of curves represented by the equation

$$\phi(x, y, c) = 0 \ldots\ldots\ldots\ldots\ldots\ldots\ldots(1).$$

Representing for brevity $\phi(x, y, c)$ by ϕ, we have on differentiating

$$\frac{d\phi}{dx}dx + \frac{d\phi}{dy}dy = 0.$$

Hence, for the intersected curves,

$$\frac{dy}{dx} = -\frac{d\phi}{dx} \div \frac{d\phi}{dy}.$$

Now representing this value by m, and the corresponding value of $\frac{dy}{dx}$ for the trajectory by m', we have, by the condition of perpendicularity, $m' = \frac{-1}{m}$. Hence for the trajectory

$$\frac{dy}{dx} = \frac{d\phi}{dy} \div \frac{d\phi}{dx},$$

or

$$\frac{d\phi}{dy}dx - \frac{d\phi}{dx}dy = 0 \ldots\ldots\ldots\ldots\ldots\ldots(2),$$

which must be true for all values of c. Hence the *differential equation of the orthogonal trajectory will be found by eliminating c between* (1) *and* (2).

Were the equation of the system of intersected curves presented in the form

$$\phi(x, y, a, b) = 0,$$

a and b being connected by a condition

$$\psi(a, b) = 0,$$

we should have to eliminate a and b between the above two equations, and the equation

$$\frac{d\phi\,(x, y, a, b)}{dy}\,dx - \frac{d\phi\,(x, y, a, b)}{dx}\,dy = 0.$$

We shall exemplify both forms of the problem.

Ex. 1. Required the orthogonal trajectory of the system of curves represented by the equation $y = cx^n$.

Here $\phi = y - cx^n$, whence by (2)

$$dx + ncx^{n-1}\,dy = 0.$$

Eliminating c,

$$xdx + nydy = 0;$$

therefore $$x^2 + ny^2 = c',$$

the equation required. We see that the trajectory will be an ellipse for all positive values of n except $n = 1$,—an ellipse, therefore, when the intersected curves are a system of common parabolas. The trajectory is a circle if $n = 1$, the intersected system then being one of straight lines passing through the origin. The trajectory is an hyperbola if n is negative.

Ex. 2. Required the orthogonal trajectory of a system of confocal ellipses.

The general equation of such a system is

$$\frac{x^2}{a^2} + \frac{y^2}{b^2} = 1,$$

a and b being connected by the condition

$$a^2 - b^2 = h^2,$$

where h is the semi-distance of the foci, and does not vary from curve to curve. Hence we have to eliminate a and b from the above equations, and the equation

$$\frac{y}{b^2}\,dx - \frac{x}{a^2}\,dy = 0;$$

16—2

the result is

$$xy\left(\frac{dy}{dx}\right)^2+(x^2-y^2-h^2)\frac{dy}{dx}-xy=0,$$

the solution of which may be deduced from that of Ex. 3, Chap. VII. Art. 10, by assuming therein $A=1$, $B=h^2$. We find

$$y^2-c^2x^2=-\frac{h^2c^2}{c^2+1},$$

and this may be reduced to the form

$$\frac{x^2}{a_1^2}-\frac{y^2}{b_1^2}=1,$$

a_1 and b_1 being connected by the condition

$$a_1^2+b_1^2=h^2.$$

Thus the trajectory is a system of hyperbolas confocal with the given system of ellipses.

6. When the trajectory is oblique, then θ being the angle which it makes with each curve of the system, and m and m' having the same significations as before,

$$m'=\frac{m+\tan\theta}{1-m\tan\theta},$$

or, substituting for m its former value $-\frac{d\phi}{dx}\div\frac{d\phi}{dy}$, and for m' its value $\frac{dy}{dx}$ as referred to the trajectory, we have on reduction

$$\frac{dy}{dx}=\frac{\frac{d\phi}{dy}\tan\theta-\frac{d\phi}{dx}}{\frac{d\phi}{dy}+\frac{d\phi}{dx}\tan\theta}\ \ldots\ldots\ldots\ldots\ldots\ldots(3),$$

an equation from which it only remains to eliminate c by means of the given equation in order to obtain the differential equation of the trajectory.

Ex. Required the general equation of the trajectories of the system of straight lines $y=ax$.

Here $\phi = y - ax$, whence by (3)

$$\frac{dy}{dx} = \frac{\tan\theta + a}{1 - a\tan\theta}$$

$$= \frac{x\tan\theta + y}{x - y\tan\theta},$$

or $$(y + x\tan\theta)\,dx + (y\tan\theta - x)\,dy = 0,$$

a homogeneous equation, an integrating factor of which being $\frac{1}{x^2+y^2}$, we have

$$\frac{ydx - xdy}{x^2+y^2} + \tan\theta\,\frac{xdx + ydy}{x^2+y^2} = 0,$$

whence integrating

$$\tan^{-1}\frac{x}{y} + \tan\theta\log(x^2+y^2)^{\frac{1}{2}} = c.$$

If we change the coordinates by assuming $x = r\cos\phi$, $y = r\sin\phi$, we get

$$r = C_1 \epsilon^{\frac{\phi}{\tan\theta}},$$

the equation of a logarithmic spiral.

The following example, which is taken from a Memoir by Mainardi (Tortolini's *Annali di Scienze Matematiche e Fisiche*, Tom. I. 251), is chiefly interesting from the mode in which the integration is effected.

Required the oblique trajectory of a system of confocal ellipses.

Representing the tangent of the angle of intersection by n, we have to eliminate a and b between the equations

$$\frac{x^2}{a^2} + \frac{y^2}{b^2} = 1, \quad a^2 - b^2 = h^2,$$

$$p = \frac{n\frac{y}{b^2} - \frac{x}{a^2}}{\frac{y}{b^2} + n\frac{x}{a^2}}.$$

The result may be expressed in the form

$$\{nx + y + (ny - x) p\} \{x - ny + (nx + y) p\} = h^2 (n - p) (1 + np).$$

To integrate this equation let us assume

$$x - ny + (nx + y) p = M (1 + np),$$

$$M \{nx + y + (ny - x) p\} = h^2 (n - p).$$

As these on multiplication reproduce the given equation, the variable M is really subject to a single condition only, and the assumption is legitimate.

Eliminating p from the last two equations, and dividing by $1 + n^2$, we have

$$(x^2 + y^2 + h^2) M = x (M^2 + h^2) \text{ .... } (a).$$

Differentiating this equation and eliminating y and p from the result by the aid of any two of the last three equations (it is evident that two only are independent), we obtain a differential equation between M and x, which is capable of expression in the form

$$\frac{nd (xM)}{\{h^2 (xM) - (xM)^2\}^{\frac{1}{2}}} + \frac{d \frac{M}{x}}{\frac{M}{x} \left(1 - \frac{M}{x}\right)^{\frac{1}{2}}} = 0.$$

Hence, by integration

$$2n \tan^{-1} \sqrt{\left(\frac{h^2}{xM} - 1\right)} + \log \frac{1 - \sqrt{\left(1 - \frac{M}{x}\right)}}{1 + \sqrt{\left(1 - \frac{M}{x}\right)}} = C,$$

in which it is only necessary to substitute for M its value in terms of x and y deduced from (a).

Curves of Pursuit.

7. The term curve of pursuit is given to the path which a point describes when moving with uniform velocity toward another point which moves with uniform velocity in a given curve.

Let x, y be the coordinates of the pursuing point, x', y' the simultaneous coordinates of the point pursued. Also let the equation of the given path of the latter be

$$f(x', y') = 0 \qquad\qquad (4).$$

Now the point pursued being always in the tangent to the path of the point which pursues, its coordinates must satisfy the equation of that tangent. Hence,

$$y' - y = \frac{dy}{dx}(x' - x) \qquad\qquad (5).$$

Lastly, the velocities of the two points being uniform, the corresponding elementary arcs will be in the constant ratio of the velocities with which they are described. Hence, if the velocity of the pursuing point be to that of the point pursued as $n : 1$, we have

$$n\sqrt{(dx'^2 + dy'^2)} = \sqrt{(dx^2 + dy^2)},$$

or, taking x as independent variable,

$$n\sqrt{\left\{\left(\frac{dx'}{dx}\right)^2 + \left(\frac{dy'}{dx}\right)^2\right\}} = \sqrt{\left\{1 + \left(\frac{dy}{dx}\right)^2\right\}} \qquad (6),$$

the sign to be given to each radical being positive or negative, according as the motion tends to increase or to diminish the corresponding arc.

From (4) and (5), when the form of the function $f(x', y')$ is determined, x' and y' may be found in terms of x, y, and $\frac{dy}{dx}$, and these values enable us to reduce (6) to an equation between x, y, $\frac{dy}{dx}$, $\frac{d^2y}{dx^2}$. It only remains to solve this differential equation of the second order. If the signs of the radicals are both changed, the motion in each curve is simply reversed, and the curve of pursuit becomes a curve of flight. But the differential equation remaining unchanged, the forms of the curves are unchanged, and only their relation inverted.

Ex. A particle which sets off from a point in the axis of x, situated at a distance a from the origin, and moves uniformly in a vertical direction parallel to the axis of y, is pursued by a particle which sets off at the same moment from the origin and travels with a velocity which is to that of the former as $n : 1$. Required the path of the latter.

The equation of the path of the first particle being $x' = a$, (5) becomes

$$y' - y = \frac{dy}{dx}(a - x),$$

whence

$$y' = y + (a - x)\frac{dy}{dx}.$$

Thus we have

$$\frac{dx'}{dx} = 0, \; \frac{dy'}{dx} = (a - x)\frac{d^2y}{dx^2},$$

and the differential equation, both radicals being positive, is

$$n(a - x)\frac{d^2y}{dx^2} = \sqrt{\left\{1 + \left(\frac{dy}{dx}\right)^2\right\}} \text{ .... } (a).$$

Hence,

$$\frac{\dfrac{d^2y}{dx^2}}{\sqrt{\left\{1 + \left(\dfrac{dy}{dx}\right)^2\right\}}} = \frac{1}{n(a - x)}.$$

Multiplying by dx and integrating

$$\frac{dy}{dx} + \sqrt{\left\{1 + \left(\frac{dy}{dx}\right)^2\right\}} = c(a - x)^{-\frac{1}{n}};$$

$$\therefore \; \frac{dy}{dx} = \frac{1}{2}\left\{c(a - x)^{-\frac{1}{n}} - \frac{1}{c}(a - x)^{\frac{1}{n}}\right\}.$$

Hence, if n be not equal to 1,

$$y = \frac{1}{2}\left\{\frac{c(a - x)^{1 - \frac{1}{n}}}{\frac{1}{n} - 1} + \frac{1}{c}\frac{(a - x)^{1 + \frac{1}{n}}}{1 + \frac{1}{n}}\right\} + c' \text{ .... } (b).$$

But if n be equal to 1, we have, on substituting C for $\frac{1}{2}\left(\frac{1}{c} - c\right)$,

$$\frac{dy}{dx} = C(x - a),$$

whence

$$y = \frac{C(x-a)^2}{2} + c' \ldots\ldots\ldots\ldots\ldots\ldots \text{(c)},$$

which represents a parabola.

8. The class of problems which we shall next consider is introduced chiefly on account of the instructive light which it throws upon the singular solutions of differential equations of the second order.

Inverse Problems in Geometry and Optics.

The problems we are about to discuss are the following: 1st, To determine the involute of a plane curve. 2ndly, To determine the form of the reflecting curve which will produce a given *caustic*, the incident rays being supposed parallel.

In both these problems we shall have occasion in a particular part of the process to solve a differential equation of the first order of the form

$$y - x\phi(p) = ff'^{-1}\phi(p) - \phi(p)\, f'^{-1}\phi(p) \ldots\ldots\ldots (7),$$

in which ϕ and f are functional symbols of given interpretation, and f'^{-1} is a functional symbol whose interpretation is inverse to that of the symbol f'. Thus, if $f(x) = \sin x$, then

$$f'(x) = \cos x, \quad f'^{-1}(x) = \cos^{-1} x.$$

It will somewhat less interrupt the theoretical observations for the sake of which the above problems are chiefly valuable, if we solve the equation (7) under its general form first.

Referring to Chap. VII. Art. 7, we see that (7) will become linear if we transform it so as to make either of the primitive variables the dependent variable, and either p or any function of p the independent variable.

Let us then assume

$$\phi(p) = v,$$

and transform the differential equation so as to make x and v the new variables.

Substituting v for $\phi(p)$ in (7), we have

$$y - xv = ff'^{-1}(v) - vf'^{-1}(v) \quad\text{.................} \quad (8).$$

Differentiating, and regarding v as independent variable,

$$\frac{dy}{dv} - x - v\frac{dx}{dv} = v\frac{d}{dv}f'^{-1}(v) - f'^{-1}(v) - v\frac{d}{dv}f'^{-1}(v)$$

$$= -f'^{-1}(v).$$

But

$$\frac{dy}{dv} = p\frac{dx}{dv} = \phi^{-1}(v)\frac{dx}{dv}.$$

Hence,

$$\{\phi^{-1}(v) - v\}\frac{dx}{dv} - x = -f'^{-1}(v),$$

or,

$$\frac{dx}{dv} + \frac{x}{v - \phi^{-1}(v)} = \frac{f'^{-1}(v)}{v - \phi^{-1}(v)}.$$

Hence, if for brevity we write

$$\int \frac{dv}{v - \phi^{-1}(v)} = \psi(v) \quad\text{......................} \quad (9),$$

we have

$$x = \epsilon^{-\psi(v)}\{C + \int \epsilon^{\psi(v)}\psi'(v) f'^{-1}(v)\, dv\}$$

$$= \epsilon^{-\psi(v)}\{C + \epsilon^{\psi(v)} f'^{-1}(v) - \int \epsilon^{\psi(v)} df'^{-1}(v)\},$$

whence

$$x - f'^{-1}(v) = \epsilon^{-\psi(v)}\{C - \int \epsilon^{\psi(v)} df'^{-1}(v)\} \quad\text{.........} \quad (10),$$

between which and (8), v must be eliminated.

If in those equations we make $f'^{-1}(v)=t$, they assume the somewhat more convenient form,

$$y-xf'(t)=f(t)-tf'(t),$$

$$x-t=\epsilon^{-\psi f'(t)}\{C-\int\epsilon^{\psi f'(t)}\,dt\},$$

and these may yet further be reduced to the form

$$x-t=\frac{y-f(t)}{f'(t)}=\frac{C-\int\epsilon^{\psi f'(t)}\,dt}{\epsilon^{\psi f'(t)}} \quad\text{..........}\quad (11).$$

From these equations it only remains to eliminate t, the forms of f and ϕ being specified, and that of ψ given by (9); and this is apparently the simplest form of the solution.

9. We shall now proceed to the special problems under consideration.

To determine the involute of a plane curve.

It is evident from the equations which present themselves in the investigation of the radius of curvature, that if x, y be the coordinates of any point in a plane curve, and x', y' those of the corresponding point in the evolute, then

$$x'=x-\frac{p(1+p^2)}{q},\quad y'=y+\frac{1+p^2}{q},$$

where $p=\dfrac{dy}{dx}$, $q=\dfrac{d^2y}{dx^2}$ (Todhunter's *Differential Calculus*, p. 295). Hence, if the equation of the evolute be

$$y'=f(x') \quad\text{........................}\quad (12),$$

we shall have on substituting therein for y' and x' the values above given,

$$y+\frac{1+p^2}{q}=f\left\{x-\frac{p(1+p^2)}{q}\right\} \quad\text{.............}\quad (13),$$

a differential equation of the second order connecting x and y, and therefore true for each point of the curve whose evolute is given. Of that evolute the curve in question is an *involute*.

Hence, if $y' = f(x')$ be the equation of a given curve, the equation of its involute will satisfy the differential equation (13).

Now suppose that nothing was known of the genesis of the above equation, and that it was required to deduce its complete primitive, and its singular solution, should such exist.

Upon examination the equation (13) will prove to be of a kind analogous to that of Chap. VII. Art. 9. If we assume

$$x - \frac{p(1+p^2)}{q} = a \qquad (14),$$

$$y + \frac{1+p^2}{q} = b \qquad (15),$$

a and b being arbitrary constants, we shall find that each of these leads by differentiation to the same differential equation of the third order, viz.

$$3pq^2 - (1+p^2)\, r = 0 \qquad (16),$$

where r stands for $\frac{d^3y}{dx^3}$. It follows hence, that a first integral of (13) will be found by eliminating q between (14) and (15), and connecting the arbitrary constants b and a by the relation $b = f(a)$. Eliminating q, we find

$$x - a + (y-b)\, p = 0 \qquad (17),$$

wherein making $b = f(a)$, we have

$$x - a + \{y - f(a)\}\, p = 0 \qquad (18),$$

for the first integral in question. Again, integrating, we have

$$(x-a)^2 + \{y - f(a)\}^2 = r^2 \qquad (19),$$

in which a and r are arbitrary constants. This is the complete primitive of (13). It is manifest from its form that it represents, not the involute of the given curve, but the circles of curvature of that involute. Indeed, that the complete primitive cannot represent the involute might have been affirmed *a priori*. The equation of the involute of a given curve cannot involve in its expression more than *one* arbitrary constant; for the only element left arbitrary in the mechanical genesis of the involute is the length of a string.

It remains to examine the singular solution of (13). This is most easily deduced by eliminating a between the first integral (18) and its derived equation with respect to a, viz. between the equations

$$x - a + \{y - f(a)\} p = 0 \ldots\ldots\ldots\ldots\ldots (20),$$

$$-1 - f'(a) p = 0 \ldots\ldots\ldots\ldots\ldots (21).$$

From the second of these we have

$$f'(a) = \frac{-1}{p},$$

$$a = f'^{-1}\left(\frac{-1}{p}\right).$$

Hence eliminating a from (20)

$$x - f'^{-1}\left(\frac{-1}{p}\right) + \left\{y - ff'^{-1}\left(\frac{-1}{p}\right)\right\} p = 0,$$

or
$$x + yp = f'^{-1}\left(\frac{-1}{p}\right) + pff'^{-1}\left(\frac{-1}{p}\right) \ldots\ldots\ldots\ldots (22),$$

which is the singular solution of (13), and the differential equation of the first order of the involute sought.

This equation is a particular case of (7). If we express it in the form

$$y - \left(\frac{-1}{p}\right) x = ff'^{-1}\left(\frac{-1}{p}\right) - \left(\frac{-1}{p}\right) f'^{-1}\left(\frac{-1}{p}\right),$$

we see that it is what (7) would become on making

$$\phi(p) = \frac{-1}{p}.$$

Hence comparing with the general solution (11) we have

$$\phi(v) = \frac{-1}{v}, \quad \phi^{-1}(v) = \frac{-1}{v},$$

$$\psi(v) = \int \frac{dv}{v + \frac{1}{v}} = \log(v^2 + 1)^{\frac{1}{2}}, \text{ by (22).}$$

Thus the system (11) becomes

$$x - t = \frac{y - f(t)}{f'(t)} = \frac{C - \int\{1 + f'(t)^2\}^{\frac{1}{2}}\,dt}{\{1 + f'(t)^2\}^{\frac{1}{2}}} \quad \ldots\ldots(23).$$

The final solution is therefore expressed in the following theorem.

Given the equation of a curve in the form $y' = f(x')$, that of its involute is found by eliminating t from the system (23).

10. Parallel rays incident, in a given direction, on a reflecting plane curve produce after reflection a caustic whose equation is given. The equation of the reflecting curve is required.

Let IP be a ray incident parallel to the axis of x on a point P in the reflecting curve SPM, Fig. 1, $PP'Q$ the reflected ray cutting the axis of x in Q and touching the caustic $S'P'M'$ in P'. Let x, y be the coordinates of P, x', y' those of P'. Let the equation of the caustic be $y' = f(x')$.

It is an easy consequence of the law of reflection that the angle PQX which the reflected ray makes with the axis of x is double of the angle PTX made by the tangent at P with the axis of x. This at once gives us the equation

$$\frac{y - y'}{x - x'} = \frac{2p}{1 - p^2},$$

where $$p = \frac{dy}{dx}.$$ Hence

$$y - y' - \frac{2p}{1 - p^2}(x - x') = 0 \quad \ldots\ldots\ldots\ldots\ldots(24).$$

As, however, (x', y') is a point at which consecutive reflected rays intersect, we are permitted to differentiate the above equation regarding x' and y' as constant while x and y vary. We thus obtain, representing $\frac{d^2y}{dx^2}$ by q,

$$p - \frac{2p}{1 - p^2} - (x - x')\,\frac{2q(1 - p^2) + 4p^2q}{(1 - p^2)^2} = 0,$$

or
$$\frac{-p-p^3}{1-p^2}-(x-x')\frac{2q(1+p^2)}{(1-p^2)^2}=0,$$

whence
$$x-x'=-\frac{p(1-p^2)}{2q},$$

and
$$x'=x+\frac{p(1-p^2)}{2q}\ldots\ldots\ldots\ldots\ldots\ldots(25).$$

Substituting this value in (24), we have
$$y-y'=\frac{2p}{1-p^2}\times-\frac{p(1-p^2)}{2q}=\frac{-p^2}{q},$$

whence
$$y'=y+\frac{p^2}{q}\ldots\ldots\ldots\ldots\ldots\ldots\ldots\ldots(26).$$

Were the equation of the reflecting curve given and that of the caustic required, it would only be necessary to substitute in (25) and (26) the values of p and q in terms of x and y derived from the former, and then by eliminating x and y from the three, to deduce the relation between x' and y'.

Conversely, to determine the reflecting curve we must eliminate x' and y' from (25), (26) and the equation of the caustic, viz. $y'=f(x')$. The result which is obtained by mere substitution is
$$y+\frac{p^2}{q}=f\left\{x+\frac{p(1-p^2)}{2q}\right\}\ldots\ldots\ldots\ldots\ldots(27),$$
a differential equation of the second order, the solution of which will determine in the fullest manner the possible relations between x and y which are consistent with the conditions of the problem.

Were this equation given and nothing known respecting its origin, we might at once infer that it is of a class analogous to those of Chap. VII. Art. 9. For writing
$$y+\frac{p^2}{q}=b,\quad x+\frac{p(1-p^2)}{2q}=a\ldots\ldots\ldots\ldots(28),$$
we find that each of these leads by differentiation to the same differential equation of the third order. For the first gives
$$3p-\frac{p^2r}{q^2}=0,$$

while the second gives

$$\frac{3}{2}-\frac{3}{2}p^2-\frac{(1-p^2)\,pr}{2q^2}=0,$$

and these lead to the same value of the differential coefficient of the third order r, viz.

$$r=\frac{3q^2}{p},$$

this constituting the essential criterion of agreement between differential equations of the third order.

Accordingly, eliminating q from (28) and afterwards making $b=f(a)$ by virtue of (27), we find

$$\frac{p}{f(a)-y}=\frac{1-p^2}{2\,(a-x)},$$

or $$y-f(a)=\frac{2p}{1-p^2}\,(x-a)\,\text{.............}\,(29),$$

which is a complete first integral of (27). We see that it agrees, and necessarily so, with (24), a only taking the place of x' and $f(a)$ that of y'.

The complete integral of (29) will be found to be

$$\{y-f(a)\}^2=4m\,(x-a)+4m^2\,\text{.............}\,(30),$$

m being an arbitrary constant. And this is the complete primitive of (27). If we substitute x' for a, which we may without loss of generality do, then $f(a)=f(x')=y'$, so that the above equation gives

$$(y-y')^2=4m\,(x-x'+m)\,\text{.............}\,(31);$$

and this is evidently the equation of a parabola whose axis is parallel to the axis of x, whose focus is upon the caustic curve, but which is in no other way limited. The complete primitive of (27) represents then a system of such parabolas.

It is plain that any such system does constitute a true solution of the problem, rays falling upon the interior arc of a parabola, and parallel to its axis, being accurately reflected to the focus.

It remains to deduce the singular solution of (27). Differentiating its first integral (29) with respect to a, we have

$$-f'(a) = -\frac{2p}{1-p^2},$$

whence $$a = f'^{-1}\left(\frac{2p}{1-p^2}\right),$$

and substituting this in (29)

$$y - ff'^{-1}\left(\frac{2p}{1-p^2}\right) = \frac{2p}{1-p^2}\left\{x - f'^{-1}\left(\frac{2p}{1-p^2}\right)\right\},$$

or $$y - \frac{2p}{1-p^2}x = ff'^{-1}\left(\frac{2p}{1-p^2}\right) - \frac{2p}{1-p^2}f'^{-1}\left(\frac{2p}{1-p^2}\right) \ldots\ldots(32).$$

This is the differential equation of the involute. Its complete integral may be deduced from the general solution in Art. 8, by making $\phi(p) = \frac{2p}{1-p^2}$, whence we have

$$\phi^{-1}(p) = \frac{-1+(1+p^2)^{\frac{1}{2}}}{p},$$

$$\psi(v) = \int\frac{dv}{v-\phi^{-1}(v)} = \int\frac{vdv}{v^2+1-(1+v^2)^{\frac{1}{2}}}$$
$$= \log\{1+\sqrt{(1+v^2)}\}.$$

Hence the system (11) becomes

$$x - t = \frac{y-f(t)}{f'(t)} = \frac{C - \int[1+\sqrt{\{1+f'(t)^2\}}]\,dt}{1+\sqrt{\{1+f'(t)^2\}}} \ldots\ldots(33),$$

from which, after the integration has been effected, t must be eliminated.

If, as before, we replace t by x', and $f(t)$ by y' and therefore $f'(t)$ by $\frac{dy'}{dx'}$, then, since we have

$$\sqrt{\{1+f'(t)^2\}}\,dt = ds'$$

where s' represents the arc of the caustic, the above system assumes the following form,

$$x - x' = \frac{y-y'}{\frac{dy'}{dx'}} = \frac{C - x' - s'}{1+\frac{ds'}{dx'}} \ldots\ldots\ldots\ldots(34),$$

from which, when s' is determined, x' and y' must be eliminated by means of the equation of the given curve.

From the above it appears that, the incident rays being parallel, the reflecting curve can always be determined when the caustic can be rectified.

We see also from the nature of the connexion between the singular solutions and the ordinary primitives of differential equations, that the reflecting curve is in reality the envelope of a system of parabolas whose axes are parallel to the direction of incident rays, whose foci are on the caustic, and whose parameters are subject to such a relation as makes that envelope to have contact of the second order with the curves out of whose differential elements it is formed. It is not merely an envelope, but an *osculating* envelope.

Analogy makes it evident that when the rays instead of being parallel issue from a given point, the reflecting curve is the osculating envelope of a system of ellipses, each of which has one focus at the radiant point, and the other on the arc of the caustic, the elliptic elements being further so conditioned as to render such osculation possible.

Lastly, it is plain that the problem of caustics in its direct and in its inverse form, as stated above, is in strict analogy with the direct and the inverse form of the problem of curvature, osculating parabolas and ellipses occupying the place and relation of osculating circles.

The above examples might also be treated by a remarkable method, the consideration of which will fitly close this chapter.

Intrinsic Equation of a Curve.

11. There are certain problems, the solution of which is much facilitated by the employment of what Dr Whewell has happily termed, the intrinsic equation of a curve, viz. the equation which expresses the relation between the length of an arc and the angle through which it bends, the latter being in more precise language the angle of deviation of the tangent from the tangent at the origin. These elements are called intrinsic because they are independent of any external lines of reference, and it will be noted that they form a system differing essentially from all systems of coordinates which begin by the defining of the position of a point, and in the application of which a curve is contemplated as a collection of points.

The conceptions of length and deviation upon which the above system is founded, might be replaced by the not less fundamental conceptions of length and curvature, the equation of the curve being then expressed in terms of its radius of curvature at the extremity of an arc and the length of that arc. Or, in place of either of these systems, we might employ that which defines a curve by the relation which connects the curvature at any point with the deviation of the tangent. Of the three elements, of length, curvature, and deviation, any two indeed will together constitute an equivalent system. Euler, in a particular class of problems, employed the combination last described. Here we shall select the one first mentioned, and shall borrow our chief illustrations of its use from the memoir of Dr Whewell (*Cambridge Philosophical Transactions*, Vol. VIII. p. 659).

Representing by s the variable length of an arc the beginning of which is assumed as origin, and by ϕ the corresponding angle of deviation, the intrinsic equation is of the form

$$s = f(\phi) \qquad (35).$$

Thus in fig. 2, $SP = s$ and $ATS = \phi$.

From this equation the ordinary equation in rectangular coordinates may be found in the following manner. Still taking the beginning of the arc as origin, let the tangent at that point be taken as the axis of x, then will the element of the curve ds be inclined at an angle ϕ to the axis x. Its projection on the axis of x will therefore be $\cos\phi ds$, and this being the differential element of the coordinate x, we have

$$dx = \cos\phi ds = \cos\phi f'(\phi)\, d\phi, \text{ by (35)}.$$

Hence
$$x = \int \cos\phi f'(\phi)\, d\phi \qquad (36),$$

and by symmetry

$$y = \int \sin\phi f'(\phi)\, d\phi \qquad (37).$$

Between these equations after integration ϕ must be eliminated; the result involving x, y and two arbitrary constants will be the equation required.

It is worth while to notice that the above result may be obtained independently of the consideration of a projection. For since $s = \int \left\{1 + \left(\frac{dy}{dx}\right)^2\right\}^{\frac{1}{2}} dx$, we have

$$\int\left\{1+\left(\frac{dy}{dx}\right)^2\right\}^{\frac{1}{2}}dx=f(\phi),$$

whence $$\left\{1+\left(\frac{dy}{dx}\right)^2\right\}^{\frac{1}{2}}dx=f'(\phi)\,d\phi\,\ldots\ldots\ldots\ldots(38).$$

But, since $\dfrac{dy}{dx}=\tan\phi$, the above becomes

$$\sec\phi dx=f'(\phi)\,d\phi,$$

$$dx=\cos\phi f'(\phi)\,d\phi,$$

$$x=\int\cos\phi f'(\phi)\,d\phi,$$

and in like manner employing for s the equivalent formula $s=\int\left\{1+\left(\dfrac{dx}{dy}\right)^2\right\}^{\frac{1}{2}}dy$, we find

$$y=\int\sin\phi f'(\phi)\,d\phi,$$

which agree with the previous expressions.

Another consequence should also be noted. From (38) we have $\left\{1+\left(\dfrac{dy}{dx}\right)^2\right\}^{\frac{1}{2}}=f'(\phi)\,\dfrac{d\phi}{dx}$.

But $$\frac{d\phi}{dx}=\frac{d}{dx}\tan^{-1}\left(\frac{dy}{dx}\right)=\frac{\dfrac{d^2y}{dx^2}}{1+\left(\dfrac{dy}{dx}\right)^2},$$ whence

$$\left\{1+\left(\frac{dy}{dx}\right)^2\right\}^{\frac{1}{2}}=f'(\phi)\,\frac{\dfrac{d^2y}{dx^2}}{1+\left(\dfrac{dy}{dx}\right)^2}.$$

Therefore $$\frac{\left\{1+\left(\dfrac{dy}{dx}\right)^2\right\}^{\frac{3}{2}}}{\dfrac{d^2y}{dx^2}}=f'(\phi).$$

Now the first member being the expression for the radius of curvature ρ of the given curve, we have

$$\rho=f'(\phi)\ldots\ldots\ldots\ldots\ldots\ldots\ldots\ldots(39).$$

Thus the radius of curvature is determined.

12. *Given the ordinary, to deduce the intrinsic equation of a curve.*

The values of s and ϕ having been first expressed in terms of the co-ordinates, it only remains to eliminate those co-ordinates between the two equations thus formed and the equation given.

Ex. To determine the intrinsic equation of the equi-angular spiral.

The polar equation of the curve being $r = C\epsilon^{m\theta}$, the arc s beginning from $\theta = 0$ is, by ordinary integration, found to be

$$s = C\frac{(m^2+1)^{\frac{1}{2}}}{m}(\epsilon^{m\theta}-1).$$

Again, as the curve cuts all its radii at the same angles the deflection of the arc between two radii vectores is equal to the angle between the radii themselves. Hence the deflection of the arc beginning with $\theta = 0$ is measured by θ. Therefore $\phi = \theta$, and the intrinsic equation becomes

$$s = C\frac{(m^2+1)^{\frac{1}{2}}}{m}(\epsilon^{m\phi}-1).$$

From this it appears that any intrinsic equation of the form

$$s = a\,(\epsilon^{m\phi}-1) \quad\ldots\ldots\ldots\ldots\ldots\ldots\ldots (40)$$

will represent an equiangular spiral.

Given the intrinsic equation of a curve, to deduce that of its evolute.

Considering the given curve as formed by the unwinding of a string from its evolute, any arc of the former may be said to *correspond* to that arc of the latter by the unwinding of the string from which it is formed. Thus if s', ϕ' represent elements of the evolute corresponding to s, ϕ in the given curve, then the origin of s' is that point of the evolute whose tangent forms the radius of curvature at the origin of s.

This premised, it is evident that we shall have

$$\phi' = \phi.$$

For the extreme differential elements of the arc of the evolute are respectively perpendicular to the corresponding extreme differential elements of an arc of the given curve. Hence the inclination of the former being equal to that of the latter, the value of ϕ is the same for both.

Secondly, any arc of the evolute is by a known property equal to the difference of the radii of curvature of the extremities of the corresponding arc of the given curve. Hence if ρ_0 represent the radius of curvature at the origin of the given curve, we shall have

$$s' = \rho - \rho_0 = f'(\phi) - f'(0), \text{ by (39)},$$

and, substituting ϕ' for ϕ,

$$s' = f'(\phi') - f'(0).$$

Dropping the accents, we may therefore affirm that if the intrinsic equation of a curve is $s = f(\phi)$, that of its evolute will be $s = f'(\phi) - f'(0)$.

Ex. The intrinsic equation of the logarithmic spiral is $s = a(\epsilon^{m\phi} - 1)$. Hence that of its evolute is

$$s = ma\epsilon^{m\phi} - ma$$
$$= ma(\epsilon^{m\phi} - 1),$$

which also denotes a logarithmic spiral.

Given the intrinsic equation of a curve in the form $s = f(\phi)$ wherein $f(\phi)$ vanishing with ϕ is supposed capable of expansion in the form

$$f(\phi) = A_1\phi + A_2\phi^2 + A_3\phi^3 + \&c. \ldots\ldots\ldots\ldots (41),$$

required the general intrinsic equation of the involute.

As to any curve there belong an infinite number of involutes depending on the different values given to that initial tangent to the curve which forms the initial radius of curvature of the involute, we shall represent the arbitrary value of that initial tangent by C.

Now if $s = F(\phi)$ be the intrinsic equation of the involute, we have by the last proposition

$$F'(\phi) - F'(0) = f(\phi).$$

But $F'(0)$, being the initial radius of curvature of the involute, is equal to C. Hence the above equation may be expressed in the form

$$\frac{dF(\phi)}{d\phi} = f(\phi) + C,$$

whence
$$F(\phi) = \int f(\phi)\, d\phi + C\phi + C',$$
$$= \frac{A_1\phi^2}{2} + \frac{A_2\phi^3}{3} \ldots + C\phi + C'.$$

Hence $F(\phi)$ vanishing with ϕ, we must have $C' = 0$. Thus the intrinsic equation of the involute, under the condition that its initial radius of curvature is a, will be

$$s = \int f(\phi)\, d\phi + a\phi \quad\ldots\ldots\ldots\ldots\ldots\ldots(42).$$

If, for distinction's sake, we represent the arc of the involute by s', the equation may be expressed in the form

$$s' = \int (a + s)\, d\phi \quad\ldots\ldots\ldots\ldots\ldots\ldots(43).$$

It is to be remembered that the lower limit of the integral is 0.

The following proposition from the memoir of Dr Whewell referred to, will illustrate the application of the above theorems.

Let any curve be evolved, and the involute evolved, and the involute of that evolved, beginning each evolution from the commencement of the curve last formed, and with a "rectilineal tail" which is of constant length for all. The curves tend continually to the form of the equiangular spiral.

Let s, s', s'', &c. be the successive curves, ϕ the angle which is the same for all, and let the tails represented in fig. 3, by AA', $A'A''$, $A''A'''$, &c. be each equal to a.

Then representing the equation of the given curve by $s = f(\phi)$, we have for the first involute the equation

$$s' = \int (a + s)\, d\phi = a\phi + \int f(\phi)\, d\phi,$$

$$s'' = \int (a + s')\, d\phi = a\phi + \frac{a\phi^2}{1\,.\,2} + \iint f(\phi)\, d\phi^2,$$

$$s''' = \int (a + s'')\, d\phi = a\phi + \frac{a\phi^2}{1\,.\,2} + \frac{a\phi^3}{1\,.\,2\,.\,3} + \iiint f(\phi)\, d\phi^3,$$

and in general

$$s^{(n)} = a\phi + \frac{a\phi^2}{1.2} + \frac{a\phi^3}{1.2.3} \ldots + \frac{a\phi^n}{1.2\ldots n} + \int^n f(\phi)\, d\phi^n \ldots (44).$$

Now giving to $f(\phi)$ the form (41), we have

$$\int^n f(\phi)\, d\phi^n = \frac{A_1\phi^{n+1}}{1.2\ldots n+1} + \frac{A_2\phi^{n+2}}{1.2\ldots n+2} + \&c.$$

We see then that the first n terms of the expression for $s^{(n)}$ in terms of ϕ are unaffected by the form of the function $f(\phi)$, while those which remain are affected with coefficients which tend to 0. Thus the limiting form of (44) becomes

$$s^{(\infty)} = a\phi + \frac{a\phi^2}{1.2} + \frac{a\phi^3}{1.2.3} \ldots + \&c.$$

$$= a\,(\epsilon^\phi - 1) \ldots\ldots\ldots\ldots\ldots\ldots\ldots\ldots (45).$$

Now this is the equation of an equiangular spiral.

EXERCISES.

1. Determine the curve whose subtangent varies as the abscissa.

2. Determine the curve whose normal varies as the square of the ordinate.

3. Shew that the curve in which the radius of curvature varies as the cube of the normal is a conic section.

4. Find a curve in which the length of the arc is in a constant ratio to the intercept cut off by the tangent from the axis of x.

5. Shew that the above is a particular case of curves of pursuit.

6. Find the orthogonal trajectory of a system of circles touching a given straight line in a given point.

7. Find the orthogonal trajectory of the system of ellipses defined by the equation $\frac{x^2}{a^2}+\frac{y^2}{b^2}=1$, b being the variable parameter.

8. Find the equation referred to polar co-ordinates of the curve in which the radius vector is equal to m times the length of the tangent.

9. Required the form of a pendent in Gothic architecture supposed to be a solid of revolution, such that the weight to be supported by each horizontal section shall be proportional to the area of that section.

10. Required the curve in which $s=ax^2$.

11. A curve is defined by this property; *viz.* that the radius of curvature at any point is a given multiple (n) of the portion of the normal intercepted between the point and the axis of abscissæ; prove that the length of any portion of the curve may be finitely expressed in terms of the ordinates of its extremities. (*Cambridge Problems*, 1849.)

12. Find a differential equation of the first order of the curve whose radius of curvature is equal to n times the normal, and shew that this is always integrable in finite terms if n be an integer.

13. Shew that if $n=2$ the curve is a cycloid, if $n=1$ a circle, if $n=-1$ a catenary.

14. The curve whose polar equation is $r^m \cos m\theta = a^m$ rolls on a fixed straight line. Assuming that straight line as the axis of x, shew that the locus of the curve described by the pole of the rolling curve will have for its equation

$$dx=\left\{\left(\frac{y}{a}\right)-1\right\}^{-\frac{1}{2}} dy.$$

(Frenet, *Recueil d'Exercises sur le Calcul Infinitésimal.*)

Note. To solve problems like the above, we observe that if RTS, Fig. 4, represent the given curve rolling on the given line OX, and APC the curve described by the pole P, then taking OX for the axis of x, and putting $OM=x$,

$MP = y$, the straight line PT joining that pole with the point of contact will be a radius vector of the given curve, but a normal of the described curve. Hence

$$r = \sqrt{\left\{1 + \left(\frac{dy}{dx}\right)^2\right\}}.y \ldots\ldots\ldots\ldots (a).$$

Again, PM is the perpendicular let fall from the pole upon the tangent of the given curve, but the ordinate y of the required curve. Hence

$$\frac{r^2 d\theta}{\sqrt{\{dr^2 + r^2 d\theta^2\}}} = y \ldots\ldots\ldots\ldots\ldots\ldots (b).$$

By means of a, b, and the equation of the given curve, eliminating r and θ, we obtain the differential equation of the curve sought.

15. In the particular case of $m = \frac{1}{2}$ the rolling curve will be a parabola, the pole its focus, and the described curve a catenary.

16. If $m = 2$, the rolling curve is an equilateral hyperbola, the pole its centre, and the described curve an elastica.

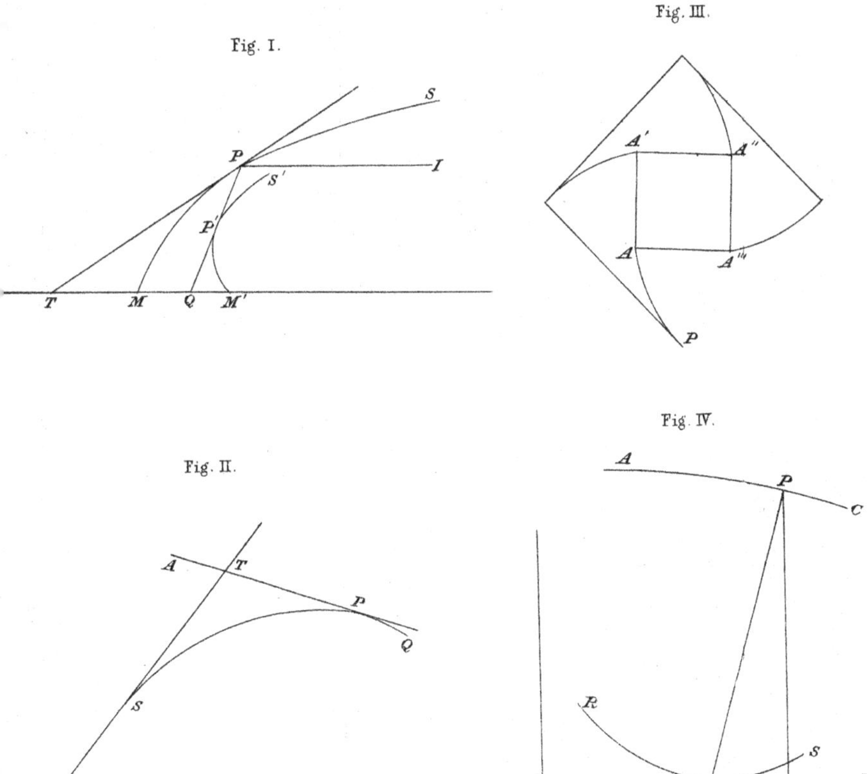
Fig. I.
S
P
I
S'
P'
T
M
Q
M'
Fig. II.
A
T
P
Q
S
Fig. III.
A'
A''
A
A'''
P
Fig. IV.
A
P
C
R
S
O
T
M
X

CHAPTER XII.

ORDINARY DIFFERENTIAL EQUATIONS WITH MORE THAN TWO VARIABLES.

1. The class of equations which we shall first consider in this Chapter, is represented by the typical form,

$$Pdx + Qdy + Rdz = 0 \quad\ldots\ldots\ldots\ldots\ldots\ldots (1),$$

P, Q and R being functions of the variables x, y, z; and it is usually termed a total differential equation of the first order with three variables.

Possibly the first observation suggested by the examination of this form will be, that it does not answer to the definition of a differential equation, as the expression of a relation involving differential coefficients, Chap. I. And certainly it does not exhibit their notation. If, however, we attempt to attach a meaning to the general form (1), we shall perceive that the idea of a limit is involved essentially. And if we study its origin, we shall see that this idea may be expressed, here as elsewhere, in the language of differential coefficients.

For (1) is not understood as implying simply that the expression,

$$P\Delta x + Q\Delta y + R\Delta z \quad\ldots\ldots\ldots\ldots\ldots\ldots\ldots (2),$$

approaches to the value 0 when the increments Δx, Δy, Δz approach that value, true though it be that the vanishing of the increments causes that expression to vanish with them. But what (1) is always understood to express is, that in the approach to the limiting state, (2) tends to vanish in consequence of the *ratios* which the increments Δx, Δy, Δz tend to assume; it is, that if we represent (2) in any of the equivalent forms

$$\frac{P\Delta x + Q\Delta y + R\Delta z}{\Delta x}\,\Delta x, \quad \frac{P\Delta x + Q\Delta y + R\Delta z}{\Delta y}\,\Delta y, \text{ \&c.}$$

the limit of the ratio expressed by the first factor of each is 0. And the problem of the integration of (1), is that of the discovery of the possible relation or relations among the primitive variables which will secure this result, supposing Δx, Δy, Δz to be so restricted as to preserve such relations unviolated.

Now whether the primitive variables are connected by *one* equation or by *two* simultaneous equations (we cannot suppose them connected by three equations without making them cease to be variable), the relation (1) is fully expressible in the language of differential coefficients. If there exist one primitive relation which, as we shall hereafter see, can only happen under particular circumstances, then

$$dz = \frac{dz}{dx}dx + \frac{dz}{dy}dy,$$

while (1) is presentable in the form

$$dz = -\frac{P}{R}dx - \frac{Q}{R}dy.$$

Hence, since dx and dy are independent, we have

$$\frac{dz}{dx} = -\frac{P}{R}, \quad \frac{dz}{dy} = -\frac{Q}{R} \quad \ldots\ldots\ldots\ldots\ldots (3),$$

a system which in the supposed case is equivalent to (1). On the other hand if, as will usually happen, two simultaneous equations connect the primitive variables, *e.g.*

$$\phi(x, y, z) = 0, \quad \psi(x, y, z) = 0 \quad \ldots\ldots\ldots\ldots\ldots (4),$$

then since we have

$$\frac{d\phi}{dx}dx + \frac{d\phi}{dy}dy + \frac{d\phi}{dz}dz = 0,$$

$$\frac{d\psi}{dx}dx + \frac{d\psi}{dy}dy + \frac{d\psi}{dz}dz = 0,$$

the elimination of dx, dy, dz between these and the original equation gives

$$P\left(\frac{d\phi}{dy}\frac{d\psi}{dz} - \frac{d\phi}{dz}\frac{d\psi}{dy}\right) + Q\left(\frac{d\phi}{dz}\frac{d\psi}{dx} - \frac{d\phi}{dx}\frac{d\psi}{dz}\right)$$
$$+ R\left(\frac{d\phi}{dx}\frac{d\psi}{dy} - \frac{d\phi}{dy}\frac{d\psi}{dx}\right) = 0 \ldots (5),$$

a result which is equivalent to (1), but is expressed in the language of partial differential coefficients. As it constitutes but a single relation between two unknown functions ϕ and ψ, one of the two may be considered arbitrary, and a particular form being given to it, we should have a partial differential equation for determining the other.

We propose indeed to discuss the equation (1) under its actual form, but it is not unimportant to shew that it constitutes no real exception to the definition of a differential equation. Treated by the methods proper to partial differential equations, the forms (3) and (5) lead to the same solutions as those investigated in this chapter.

2. The foregoing remarks admit of geometrical illustrations.

If x, y, z and $x + \Delta x$, $y + \Delta y$, $z + \Delta z$ are the coordinates of two points, then will the value of the expression $P\Delta x + Q\Delta y + R\Delta z$, where P, Q, R are given functions of x, y, z, depend solely upon the positions of the points.

If we suppose the second point to approach the first *along any path*, the value of the above expression will approach to 0 in consequence of the quantities Δx, Δy, Δz approaching to 0, and independently of the *ratios* which they assume in vanishing. But this is not in accordance with the understood meaning of the equation (1).

The increments therefore not being independent, either they are connected by one relation, in which case one point being given the other must lie on the surface which that relation determines, and its approach to the first must be made along that surface, but is in no other way restricted; or the increments are connected by two relations, and then, the first point being given, the second must be on the line determined by those relations, and its approach to the first must be made along that line, and therefore in a definite path.

3. These considerations suggest to us the following questions for analysis, viz.:

1st. Under what circumstances is the solution of the equation $Pdx + Qdy + Rdz = 0$, expressed by a single relation between the primitive variables—a relation which with the arbitrary constant of integration will represent a family of surfaces;—and how is such a relation to be determined?

2ndly. How is the solution to be obtained when the above condition is not satisfied?

These questions we shall next consider.

The equation $Pdx + Qdy + Rdz = 0$, derivable from a single primitive.

From the given equation, we have

$$dz = -\frac{P}{R}dx - \frac{Q}{R}dy \quad \text{.................... (6).}$$

But the existence of a single primitive involves the supposition that z is a function of x and y, and therefore that we have

$$\frac{dz}{dx} = -\frac{P}{R}, \quad \frac{dz}{dy} = -\frac{Q}{R} \quad \text{.................. (7).}$$

Hence, if $\frac{P}{R}$ and $\frac{Q}{R}$ do not contain z, we have by the property of differential coefficients,

$$\frac{d}{dy}\frac{P}{R} = \frac{d}{dx}\frac{Q}{R}.$$

Should however $\frac{P}{R}$ and $\frac{Q}{R}$ both or either of them contain z, then, because we can still regard them as *ultimately* functions of x and y, for z is such by hypothesis, we must change the above into

$$\frac{d}{dy}\frac{P}{R} + \frac{dz}{dy}\frac{d}{dz}\frac{P}{R} = \frac{d}{dx}\frac{Q}{R} + \frac{dz}{dx}\frac{d}{dz}\frac{Q}{R}.$$

Lastly, substituting here for $\frac{dz}{dy}$ and $\frac{dz}{dx}$ their values given in (7), effecting the differentiations, and reducing, we have

$$P\left(\frac{dQ}{dz} - \frac{dR}{dy}\right) + Q\left(\frac{dR}{dx} - \frac{dP}{dz}\right) + R\left(\frac{dP}{dy} - \frac{dQ}{dx}\right) = 0 \quad \text{.... (8),}$$

an equation of condition which, when identically satisfied,

indicates that the proposed equation admits of a single primitive.

4. *To deduce the complete primitive of the differential equations* $Pdx + Qdy + Rdz = 0$ *when the equation of condition* (8) *is satisfied.*

The supposed primitive involving all the variables x, y, z, it is evident that if we differentiate it on the hypothesis that z is constant, we shall arrive at a result equivalent to $Pdx + Qdy = 0$. It is also evident that if the primitive contained a function of z for one of its terms, that term, whatever the form of the function might be, would disappear in the differentiation.

Conversely then if we integrate the equation

$$Pdx + Qdy = 0 \quad\text{..........................}(9),$$

regarding z as constant, and adding in the place of an arbitrary constant an arbitrary function of z, we shall arrive at a result which will necessarily *include* the complete primitive, and in which it will only remain necessary to determine what form must be given to the arbitrary function of z.

Thus, if the integrating factor of (9) be μ, and if, assuming z constant, we write

$$\mu\,(Pdx + Qdy) = \frac{dV}{dx}\,dx + \frac{dV}{dy}\,dy,$$

then will the complete primitive be of the form

$$V = \phi(z) \quad\text{..........................}(10),$$

in which it only remains to determine $\phi(z)$. And this will be done by differentiating with respect to all the variables and comparing with the given equation.

Differentiating (10) then with respect to x, y, z, and transposing we have

$$\frac{dV}{dx}\,dx + \frac{dV}{dy}\,dy + \left\{\frac{dV}{dz} - \frac{d\phi(z)}{dz}\right\} dz = 0,$$

whence

$$\mu\,(Pdx + Qdy) + \left\{\frac{dV}{dz} - \frac{d\phi(z)}{dz}\right\} dz = 0.$$

Now by the given equation, $Pdx+Qdy=-Rdz$. Substituting, and rejecting the common factor dz, we have

$$-\mu R+\frac{dV}{dz}-\frac{d\phi(z)}{dz}=0,$$

whence

$$\frac{d\phi(z)}{dz}=\frac{dV}{dz}-\mu R\text{.................}(11),$$

the second member of which must, on the hypothesis that a single primitive exists, be reducible to a function of z by means of (10). The solution of the equation thus reduced will determine $\phi(z)$, the value of which substituted in (10) will give the complete primitive.

Although we are fully entitled to affirm that the equation determining $\phi(z)$ must, whenever a single primitive exists, be reducible to a form not involving x and y; it may be proper to verify this conclusion *a posteriori*.

Let us then inquire under what condition the function $\frac{dV}{dz}-\mu R$, can be freed from both x and y by means of the equation $V=\phi(z)$. Evidently this can only be the case when $\frac{dV}{dz}-\mu R$ and V are so related that, considered with respect to x and y alone, the one is a function of the other. Thus we have by the equation of condition (Chap. IV. Art. 3)

$$\frac{dV}{dx}\frac{d}{dy}\left(\frac{dV}{dz}-\mu R\right)-\frac{dV}{dy}\frac{d}{dx}\left(\frac{dV}{dz}-\mu R\right)=0,$$

or

$$\frac{dV}{dx}\frac{d^2V}{dz\,dy}-\frac{dV}{dy}\frac{d^2V}{dz\,dx}+\mu\left(\frac{dR}{dx}\frac{dV}{dy}-\frac{dR}{dy}\frac{dV}{dx}\right)$$

$$+R\left(\frac{dV}{dy}\frac{d\mu}{dx}-\frac{dV}{dx}\frac{d\mu}{dy}\right)=0\text{.........}(12).$$

Now since $\frac{dV}{dx}=\mu P,\quad \frac{dV}{dy}=\mu Q$, we have

$$\frac{dV}{dx}\frac{d^2V}{dz\,dy}-\frac{dV}{dy}\frac{d^2V}{dz\,dx}=\mu P\frac{d}{dz}(\mu Q)-\mu Q\frac{d}{dz}(\mu P)$$

$$=\mu^2\left(P\frac{dQ}{dz}-Q\frac{dP}{dz}\right)\dots\dots\dots\dots\dots(13).$$

Thus also

$$\mu\left(\frac{dR}{dx}\frac{dV}{dy}-\frac{dR}{dy}\frac{dV}{dx}\right)=\mu^2\left(Q\frac{dR}{dx}-P\frac{dR}{dy}\right)\dots\dots(14).$$

Lastly,

$$R\left(\frac{dV}{dy}\frac{d\mu}{dx}-\frac{dV}{dx}\frac{d\mu}{dy}\right)=\mu R\left(Q\frac{d\mu}{dx}-P\frac{d\mu}{dy}\right)\dots\dots.(15).$$

But since μ is the integrating factor of $Pdx+Qdy$ we have by Chap. IV.,

$$Q\frac{d\mu}{dx}-P\frac{d\mu}{dy}=\mu\left(\frac{dP}{dy}-\frac{dQ}{dx}\right),$$

which reduces (15) to the form

$$R\left(\frac{dV}{dy}\frac{d\mu}{dx}-\frac{dV}{dx}\frac{d\mu}{dy}\right)=\mu^2R\left(\frac{dP}{dy}-\frac{dQ}{dx}\right)\dots\dots\dots(16)$$

Substituting these values in (12) and rejecting the common factor μ^2, there results

$$P\frac{dQ}{dz}-Q\frac{dP}{dz}+Q\frac{dR}{dx}-P\frac{dR}{dy}+R\frac{dP}{dy}-R\frac{dQ}{dx}=0,$$

or

$$P\left(\frac{dQ}{dz}-\frac{dR}{dy}\right)+Q\left(\frac{dR}{dx}-\frac{dP}{dz}\right)+R\left(\frac{dP}{dy}-\frac{dQ}{dx}\right)=0\dots(17);$$

and this is identical with the equation of condition (8). The conclusion is therefore established.

It follows also that it is not necessary in any proposed case to apply directly the above equation of condition. It is *implicitly* involved in the very process of solution.

5. The results of the above investigation are contained in the following Rule.

RULE. *Integrate the proposed equation on the hypothesis that one of the variables is constant and its differential therefore equal to* 0, *adding an arbitrary function of that variable in the place of an arbitrary constant. Then differentiating with respect to all the variables, determine the arbitrary function by the condition that the result of such differentiation shall be equivalent to the equation given. The equation expressing such condition will, if a single primitive exist, be reducible by previous results to a form in which no other variable than the one involved in the arbitrary function will remain.*

Ex. 1. Given $(y+a)^2\,dx + z\,dy - (y+a)\,dz = 0$.

Here $P = (y+a)^2$, $Q = z$, $R = -y-a$, values which identically satisfy the condition (8). The equation therefore admits of a single complete primitive.

Regarding z as constant we have first to integrate the equation

$$(y+a)^2\,dx + z\,dy = 0.$$

Dividing by $(y+a)^2$, we have

$$dx + \frac{z\,dy}{(y+a)^2} = 0,$$

the solution of which is

$$x - \frac{z}{y+a} = \phi(z),$$

$\phi(z)$ being an arbitrary function of z introduced in the place of an arbitrary constant.

Now, differentiating with respect to all the variables, we have

$$dx + \frac{z}{(y+a)^2}dy - \left\{\frac{1}{y+a} + \frac{d\phi(z)}{dz}\right\}dz = 0,$$

or $$(y+a)^2 dx + z\,dy - \left\{y + a + (y+a)^2\frac{d\phi(z)}{dz}\right\}dz,$$

which agrees with the equation given, if we have

$$-(y+a) = -\left\{y + a + (y+a)^2\frac{d\phi(z)}{dz}\right\},$$

or $$\frac{d\phi(z)}{dz} = 0.$$

Here then $\phi(z) = c$ and the complete primitive is

$$x - \frac{z}{y+a} = c \quad\ldots\ldots\ldots\ldots\ldots\ldots\ldots\ldots (a).$$

If we commence by regarding y as constant we obtain by a first integration

$$z - (y+a)\,x = \phi(y),$$

whence, differentiating and comparing with the given equation,

$$x + \frac{d\phi(y)}{dy} = \frac{z}{y+a}.$$

This equation involves both x and y, but it is reducible by the previous one to the form

$$\frac{d\phi(y)}{dy} = \frac{\phi(y)}{y+a},$$

or

$$\frac{d\phi(y)}{\phi y} = \frac{dy}{y+a},$$

of which the integral may be expressed in the form

$$\phi(y) = b\,(y+a),$$

b being an arbitrary constant. Hence, finally

$$z = (y+a)\,x + b\,(y+a)$$

$$= (y+a)(x+b),$$

and this is equivalent to the former result. (a).

Ex. 2. Given $zdz + (x-a)\,dx = \{h^2 - z^2 - (x-a)^2\}^{\frac{1}{2}}\,dy$.

Integrating as if y were constant we have

$$z^2 + (x-a)^2 = \phi(y) \quad\ldots\ldots\ldots\ldots\ldots\ldots (a).$$

Differentiating and comparing with the given equation

$$\frac{1}{2}\frac{d\phi(y)}{dy} = \{h^2 - z^2 - (x-a)^2\}^{\frac{1}{2}}.$$

$$= \{h^2 - \phi(y)\}^{\frac{1}{2}}, \text{ by } (a).$$

Hence,

$$\frac{-d\phi(y)}{2\,\{h^2 - \phi(y)\}^{\frac{1}{2}}} = -\,dy.$$

Therefore integrating

$$\{h^2 - \phi(y)\}^{\frac{1}{2}} = -y + b,$$

b being an arbitrary constant. Hence determining $\phi(y)$, and substituting in (a), we have finally

$$z^2 + (x-a)^2 + (y-b)^2 = h^2,$$

where b is arbitrary.

Homogeneous Equations.

6. When the equation $Pdx + Qdy + Rdz = 0$ is homogeneous with respect to x, y, z, its solution will be facilitated by a transformation similar to that employed for homogeneous equations with two variables.

Assuming $x = uz$, $y = vz$, we obtain by substitution a result of the form

$$L\frac{dz}{z} = Mdu + Ndv \quad\text{..................(18)}.$$

If L be equal to 0 this simply gives

$$Mdu + Ndv = 0,$$

which can always be made integrable by a factor. If L be not equal to 0 we have

$$\frac{dz}{z} = \frac{M}{L}du + \frac{N}{L}dv;$$

and here the first member being an exact differential the second will be such also if a complete primitive exist. After integration, u and v must be replaced by their values $\frac{x}{z}$, $\frac{y}{z}$.

Ex. 3. Given $(ay - bz)\,dx + (cz - ax)\,dy + (bx - cy)\,dz = 0$.

This equation satisfies the equation of condition (8).

Assuming $x = uz$, $y = vz$ it becomes simply

$$(av - b)\,du - (au - c)\,dv = 0,$$

or

$$\frac{du}{au - c} = \frac{dv}{av - b},$$

the solution of which is

$$\frac{au-c}{av-b}=C,$$

whence the complete primitive sought will be

$$\frac{ax-cz}{ay-bz}=C.$$

Ex. 4. Given

$$(y^2+yz+z^2)\,dx+(x^2+xz+z^2)\,dy+(x^2+xy+y^2)\,dz=0.$$

Assuming $x=uz$, $y=vz$, we have on reduction

$$\frac{dz}{z}=-\frac{(v^2+v+1)\,du+(u^2+u+1)\,dv}{(u+v+1)\,(uv+u+v)},$$

or

$$\frac{dz}{z}=\frac{du+dv}{u+v+1}-\frac{(v+1)\,du+(u+1)\,dv}{uv+u+v},$$

whence integrating

$$\log z=\log\frac{u+v+1}{uv+u+v}+C.$$

Finally we have

$$\frac{xy+xz+yz}{x+y+z}=C',$$

for the complete primitive.

The last two equations might have been integrated without preliminary transformation. (Lacroix, Tom. II. pp. 507—10).

Integrating factors.

7. The equation $Pdx+Qdy+Rdz=0$ can also, when there exists a single complete primitive, be integrated by means of a factor.

If μ be that factor, then, since the expression

$$\mu Pdx+\mu\,Qdy+\mu\,Rdz$$

must be an exact differential, we must have

$$\frac{d(\mu Q)}{dz} = \frac{d(\mu R)}{dy}, \quad \frac{d(\mu R)}{dx} = \frac{d(\mu P)}{dz},$$

$$\frac{d(\mu P)}{dy} = \frac{d(\mu Q)}{dx},$$

equations to which we may give the forms

$$Q\frac{d\mu}{dz} - R\frac{d\mu}{dy} + \mu\left(\frac{dQ}{dz} - \frac{dR}{dy}\right) = 0,$$

$$R\frac{d\mu}{dx} - P\frac{d\mu}{dz} + \mu\left(\frac{dR}{dx} - \frac{dP}{dz}\right) = 0,$$

$$P\frac{d\mu}{dy} - Q\frac{d\mu}{dx} + \mu\left(\frac{dP}{dy} - \frac{dQ}{dx}\right) = 0.$$

Multiplying these equations by P, Q, and R, respectively, adding, and dividing by μ, we have

$$P\left(\frac{dQ}{dz} - \frac{dR}{dy}\right) + Q\left(\frac{dR}{dx} - \frac{dP}{dz}\right) + R\left(\frac{dP}{dy} - \frac{dQ}{dx}\right) = 0 \ldots (19),$$

the same equation of condition which was before obtained.

When this equation is satisfied a particular form of the factor μ will frequently suggest itself.

In Ex. 3 the functions $\frac{1}{(ay-bz)^2}$, $\frac{1}{(cz-ax)^2}$, $\frac{1}{(bx-ay)^2}$ are integrating factors. In Ex. 4 the functions $\frac{1}{(x+y+z)^2}$ and $\frac{1}{(xy+xz+yz)^2}$ are integrating factors.

Equations not derivable from a single primitive.

8. To solve the equation $Pdx + Qdy + Rdz = 0$, when the equation of condition (8) is not satisfied.

In this case the solution consists of two simultaneous equations between x, y, z, one of which is perfectly arbitrary in form.

For representing an assumed arbitrary equation in the form

$$f(x, y, z) = 0 \text{.......................} (20),$$

and differentiating, we have

$$\frac{df(x, y, z)}{dx} dx + \frac{df(x, y, z)}{dy} dy + \frac{df(x, y, z)}{dz} dz = 0.$$

Now these two equations enabling us, when the form of $f(x, y, z,)$ is specified, to eliminate one of the variables and its differential, e.g. z and dz, from the equation given, permit us to reduce it to the form

$$Mdx + Ndy = 0,$$

M and N being functions of x and y. Solving this, we obtain an equation involving an arbitrary constant, and this equation together with (20) will constitute a solution. By giving different forms to $f(x, y, z)$ every possible solution may be obtained. What a solution thus found represents in geometrical construction is the drawing, on a particular surface, of a family of lines, each of which satisfies at every point the condition $Pdx + Qdy + Rdz = 0$ Now dx, dy, dz are proportional to the directing cosines of the tangent line. Hence the geometrical problem may be represented as that of drawing on a given surface a family of lines, in each of which the directing cosines $\cos\phi$, $\cos\psi$, $\cos\chi$ at any point shall satisfy the condition

$$P\cos\phi + Q\cos\psi + R\cos\chi = 0 \text{...........} (21).$$

Ex. Required the most general solution of the equation

$$xdx + ydy + c\left(1 - \frac{x^2}{a^2} - \frac{y^2}{b^2}\right)^{\frac{1}{2}} dz = 0 \text{...........} (a),$$

which is consistent with the assumption that it shall represent a series of lines traced upon the ellipsoid whose equation is

$$\frac{x^2}{a^2} + \frac{y^2}{b^2} + \frac{z^2}{c^2} = 1 \text{ .... } (b).$$

It will be found that (a) does not satisfy the equation of condition (8).

Differentiating (b), we have

$$\frac{x\,dx}{a^2}+\frac{y\,dy}{b^2}+\frac{z\,dz}{c^2}=0,$$

whence

$$dz=-\frac{c^2}{z}\left(\frac{x\,dx}{a^2}+\frac{y\,dy}{b^2}\right)$$

$$=\frac{-c\left(\frac{x\,dx}{a^2}+\frac{y\,dy}{b^2}\right)}{\left(1-\frac{x^2}{a^2}-\frac{y^2}{b^2}\right)^{\frac{1}{2}}}.$$

and this reduces (a) to

$$x\,dx+y\,dy-c^2\left(\frac{x\,dx}{a^2}+\frac{y\,dy}{b^2}\right)=0\text{..........}(c),$$

the integral of which is

$$\left(1-\frac{c^2}{a^2}\right)x^2+\left(1-\frac{c^2}{b^2}\right)y^2=C\text{ ....}(d),$$

indicating that the projections of the proposed family of lines will be a certain series of central conic sections.

If $a=b=c=1$ the proposed equation admits of a single primitive, viz. $x^2+y^2+z^2=1$. And any line traced on the surface of which this is the equation will satisfy the differential equation; for the equation (b) by which the lines are ordinarily determined is now reduced to an identity.

The above method of solution is due to Newton. Monge has however remarked that the general solution may be expressed by the equations (10) and (11) of Art. 8 viz. by the simultaneous system

$$V=\phi(z)\text{ ....}(22),$$

$$\frac{dV}{dz}-\mu R=\phi'(z)\text{......................}(23),$$

where μ is the integrating factor, and V the corresponding integral of the expression $P\,dx+Q\,dy$. It is indeed shewn in that Article that (22) does satisfy the differential equation provided that the condition (23) is satisfied. But there is no *practical* advantage in the employment of Monge's form. Applied to the problem of drawing on a given surface lines satisfying the condition expressed by the differential equation, it makes the determination of the arbitrary function $\phi(z)$ itself dependent on the solution of a differential equation.

Thus in the example last considered we have, on giving to μ the value 2,

$$V=x^2+y^2, \quad R=c\left(1-\frac{x^2}{a^2}-\frac{y^2}{b^2}\right)^{\frac{1}{2}},$$

so that the *general* solution assumes the form

$$x^2+y^2=\phi(z)$$

$$-2c\left(1-\frac{x^2}{a^2}-\frac{y^2}{b^2}\right)^{\frac{1}{2}}=\phi'(z).$$

To apply this to the problem of drawing lines satisfying the conditions of the problem on the ellipsoid

$$\frac{x^2}{a^2}+\frac{y^2}{b^2}+\frac{z^2}{c^2}=1 \quad \text{.......................(e)},$$

it is necessary from the above three equations to eliminate x and y. From the second and third which here suffice, we have

$$-2z=\phi'(z),$$

whence $$\phi(z)=-z^2-C.$$

Therefore $$x^2+y^2+z^2=C \quad \text{....................}(f).$$

The particular solution sought is therefore expressed by the equations (e) and (f), which are together equivalent to the previous solution expressed by (b) and (d).

Total differential equations containing more than three variables.

9. It will suffice to make a few observations on the equation with four variables

$$Pdx+Qdy+Rdz+Tdt=0 \quad \text{..............(24)},$$

and to direct attention to the general analogy.

Writing the above equation in the form

$$dt = -\frac{P}{T}dx - \frac{Q}{T}dy - \frac{R}{T}dz \text{..............(25)},$$

it is evident that, in order that it should be derivable from a complete primitive, we must have

$$\left(\frac{d}{dx}\right)\frac{Q}{T} = \left(\frac{d}{dy}\right)\frac{P}{T}, \quad \left(\frac{d}{dy}\right)\frac{R}{T} = \left(\frac{d}{dz}\right)\frac{Q}{T}, \quad \left(\frac{d}{dz}\right)\frac{P}{T} = \left(\frac{d}{dx}\right)\frac{R}{T},$$

where $\left(\frac{d}{dx}\right)$ refers to x not only as appearing independently, but also as implicitly involved in t; and so on for the rest.

Effecting the differentiations, and substituting for $\frac{dt}{dx}$, $\frac{dt}{dy}$, $\frac{dt}{dz}$ their values implied in (25), we have

$$\left.\begin{aligned} T\left(\frac{dQ}{dx} - \frac{dP}{dy}\right) + P\left(\frac{dT}{dy} - \frac{dQ}{dt}\right) + Q\left(\frac{dP}{dt} - \frac{dT}{dx}\right) &= 0 \\ T\left(\frac{dR}{dy} - \frac{dQ}{dz}\right) + Q\left(\frac{dT}{dz} - \frac{dR}{dt}\right) + R\left(\frac{dQ}{dt} - \frac{dT}{dy}\right) &= 0 \\ T\left(\frac{dP}{dz} - \frac{dR}{dx}\right) + R\left(\frac{dT}{dx} - \frac{dP}{dt}\right) + P\left(\frac{dR}{dt} - \frac{dT}{dz}\right) &= 0 \end{aligned}\right\} \text{...(26)},$$

which are the equations of condition of existence of a single complete primitive.

We are justified by the above analysis in affirming that these equations of condition are sufficient; though symmetry makes it evident that we must also have the equation

$$\mathfrak{D} \equiv P\left(\frac{dQ}{dz} - \frac{dR}{dy}\right) + Q\left(\frac{dR}{dx} - \frac{dP}{dz}\right) + R\left(\frac{dP}{dy} - \frac{dQ}{dx}\right) = 0 \text{...(27)}.$$

Jacobi has however independently proved that this equation is involved in the others. (*Mathematische Werke*, Vol. I. p. 149.)

It is obvious that when there exist n variables, the number of independent equations of condition is $\frac{(n-1)(n-2)}{2}$, being the number of ways of equating two partial differential coefficients in a system in which $n-1$ are contained.

The solution of any such equation may be effected by an extension of the method adopted for equations with three variables. We must integrate as if all but two of the variables were constant, adding, in the place of an arbitrary constant, an arbitrary function of the variables which remain. This function we must determine by differentiating with respect to all the variables, and comparing with the equation given. If a single primitive exist, such determination will be possible. If a single primitive do not exist, we must, following the analogy of the corresponding case for three variables, endeavour to express the solution by a system of simultaneous equations. And such is indeed its general form. Pfaff, in a memoir published by the Berlin Academy 1814—15, has shewn that, according as the number of variables is $2n$ or $2n+1$, the number of integral equations is n or $n+1$ *at most*. His method, which is remarkable, consists of alternate integrations and transformations. For important commentaries and additions see Jacobi, (*Werke*, Tom. I. p. 140), and Raabe, (*Crelle*, Tom. XIV. p. 123).

Ex. Given $(2x+y^2+2xy_2-y_1)dx+2xy\,dy-xdy_1+x^2dy_2=0$.

If we suppose the variables y_1, y_2, constant, we have to integrate

$$(2x+y^2+2xy_2-y_1)\,dx+2xy\,dy=0,$$

which, on substituting an arbitrary function of y_1, y_2 represented by ϕ, for an arbitrary constant, gives

$$x^2+xy^2+x^2y_2-xy_1=\phi.$$

Differentiating with respect to all the variables, we have

$$(2x+y^2+2xy_2-y_1)\,dx+2xy\,dy-xdy_1+x^2dy_2$$
$$=\frac{d\phi}{dy_1}dy_1+\frac{d\phi}{dy_2}dy_2.$$

Comparing this with the given equation, we have

$$\frac{d\phi}{dy_1}dy_1 + \frac{d\phi}{dy_2}dy_2 = 0,$$

whence $\phi = c$ and the solution is

$$x^2 + xy^2 + x^2y_2 - xy_1 = c \text{ .... } (a).$$

Had we begun by making x and y constant, we should have had as the result of the first integration,

$$x^2y_2 - xy_1 = \phi \text{ .... } (b)$$

ϕ denoting a function of x and y. Differentiating with respect to all the variables and comparing with the given equation, we should find

$$d\phi = -(2x + y^2)\,dx - 2xy\,dy,$$

whence, $\phi = -x^2 - xy^2 + c,$

the substitution of which in (b) reproduces the former solution (a).

Equations of an order higher than the first.

10. When an equation of the form,

$$Adx^2 + Bdy^2 + Cdz^2 + 2Ddy\,dz + 2Edx\,dz + 2Fdx\,dy = 0 \text{ ... } (28),$$

is resolvable into two equations each of the form

$$Pdx + Qdy + Rdz = 0,$$

the solution of either of these obtained by previous methods, will be a particular solution of (28), and the two solutions taken disjunctively will constitute the complete solution, which is therefore expressed by the *product* of the equations of these solutions, each reduced to the form $V = 0$.

The condition under which (28) is resolvable as above, is expressed by the equation,

$$ABC + 2DEF - AD^2 - BE^2 - CF^2 = 0 \text{ .... } (29).$$

This is shewn by solving (28) with respect to dx, and assuming the quantity under the radical to be a complete square.

Thus, the equation $x^2dx^2 + y^2dy^2 - z^2dz^2 + 2xydx\,dy = 0$, which will be found to satisfy the above condition, is resolvable into the two equations,

$$xdx + ydy + zdz = 0, \qquad xdx + ydy - zdz = 0,$$

whence, $x^2 + y^2 + z^2 = c \ldots (a)$, $\qquad x^2 + y^2 - z^2 = c' \ldots\ldots (b)$.

Geometrically the solution is expressed by lines drawn in any manner on the surface, either of the sphere (a), or of the hyperboloid (b).

When the condition (29) is not satisfied, the proposed equation does not admit of a single primitive, or of any *disjunctive* system of primitives. But it does in general admit of a solution expressed by a system of *simultaneous* equations. Thus, if we integrate the equation $dz^2 = m^2(dx^2 + dy^2)$, supposing x constant, we find $z = my + C$, or, replacing C by a function of x,

$$z = my + \phi(x) \ldots\ldots\ldots\ldots\ldots\ldots\ldots\ldots (c).$$

On substitution and integration, we find that this will satisfy the proposed equation if we have

$$2y = m\int \frac{dx}{\phi'(x)} - \frac{1}{m}\phi(x) + c \ldots\ldots\ldots\ldots (d),$$

the system (c) (d) will therefore constitute a solution of the equation given. We enter not into the question whether it is the most general solution or not, proposing merely to exemplify the kind of solution of which the equation admits.

To this we may add that all equations which do not satisfy the conditions of integrability, though they may present themselves in the form of *ordinary*, have a far more intimate connexion with *partial* differential equations; and that this connexion affords the best clue to the solution of their theoretical difficulties.

EXERCISES.

1. $\dfrac{dx}{x-a}+\dfrac{dy}{y-b}+\dfrac{dz}{z-c}=0.$

2. $(x-3y-z)\,dx+(2y-3x)\,dy+(z-x)\,dz=0.$

3. $(y+z)\,dx+(z+x)\,dy+(x+y)\,dz=0.$

4. Shew that an integrating factor of the last equation is $\dfrac{1}{(x+y+z)^2}$, and deduce a general expression for such factors.

5. $(y+z)\,dx+dy+dz=0.$

6. $ay^2z^2dx+bz^2x^2dy+cx^2y^2dz=0.$

7. $(x^2y-y^3-y^2z)\,dx+(xy^2-x^2z-x^3)\,dy+(xy^2+x^2y)\,dz=0.$

8. $(2x^2+2xy+2xz^2+1)\,dx+dy+2zdz=0.$

9. $(2x+y^2+2yz)\,dx+2xy\,dy-dw+x^2dz=0.$

10. Is the equation $(1+2m)\,xdx+y\,(1-x)\,dy+zdz=0$ derivable from a single primitive of the form $\phi\,(x,\,y,\,z)=c$?

11. Shew that any system of lines described on the surface of the sphere $x^2+y^2+z^2=r^2$, and satisfying the above equation, would be projected on the plane xy in parabolas.

12. Shew that Monge's method would, if we integrate first with respect to x and z, present the solution of the equation of Ex. 10, in the form

$$(1+2m)\,x^2+z^2=\phi\,(y),\quad 2y\,(1-x)=-\phi'\,(y).$$

13. Applying this form to the problem of Ex. 11, form and solve the differential equation for the determination of $\phi\,(y)$, and shew that it leads to the result stated in that Example.

14. Find the equation of the projections of the same system of curves on the plane yz.

CHAPTER XIII.

SIMULTANEOUS DIFFERENTIAL EQUATIONS.

1. We have hitherto considered only single differential equations. We have now to treat of systems of differential equations.

Of such by far the most important class is that in which one of the variables is independent and the others are dependent upon it, the number of equations in the system being equal to the number of dependent variables. Thus in the chief problem of physical astronomy—the problem of the motion of a system of material bodies abandoned to their mutual attractions—there is but one independent variable, the time; the dependent variables are the coördinates, which, varying with the time, determine the varying positions of the several members of the material system; while, lastly, the number of equations being equal to the number of coordinates involved, the dependence of the latter upon the time is made determinate.

Such a system of equations may properly be called a determinate system.

We propose in this chapter to treat only of systems of equations of the above class. And in the first instance we shall speak of simultaneous differential equations of the first order and degree, beginning with particular examples, and proceeding to the consideration of their general theory.

Particular Illustrations.

2. The simplest class of examples is that in which the equations of the given system are separately integrable.

Ex. 1. Given $ldx + mdy + ndz = 0$, $xdx + ydy + zdz = 0$.

Integrating separately, we have

$$lx + my + nz = c, \quad x^2 + y^2 + z^2 = c';$$

and these equations expressing the complete solution of the given system may be said to constitute the primitive system.

Another class of examples is that in which, while the equations of the given system are not all separately integrable, they admit of being so combined as to produce an equivalent system of equations which are separately integrable.

Ex. 2. Given $\frac{dx}{dt} + \frac{2x}{t} = 1$, $\frac{dy}{dt} = x + y + \frac{2x}{t} - 1$.

Here the first equation alone is separately integrable, and gives

$$x = \frac{t}{3} + \frac{c}{t^2} \quad\text{.........................}\quad (a).$$

Also by addition of the given equations, we have

$$\frac{dx + dy}{dt} = x + y;$$

$$\therefore \frac{dx + dy}{x + y} = dt,$$

$$\log (x + y) = t + c' \quad\text{.......................}\quad (b).$$

The primitive system is therefore expressed by (a) and (b).

In both the above examples we see that the number of equations of the solution is equal to that of the equations of the system given, and that each equation of the solution involves a distinct arbitrary constant. And it is evident that this must be the case whenever we can combine the given equations into an equivalent system of integrable equations of the first order. But as we have not proved that such combination is possible, the following question becomes important, viz. what is the nature of the solution of a system of simultaneous equations of the first order and degree?

This question will be considered in the next section.

General theory of simultaneous equations of the first order and degree.

3. We shall seek first to establish the general theory of a system composed of two equations between three variables, and therefore of the form

$$Pdx + Qdy + Rdz = 0,$$
$$P'dx + Q'dy + R'dz = 0 \quad\text{.................}\quad (1),$$

the coefficients P, P', &c. being functions of the variables, or constants.

We design to consider the above system first, and with the greater care, because there is scarcely any part of the general theory which it does not serve to exemplify.

PROP. *The solution of the system* (1) *can always be made to depend upon that of an ordinary differential equation of the second order between two of the primitive variables, and it always consists of two equations involving two arbitrary constants.*

By algebraic solution of the system (1) we have

$$dy = \frac{RP' - PR'}{QR' - RQ'}\, dx, \quad dz = \frac{PQ' - QP'}{QR' - RQ'}\, dx \dots\dots\dots (2).$$

As the coefficients of dx in the second members of these equations are functions of x, y, z we may express the reduced system in the form

$$dy = \phi(x, y, z)\, dx, \quad dz = \psi(x, y, z)\, dx,$$

whence, regarding x as independent variable,

$$\frac{dy}{dx} = \phi(x, y, z) \dots\dots\dots\dots\dots\dots\dots\dots (3),$$

$$\frac{dz}{dx} = \psi(x, y, z) \dots\dots\dots\dots\dots\dots\dots\dots (4).$$

Thus the given system enables us to express $\frac{dy}{dx}$ and $\frac{dz}{dx}$ by known functions of x, y, z.

Now differentiating (3), still on the assumption that x is the independent variable and representing for brevity $\phi(x, y, z)$ by ϕ, $\psi(x, y, z)$ by ψ, we have

$$\frac{d^2y}{dx^2} = \frac{d\phi}{dx} + \frac{d\phi}{dy}\frac{dy}{dx} + \frac{d\phi}{dz}\frac{dz}{dx},$$

or substituting for $\frac{dz}{dx}$ its value given by (4),

$$\frac{d^2y}{dx^2} = \frac{d\phi}{dx} + \frac{d\phi}{dy}\frac{dy}{dx} + \psi\frac{d\phi}{dz} \dots\dots\dots\dots\dots\dots (5).$$

This equation involves $\frac{dy}{dx}$ and $\frac{d^2y}{dx^2}$ together with the quantities $\frac{d\phi}{dx}$, $\frac{d\phi}{dy}$, $\frac{d\phi}{dz}$ and ψ, which are known functions of x, y, and z. Hence eliminating z by means of (3) we have a final equation involving $\frac{dy}{dx}$, $\frac{d^2y}{dx^2}$, x, and y. The complete primitive of this differential equation of the second order will enable us to express y as a function of x and two arbitrary constants. Suppose the value thus obtained for y to be

$$y = \chi(x, c_1, c_2) \text{..........................(6)}.$$

Then we have by virtue of (3)

$$\phi(x, y, z) = \frac{d\chi(x, c_1, c_2)}{dx} \text{..................(7)}.$$

These two equations involving two arbitrary constants contain the complete solution of the system given.

4. It is important to observe that the system (2) may be expressed in the symmetrical form

$$\frac{dx}{QR' - RQ'} = \frac{dy}{RP' - PR'} = \frac{dz}{PQ' - QP'}.$$

If we represent the denominators of the above reduced system by X, Y, Z, it becomes

$$\frac{dx}{X} = \frac{dy}{Y} = \frac{dz}{Z} \text{..........................(8)}.$$

This, then, may be regarded as the symmetrical form of a system composed of two differential equations of the first order.

Again, the complete solution of such a system as is expressed by (6) and (7) consists of two equations connecting the variables x, y, z with two arbitrary constants. If we solve these equations with respect to the constants, the solution assumes the form

$$\phi_1(x, y, z) = c_1 \quad \phi_2(x, y, z) = c_2 \text{............(9)}.$$

Thus a system of two differential equations of the first order may, without loss of generality, be presented in the symmetrical form (8), and its complete solution in the symmetrical form (9).

Ex. 1. Given

$$(5y+9z)\,dx+dy+dz=0,\quad (4y+3z)\,dx+2dy-dz=0.$$

Here we find by algebraic solution

$$\frac{dy}{dx}=-3y-4z \quad\ldots\ldots\ldots\ldots\ldots\ldots\ldots\ldots(a),$$

$$\frac{dz}{dx}=-2y-5z \quad\ldots\ldots\ldots\ldots\ldots\ldots\ldots\ldots(b),$$

whence

$$\frac{d^2y}{dx^2}=-3\frac{dy}{dx}-4\frac{dz}{dx}$$

$$=-3\frac{dy}{dx}+8y+20z, \text{ by } (b).$$

Eliminating z by (a), we have on reduction

$$\frac{d^2y}{dx^2}+8\frac{dy}{dx}+7y=0,$$

a linear equation with constant coefficients whose complete primitive is

$$y=C_1\epsilon^{-x}+C_2\epsilon^{-7x} \quad\ldots\ldots\ldots\ldots\ldots\ldots\ldots(c).$$

Equating the value of $\frac{dy}{dx}$ hence determined with that given in (a) we have

$$3y+4z=C_1\epsilon^{-x}+7\,C_2\epsilon^{-7x} \quad\ldots\ldots\ldots\ldots\ldots(d).$$

The complete solution is therefore expressed by (c) and (d).

Theoretically it is of no consequence which of the primitive variables we assume as independent. But practically the question is of some importance as affecting the character of the final differential equation.

Ex. 2. Given $\frac{dx}{dt}-3x+y=0,\ \frac{dy}{dt}-x-y=0.$

Differentiating the first equation we have

$$\frac{d^2x}{dt^2}-3\frac{dx}{dt}+\frac{dy}{dt}=0,$$

from which eliminating $\frac{dy}{dt}$ by the second equation we have

$$\frac{d^2x}{dt^2}-3\frac{dx}{dt}+x+y=0,$$

Hence eliminating y by the first equation

$$\frac{d^2x}{dt^2} - 4\frac{dx}{dt} + 4x = 0.$$

Integrating

$$x = (C + C't)\epsilon^{2t},$$

and this value of x substituted in the first equation gives

$$y = (C - C' + C't)\epsilon^{2t}.$$

The last two equations constitute the primitive system.

We choose next an example in which the given system involves functions of the independent variable in the second members.

Ex. 3. Given $\dfrac{dx}{dt} + 5x - 2y = \epsilon^t$, $\dfrac{dy}{dt} - x + 6y = \epsilon^{2t}$.

Here, differentiating the first equation, we have

$$\frac{d^2x}{dt^2} + 5\frac{dx}{dt} - 2\frac{dy}{dt} = \epsilon^t.$$

Eliminating $\dfrac{dy}{dt}$ by the second equation of the given system, we have

$$\frac{d^2x}{dt^2} + 5\frac{dx}{dt} - 2x + 12y = \epsilon^t + 2\epsilon^{2t}.$$

And, eliminating y by means of the first equation of the system,

$$\frac{d^2x}{dt^2} + 11\frac{dx}{dt} + 28x = 7\epsilon^t + 2\epsilon^{2t},$$

a linear differential equation of the second order whose solution is

$$x = C_1\epsilon^{-4t} + C_2\epsilon^{-7t} + \frac{7}{40}\epsilon^t + \frac{1}{27}\epsilon^{2t}.$$

Hence, by the first of the given equations,

$$2y - 5x + \epsilon^t = -4C_1\epsilon^{-4t} - 7C_2\epsilon^{-7t} + \frac{7}{40}\epsilon^t + \frac{2}{27}\epsilon^{2t}.$$

The last two equations are the complete primitives of the system given.

5. The above theory may be extended to all systems which are composed of n differential equations of the first order and degree connecting $n+1$ variables.

Assume x (independent) and $x_1, x_2, \ldots x_n$ (dependent) as the variables of the system. Then there exist n differential equations of the form

$$Pdx + P_1 dx_1 + P_2 dx_2 \ldots + P_n dx_n = 0 \ldots\ldots\ldots (10),$$

P, P_1, &c. being functions of the variables. These equations exactly suffice to determine the ratios of the differentials dx, $dx_1, \ldots dx_n$, and thus assume the symmetrical form

$$\frac{dx}{X} = \frac{dx_1}{X_1} = \frac{dx_2}{X_2} \ldots = \frac{dx_n}{X_n} \ldots\ldots\ldots\ldots\ldots (11),$$

X, X_1, &c. being determinate functions of the variables.

This premised, the solution of the system (11) depends upon the solution of a single differential equation of the nth order connecting two of the variables.

Let us select for the two x and x_1.

Now (11) gives

$$\frac{dx_1}{dx} = +\frac{X_1}{X}, \quad \frac{dx_2}{dx} = \frac{+X_2}{X}, \quad \ldots \frac{dx_n}{dx} = \frac{+X_n}{X} \ldots\ldots\ldots (12).$$

Differentiate the first of these $n-1$ times in succession, regarding x as independent variable and continually substituting for $\frac{dx_2}{dx}, \ldots \frac{dx_n}{dx}$ their values as given by the $n-1$ last equations of the above system. We thus obtain, including the equation operated upon, n equations connecting

$$\frac{dx_1}{dx}, \frac{d^2x_1}{dx^2} \ldots \frac{d^nx_1}{dx^n}$$

with the primitive variables and therefore enabling us, 1st, to express the above n differential coefficients in terms of those variables, 2ndly, by elimination of the $n-1$ variables, x_2, x_3, $\ldots x_n$ to deduce a single equation of the form

$$F\left(x, x_1, \frac{dx_1}{dx}, \frac{d^2x_1}{dx^2} \ldots \frac{d^nx_1}{dx_n}\right) = 0 \ldots\ldots\ldots\ldots (13).$$

Now this being a differential equation of the nth order, there exist, Chap. IX. Art. 1, n first integrals involving n distinct arbitrary constants and capable of expression in the form

$$\left.\begin{aligned} F_1\left(x, x_1, \frac{dx_1}{dx}, \frac{dx_1^2}{dx^2} \cdots \frac{dx_1^{n-1}}{dx_{n-1}}\right) &= C_1 \\ F_2\left(x, x_1, \frac{dx_1}{dx}, \frac{d^2x_1}{dx^2} \cdots \frac{d^{n-1}x_1}{dx_{n-1}}\right) &= C_2 \\ F_n\left(x, x_1, \frac{dx_1}{dx}, \frac{d^2x_1}{dx^2} \cdots \frac{d^{n-1}x_1}{dx^{n-1}}\right) &= C_n \end{aligned}\right\} \text{.........(14).}$$

If in this system we substitute for $\frac{dx_1}{dx}, \frac{d^2x_1}{dx^2} \cdots \frac{d^{n-1}x_1}{dx^{n-1}}$ their values in terms of the primitive variables above referred to, we shall obtain a system of n equations of the form

$$\left.\begin{aligned} \phi_1(x, x_1, x_2 \dots x_n) &= C_1 \\ \phi_2(x, x_1, x_2 \dots x_n) &= C_2 \\ \phi_n(x, x_1, x_2 \dots x_n) &= C_n \end{aligned}\right\} \text{.......................(15).}$$

This is the primitive system sought.

And thus the following Propositions are established, viz. 1st, that a system of differential equations of the first order connecting $n+1$ variables is expressible in the symmetrical form (11). 2ndly, that its complete solution depends on that of an ordinary differential equation of the nth order (13). 3rdly, that that solution consists of n equations connecting the primitive variables with n arbitrary constants and theoretically expressible in the form (15).

These very important propositions were first established by Lagrange, but the above demonstration of them is taken from a memoir by Jacobi*.

It is not necessary, as is evident from the examples already given, actually to determine the n first integrals of the differential equation (13). The complete primitive and the successive equations obtained from it by differentiation enable us to accomplish the same object. Neither is it always necessary to

* *Ueber die Integration der partiellen Differential-Gleichungen erster ordnung. Crelle,* Tom. II. p. 317.

proceed to differential equations of an order higher than the first. This point will be illustrated in the following sections.

Linear equations of the first order with constant coefficients.

6. The characters here mentioned have reference only to the dependent variables which are the true unknown quantities of the system. Thus the equation

$$a\frac{dx}{dt}+b\frac{dy}{dt}+cx+ey=\phi(t)$$

would be described as linear and with constant coefficients.

The solution of any system of n such equations is by the foregoing general method reducible to that of an ordinary linear differential equation of the nth order with constant coefficients. And this method is in the two following respects the best of all, viz. 1st, because of its fundamental character, 2ndly, because it leads directly to the expression of the values of the dependent variables.

The solution of such a system may however also be effected by the method of indeterminate multipliers, and this we propose here to exemplify. Its advantage is that it generally presents the equations of the solution under a common type, so that their discovery is made to depend upon the discovery of a single general form.

Ex. Given $\dfrac{dx}{dt}=ax+by+c, \qquad \dfrac{dy}{dt}=a'x+b'y+c'.$

Multiplying the second equation by an indeterminate quantity m, and adding to the first, we have

$$\frac{dx+mdy}{dt}=(a+ma')\,x+(b+mb')\,y+c+mc'$$

$$=(a+ma')\left(x+\frac{b+mb'}{a+ma'}y+\frac{c+mc'}{a+ma'}\right)$$

$$=(a+ma')\left(x+my+\frac{c+mc'}{a+ma'}\right)\ldots\ldots\ldots\ldots(a),$$

provided that we determine m so as to satisfy the condition

$$m = \frac{b + mb'}{a + ma'},$$

or

$$a'm^2 + (a - b')\,m - b = 0 \quad\text{.................}(b).$$

Now (a) gives

$$\frac{dx + mdy}{x + my + \dfrac{c + mc'}{a + ma'}} = (a + ma')\,dt,$$

whence on integration

$$\log\left(x + my + \frac{c + mc'}{a + ma'}\right) = (a + ma')\,t + C \quad\text{......}(c).$$

In this equation it only remains to substitute in succession the two values of m furnished by (b). The two resulting equations, in which the arbitrary constants must of course be supposed different, will express the complete solution of the problem.

When the values of m are equal, the form (c) furnishes directly only a single equation of the complete solution. We may deduce the other equation, either by the method of limits (assuming the law of continuity), or by eliminating x from the given system by means of (c), and then forming a new differential equation between y and t. It seems preferable however to employ the general method of Art. 5, by which all difficulties connected with the presence of equal or imaginary roots are referred to the corresponding cases of ordinary differential equations.

7. Simultaneous equations are so often presented under the symmetrical form (11) that the appropriate mode of treatment deserves to be carefully studied, especially as it possesses the superiority, always in point of elegance, and frequently in point of convenience, over other processes.

It is known that each member of a system of equal fractions is equal to the fraction which would be formed by dividing any linear homogeneous function of their numerators by the same function of their denominators. Hence if we have a system of equations of the form

$$\frac{dx_1}{X_1}=\frac{dx_2}{X_2}\cdots\cdots=\frac{dx_n}{X_n}=\frac{dt}{T}\cdots\cdots\cdots\cdots\cdots\cdots \text{(16)},$$

in which we suppose t the independent variable, and T a function of t only, then we shall have

$$\frac{dt}{T}=\frac{dx_1+m\,dx_2\ldots+r\,dx_n}{X_1+mX_2\ldots+rX_n}\cdots\cdots\cdots\cdots\cdots \text{(17)}.$$

Hence, should the first member be an exact differential, the inquiry is suggested whether the multipliers m , ... r cannot be so determined, whether as functions of the variables or as constants, as to render the second member such also. Now when the system of equations is linear and with constant coefficients this can always be effected. It may be observed that the *character* of the system is as manifest from inspection of the symmetrical form (16) as of the ordinary form. If the system be linear and with constant coefficients the denominators X_1, X_2,...X_n will, when considered with respect to the dependent variables x_1, x_2,...x_n, be linear and with constant coefficients.

In the employment of this method it is often of great advantage to introduce a new independent variable, and to consider all the variables of the given system as dependent upon it. We are thus enabled to secure the condition above adverted to, of having one member of the symmetrical system an exact differential.

Ex. Given $\dfrac{dx}{ax+by+c}=\dfrac{dy}{a'x+b'y+c'}$.

Let us introduce a new variable t so as to give to the system the form

$$\frac{dx}{ax+by+c}=\frac{dy}{a'x+b'y+c'}=\frac{dt}{t}\ \cdots\cdots\cdots\cdots \text{(a)}.$$

Here the third member being an exact differential, we shall write

$$\frac{dt}{t}=\frac{dx+mdy}{ax+by+c+m\,(a'x+b'y+c')}$$

$$=\frac{dx+mdy}{(a+ma')\,x+(b+mb')\,y+c+mc'}$$

$$=\frac{1}{a+ma'}\,\frac{(a+ma')\,dx+(a+ma')\,m\,dy}{(a+ma')\,x+(b+mb')\,y+c+mc'}.$$

The second member of this equation will be an exact differential if we have

$$(a+ma')\,m=b+mb' \;\dots\dots\dots\dots\dots\dots\; (b),$$

the integral corresponding to each value of m thus determined being of the form

$$\log t+C=\frac{1}{a+ma'}\log\{(a+ma')\,x+(b+mb')\,y+c+mc'\},$$

or

$$C't=\{ax+by+c+m\,(a'x+b'y+c')\}^{\frac{1}{a+ma'}}.$$

If the roots of the quadratic (b) are m_1 and m_2, we thus find

$$\left.\begin{aligned}C_1t&=\{ax+by+c+m_1\,(a'x+b'y+c')\}^{\frac{1}{a+m_1a'}}\\ C_2t&=\{ax+by+c+m_2\,(a'x+b'y+c')\}^{\frac{1}{a+m_2a'}}\end{aligned}\right\}\dots\dots\dots (c),$$

for the primitive equations of the system (a). Those of the given system will be obtained by eliminating t. The result assumes the remarkable form

$$\frac{\{ax+by+c+m_1\,(a'x+b'y+c')\}^{\frac{1}{a+m_1a'}}}{\{ax+by+c+m_2\,(a'x+b'y+c')\}^{\frac{1}{a+m_2a'}}}=C\;\dots\dots\; (d).$$

Ex. 2. Given $\dfrac{dx}{X}=\dfrac{dy}{Y}=\dfrac{dz}{Z}$, where

$$\left.\begin{aligned}X&=ax+by+cz+d\\ Y&=a'x+b'y+c'z+d'\\ Z&=a''x+b''y+c''z+d''\end{aligned}\right\}\dots\dots\dots\dots\dots\dots (a).$$

Introducing a new variable t, so as to give to the system the more complete form

$$\frac{dt}{t}=\frac{dx}{X}=\frac{dy}{Y}=\frac{dz}{Z} \quad\text{.....................}\quad (b),$$

we have

$$\frac{dt}{t}=\frac{ldx+mdy+ndz}{lX+mY+nZ}$$

$$=\frac{ldx+mdy+ndz}{\lambda\,(lx+my+nz+r)} \quad\text{.............}\quad (c).$$

Provided that we assume

$$\left.\begin{array}{l} al+a'm+a''n=\lambda l \\ bl+b'm+b''n=\lambda m \\ cl+c'm+c''n=\lambda n \\ dl+d'm+d''n=\lambda r \end{array}\right\} \quad\text{..................}\quad (d).$$

The first three of these may be written in the form

$$\left.\begin{array}{l} (a-\lambda)\,l+a'm+a''n=0 \\ bl+(b'-\lambda)\,m+b''n=0 \\ cl+c'm+(c''-\lambda)\,n=0 \end{array}\right\} \quad\text{..................}\quad (e),$$

whence eliminating l, m, n we have the well known cubic

$$(a-\lambda)\,(b'-\lambda)\,(c''-\lambda)-b''c'\,(a-\lambda)$$
$$-ca''\,(b'-\lambda)-ba'\,(c''-\lambda)+a'b''c+a''bc'=0 \;\ldots\; (f).$$

Now let the values of λ hence found be λ_1, λ_2, λ_3, and the corresponding values of l, m, n, r, be l_1, m_1, n_1, r_1, l_2, m_2, &c. then integrating (c) we shall have the system

$$c_1t=(l_1x+m_1y+n_1z+r_1)^{\frac{1}{\lambda_1}},$$

$$c_2t=(l_2x+m_2y+n_2z+r_2)^{\frac{1}{\lambda_2}},$$

$$c_3t=(l_3x+m_3y+n_3z+r_3)^{\frac{1}{\lambda_3}}.$$

Hence eliminating t by equating its values, we find as the general solution of the original system of equations

$$(l_1x+m_1y+n_1z+r_1)^{\frac{1}{\lambda_1}}=C\,(l_2x+m_2y+n_2z+r_2)^{\frac{1}{\lambda_2}}$$
$$=C'\,(l_3x+m_3y+n_3z+r_3)^{\frac{1}{\lambda_3}}\ldots(g).$$

In the same way we may integrate the general system

$$\frac{dx_1}{X_1}=\frac{dx_2}{X_2}\ldots=\frac{dx_n}{X_n},$$

where X_1, X_2, ... X_n are any linear functions of the variables.

8. From the above results the solutions of various symmetrical systems in which the denominators are not linear may be deduced. The most remarkable of such deductions is the following.

Suppose that in the system

$$\frac{dx'}{ax'+by'+cz'}=\frac{dy'}{a'x'+b'y'+c'z'}=\frac{dz'}{a''x'+b''y'+c''z'}\ldots(a),$$

the solution of which is known from what precedes, we substitute

$$x'=xz',\quad y'=yz',$$

x and y being new variables introduced in the place of x' and y'. The result is

$$\frac{z\,dx+x\,dz}{ax+by+c}=\frac{z\,dy+y\,dz}{a'x+b'y+c}=\frac{dz}{a''x+b''y+c''},$$

to which we may obviously give the form

$$\frac{z\,dx}{ax+by+c-x\,(a''x+b''y+c'')}=\frac{z\,dy}{a'x+b'y+c'-y\,(a''x+b''y+c'')}$$
$$=\frac{dz}{a''x+b''y+c''}.$$

Dividing the first equation of this system by z, we have

$$\frac{dx}{ax+by+c-x(a''x+b''y+c'')}=\frac{dy}{a'x+b'y+c'-y(a''x+b''y+c'')}\ldots(b).$$

Now this on clearing of fractions will be found to be of the same form as Jacobi's equations (*Crelle*, Tom. XXIV. p. 1), whose solution on other grounds has been explained, Chap. V. Art. 8.

We see that the solution of (b) is deducible from that of the system (a) by changing x' into xz', y' into yz', and eliminating z.

And just in this way the solution of any symmetrical non-linear system of the form

$$\frac{dx_1}{X_1 - x_1X} = \frac{dx_2}{X_2 - x_2X} \dots = \frac{dx_n}{X_n - x_nX} \dots\dots (18),$$

in which X, X_1, X_2,...X_n are linear functions of the variables x_1, x_2, ... x_n may be made to flow from that of a symmetrical system of the form

$$\frac{dx_1}{X_1} = \frac{dx_2}{X_2} \dots = \frac{dx_{n+1}}{X_{n+1}} \dots\dots\dots\dots\dots\dots (19),$$

in which X_1, X_2,...X_{n+1} are linear *homogeneous* functions of the variables x_1, x_2,...x_{n+1}. The general solution of the system (18) seems to have been first obtained by Hesse (*Crelle*, Tom. XXV. p. 171).

9. Lastly, certain systems of linear equations which have not constant coefficients may be solved by the above method.

Thus the solution of the equations

$$\left.\begin{aligned} \frac{dx}{dt} + T(ax + by) &= T_1 \\ \frac{dy}{dt} + T(a'x + b'y) &= T_2 \end{aligned}\right\} \dots\dots\dots\dots\dots (a),$$

where T, T_1, T_2 are functions of the independent variable, may be reduced to that of an ordinary linear differential equation of the first order.

For proceeding as before, we find

$$\frac{d(x + my)}{dt} + \lambda T(x + my) = T_1 + mT_2 \dots\dots\dots (b),$$

provided that λ and m be determined by the conditions

$$\lambda = a + ma', \quad \lambda m = b + mb' \dots\dots\dots\dots\dots (c).$$

Hence eliminating λ, we have

$$m(a + ma') = b + mb' \text{ .... } (d),$$

which gives two values for m. Integrating (b) regarded as a linear equation of the first order between $x + my$ and t, and substituting for λ its value in terms of m given by the first equation of the system (c), we have

$$x + my = \epsilon^{-(a+ma')\int T dt} \{C + \int \epsilon^{(a+ma')\int T dt} (T_1 + mT_2)\, dt\} \text{......} (e),$$

in which it remains to substitute for m its values given by (d).

Ex. Given $\dfrac{dx}{dt} + \dfrac{2z}{t}(x - y) = 1, \quad \dfrac{dy}{dt} + \dfrac{1}{t}(x + 5y) = t.$

The solution is

$$x + y = \frac{1}{t^3}\left(C_1 + \frac{t^4}{4} + \frac{t^5}{5}\right), \quad x + 2y = \frac{1}{t^4}\left(C_2 + \frac{t^5}{5} + \frac{t^6}{3}\right).$$

If in the system (a) we make $T = 1$, it becomes a system of equations with constant coefficients but possessed of second members.

The general system analogous to (a) when the number of variables is increased, may be solved by the same method. It may be well to notice that the equivalent symmetrical form is

$$\frac{dx_1}{X_1 + T_1} = \frac{dx_2}{X_2 + T_2} \dots = \frac{dx_n}{X_n + T_n} = \frac{dt}{T} \text{ ....} (20),$$

where $X_1, X_2, \dots X_n$ are linear homogeneous functions of the dependent variables, and $T, T_1, \dots T_n$ are functions of t. Treated under this form, it is obvious that its solution will be made to depend upon that of a linear differential equation of the first order, and an auxiliary algebraic equation of the nth degree.

Equations of an order higher than the first.

10. Any system of simultaneous equations of an order higher than the first is reducible to a system of the first order.

And this reduction though not always necessary for the purpose of solution is theoretically important, because it enables us to predicate what *kind* of solution is possible.

To effect this reduction it is only necessary to regard as a new variable and to express as such by a new symbol, each differential coefficient, except the highest, of each dependent variable in the given equations. The transformed equations will thus be of the first order, and the connecting relations of the first order also; and the two together will constitute a system of simultaneous equations of the first order.

Ex. Given the dynamical system

$$\frac{d^2x}{dt^2}=X, \quad \frac{d^2y}{dt^2}=Y, \quad \frac{d^2z}{dt^2}=Z,$$

where X, Y, Z are functions of the variables.

Here if we assume

$$\frac{dx}{dt}=x', \quad \frac{dy}{dt}=y', \quad \frac{dz}{dt}=z',$$

the given system assumes the form

$$\frac{dx'}{dt}=X, \quad \frac{dy'}{dt}=Y, \quad \frac{dz'}{dt}=Z.$$

Thus we have in the whole six equations of the first order between the six dependent variables x, y, z, x', y', z', and the independent variable t.

The complete solution of the latter system will therefore consist of six equations connecting the above system of variables with six arbitrary constants.

If from these six equations we eliminate the three new variables x', y', z', we obtain three equations connecting the original variables x, y, z, t with the above-mentioned six arbitrary constants.

And thus it might be shewn that the complete solution of any system of three differential equations of the second order between four variables will be expressed by three primitive equations connecting these variables with six arbitrary constants.

And still more generally, the complete solution of a system of n differential equations containing $n+1$ variables of which one is independent will consist of n equations connecting those variables with a number of constants equal to the sum of the indices of order of the several highest differential coefficients.

For let t be the independent and x one of the dependent variables, and let the highest differential coefficient of x which presents itself be $\frac{d^n x}{dt^n}$. Then in the reduction of the system of given equations to a system of equations of the first order it is necessary to introduce $n-1$ new variables connected with x by the relations

$$\frac{dx}{dt} = x_1, \quad \frac{dx_1}{dt} = x_2 \ldots \frac{dx_{n-2}}{dt} = x_{n-1}.$$

Thus the number of variables in the transformed system corresponding to x and its differential coefficients will be n, and as a similar remark applies to all the other variables, it appears that the total number of variables of the transformed system will be equal to the sum of the indices of the orders of the highest differential coefficients of the several dependent variables in the system given. Such then will be the number of equations of the transformed system, and such the number of constants introduced by their complete integration. Art. 5.

It is also evident that if from the equations by which the complete solution is expressed we eliminate all the new variables there will remain a number of equations equal in number to the original equations, and connecting the primitive variables with the constants above mentioned. Thus the proposition is established.

The transformation above employed is further important, because in the highest class of researches on theoretical dynamics it is always supposed that the differential equations of motion are reduced to a system of simultaneous equations of the first order.

At the same time it is not necessary for ordinary purposes to effect this reduction. Differentiation and elimination always enable us to arrive at a differential equation, higher in order, between two of the variables. The method of indeterminate multipliers may also be sometimes used with advantage. No *general* rule can however be given.

Ex. 1. Given $\frac{d^2x}{dt^2} = ax + by$, $\frac{d^2y}{dt^2} = a'x + b'y$.

1st method. Differentiating the first equation twice with respect to t, we have

$$\frac{d^4x}{dt^4} = a\frac{d^2x}{dt^2} + b\frac{d^2y}{dt^2}.$$

Eliminating y and $\frac{d^2y}{dt^2}$ from the above three equations, we have

$$\frac{d^4x}{dt^4} - (a + b')\frac{d^2x}{dt^2} + (ab' - a'b)x = 0 \text{ .... } (a).$$

The complete integral of this linear equation with constant coefficients will determine x, whence y is given by the formula

$$y = \frac{1}{b}\left(\frac{d^2x}{dt^2} - ax\right).$$

2nd method. From the given equations we find

$$\frac{d^2x}{dt^2} + m\frac{d^2y}{dt^2} = (a + ma')x + (b + mb')y$$

$$= (a + ma')\left(x + \frac{b + mb'}{a + ma'}y\right).$$

Let $x + my = u$, then provided that we determine m by the condition

$$m = \frac{b + mb'}{a + ma'} \text{ .... } (b),$$

we shall have

$$\frac{d^2u}{dt^2} = (a + ma')u,$$

whence $$u = C_1\epsilon^{(a+ma')^{\frac{1}{2}}t} + C_2\epsilon^{-(a+ma')^{\frac{1}{2}}t}.$$

Let m_1, m_2 be the values of m given by (b), then the complete primitive system is

$$x + m_1 y = C_1 \epsilon^{(a+m_1 a')^{\frac{1}{2}} t} + C_2 \epsilon^{-(a+m_1 a')^{\frac{1}{2}} t},$$

$$x + m_2 y = C_3 \epsilon^{(a+m_2 a')^{\frac{1}{2}} t} + C_4 \epsilon^{-(a+m_2 a')^{\frac{1}{2}} t},$$

and this is really equivalent to the previous solution, though more symmetrical.

Ex. 2. The approximate equations for the horizontal motion of a pendulum when the influence of the earth's rotation is taken into account* are

$$\left.\begin{aligned} \frac{d^2x}{dt^2} - 2r\frac{dy}{dt} + \frac{gx}{l} = 0 \\ \frac{d^2y}{dt^2} + 2r\frac{dx}{dt} + \frac{gy}{l} = 0 \end{aligned}\right\} \ldots\ldots\ldots\ldots\ldots\ldots (a),$$

l, g, and r being constants representing the length of the pendulum, the force of gravity, and the vertical component of the force resulting from the earth's rotation, respectively.

As the equations have constant coefficients they admit of complete integration. If we differentiate so as to enable us to eliminate y, $\frac{dy}{dx}$ and $\frac{d^2y}{dx^2}$, we find as the result

$$\frac{d^4x}{dt^4} + 2\left(2r^2 + \frac{g}{l}\right)\frac{d^2x}{dt^2} + \frac{g^2}{l^2}x = 0 \ldots\ldots\ldots\ldots (b),$$

the complete solution of which is of the form

$$x = A\cos(m_1 t + \alpha) + B\cos(m_2 t + \beta) \ldots\ldots\ldots (c),$$

where A, α, B, β are arbitrary constants, and m_1^2, m_2^2 are the two roots with signs changed of the equation

$$t^2 - 2\left(2r^2 + \frac{g}{l}\right)t + \frac{g^2}{l^2} = 0.$$

* Jullien, *Problèmes de Mécanique Rationnelle*, Tom. II. p. 233.

From the above value of x that of y may be obtained by means of the formula

$$y = -\frac{l}{2rg}\frac{d^3x}{dt^3} - \frac{l}{g}\left(2r + \frac{g}{2rl}\right)\frac{dx}{dt} \dots\dots\dots\dots (d),$$

which is readily deduced from the given equations.

The above system may also be solved by assuming

$$\left.\begin{aligned} x &= x' \cos rt + y' \sin rt \\ y &= -x' \sin rt + y' \cos rt \end{aligned}\right\} \dots\dots\dots\dots\dots\dots (e).$$

The transformed equations are

$$\frac{d^2x'}{dt^2} + \frac{g}{l}x' = 0, \quad \frac{d^2y'}{dt^2} + \frac{g}{l}y' = 0,$$

whence we find

$$\left.\begin{aligned} x' &= A \cos \frac{g^{\frac{1}{2}}t}{l^{\frac{1}{2}}} + B \sin \frac{g^{\frac{1}{2}}t}{l^{\frac{1}{2}}} \\ y' &= A' \cos \frac{g^{\frac{1}{2}}t}{l^{\frac{1}{2}}} + B' \sin \frac{g^{\frac{1}{2}}t}{l^{\frac{1}{2}}} \end{aligned}\right\} \dots\dots\dots\dots\dots (f).$$

11. In problems connected with central forces particular forms of the following system of equations present themselves, viz.

$$\frac{d^2x}{dt^2} = \frac{dR}{dx}, \quad \frac{d^2y}{dt^2} = \frac{dR}{dy}, \quad \frac{d^2z}{dt^2} = \frac{dR}{dz} \dots\dots\dots\dots (a),$$

where R is a given function of the quantity $\sqrt{(x^2 + y^2 + z^2)}$ or r. Multiplying the above equations by dx, dy, dz respectively, and integrating, we have

$$\frac{1}{2}\left\{\left(\frac{dx}{dt}\right)^2 + \left(\frac{dy}{dt}\right)^2 + \left(\frac{dz}{dt}\right)^2\right\} = R + B \dots\dots\dots\dots (b),$$

B being an arbitrary constant.

Again, since $\dfrac{dR}{dx} = \dfrac{dR}{dr}\dfrac{dr}{dx} = \dfrac{x}{r}\dfrac{dR}{dr}$, &c. the given system of equations may be expressed in the form

$$\frac{d^2x}{dt^2} = \frac{x}{r}\frac{dR}{dr}, \quad \frac{d^2y}{dt^2} = \frac{y}{r}\frac{dR}{dr}, \quad \frac{d^2z}{dt^2} = \frac{z}{r}\frac{dR}{dr}.$$

Now if from each pair of equations we eliminate $\frac{dR}{dr}$, we obtain

$$x\frac{d^2y}{dt^2}-y\frac{d^2x}{dt^2}=0,\quad y\frac{d^2z}{dt^2}-z\frac{d^2y}{dt^2}=0,\quad z\frac{d^2x}{dt^2}-x\frac{d^2z}{dt^2}=0,$$

of which it is evident that two only are independent. Integrating these, we have

$$x\frac{dy}{dt}-y\frac{dx}{dt}=C_1,\quad y\frac{dz}{dt}-z\frac{dy}{dt}=C_2,\quad z\frac{dx}{dt}-x\frac{dz}{dt}=C_3,$$

C_1, C_2, C_3 being constants.

Squaring the last three equations and adding, we obtain a result which may be expressed in the form

$$(x^2+y^2+z^2)\left\{\left(\frac{dx}{dt}\right)^2+\left(\frac{dy}{dt}\right)^2+\left(\frac{dz}{dt}\right)^2\right\}-\left(x\frac{dx}{dt}+y\frac{dy}{dt}+z\frac{dz}{dt}\right)^2$$
$$=C_1^2+C_2^2+C_3^2=A^2,$$

or, by virtue of (b) and of the known value of r,

$$2r^2(R+B)-\left(r\frac{dr}{dt}\right)^2=A^2 \quad\text{.................}(c),$$

whence
$$dt=\frac{rdr}{\sqrt{\{2r^2(R+B)-A^2\}}}\quad\text{..............}(d),$$

$$t+\alpha=\int\frac{rdr}{\sqrt{\{2r^2(R+B)-A^2\}}}\quad\text{............}(e).$$

Again, it is evident that by means of (c) we can eliminate R from each equation of the system (a). For (c) gives

$$R=-B+\frac{1}{2}\left\{\frac{A^2}{r^2}+\left(\frac{dr}{dt}\right)^2\right\}.$$

Substituting which in the first of the given equations, we have

$$\frac{d^2x}{dt^2}=\frac{x}{r}\left(-\frac{A^2}{r^3}+\frac{dr}{dt}\,\frac{d}{dr}\,\frac{dr}{dt}\right)$$
$$=\frac{x}{r}\left(-\frac{A^2}{r^3}+\frac{d^2r}{dt^2}\right).$$

Hence $$r\frac{d^2x}{dt^2}-x\frac{d^2r}{dt^2}+\frac{A^2x}{r^3}=0,$$

or $$\frac{d}{dt}r^2\frac{d}{dt}\frac{x}{r}+\frac{A^2x}{r^3}=0;$$

$$\therefore\ r^2\frac{d}{dt}r^2\frac{d}{dt}\left(\frac{x}{r}\right)+A^2\frac{x}{r}=0.$$

If we now assume $\frac{Adt}{r^2}=d\phi$, the above becomes

$$\frac{d^2\left(\frac{x}{r}\right)}{d\phi^2}+\frac{x}{r}=0,$$

whence $$\frac{x}{r}=a_1\cos\phi+b_1\sin\phi\ldots\ldots\ldots\ldots(f).$$

In like manner, we find

$$\frac{y}{r}=a_2\cos\phi+b_2\sin\phi\ldots\ldots\ldots\ldots(g),$$

$$\frac{z}{r}=a_3\cos\phi+b_3\sin\phi\ldots\ldots\ldots\ldots(h),$$

in which we must substitute for ϕ its value, viz.

$$\phi=\int\frac{Adt}{r^2}=\int\frac{Adr}{r\sqrt{\{(2r^2(R+B)+A^2)\}}}\ldots\ldots(i).$$

To this expression it would be superfluous to annex an arbitrary constant before that substitution. For each of the second members of (f), (g), (h) is expressible in the form $C\cos(\phi+C')$, in which ϕ is already provided with an arbitrary constant.

The solution is therefore expressed by means of (e) and (i), which determine r and the auxiliary ϕ as functions of t, and by (f), (g), (h), which then enable us to express x, y, z as functions of t. As we have however made no attempt to preserve independence in the series of results, the constants will not be independent. If we add the squares of (f), (g), (h), we shall have

$$1 = (a_1^2 + a_2^2 + a_3^2)\cos^2\phi + 2(a_1b_1 + a_2b_2 + a_3b_3)\sin\phi\cos\phi + (b_1^2 + b_2^2 + b_3^2)\sin^2\phi,$$

which involves the relations among the constants

$$a_1^2 + a_2^2 + a_3^2 = 1, \quad b_1^2 + b_2^2 + b_3^2 = 1, \quad a_1b_1 + a_2b_2 + a_3b_3 = 0 \ldots (k).$$

The six constants in (f), (g), (h), thus limited supply the place of only three arbitrary constants, and there being three also involved in (e), the total number is six, as it ought to be.

In the same way we may integrate the more general system

$$\frac{d^2x_1}{dt^2} = \frac{dR}{dx_1}, \quad \frac{d^2x_2}{dt^2} = \frac{dR}{dx_2}, \ldots \quad \frac{d^2x_n}{dt^2} = \frac{dR}{dx_n},$$

where R is a function of $\sqrt{(x_1^2 + x_2^2 \ldots + x_n^2)}$. The results, which have no application in our astronomy, are of the form which the above analysis would suggest. Biuet, to whom the method is due, has applied it to the problem of elliptic motion. (Liouville, Tom. II. p. 457). For all practical ends the employment of polar co-ordinates, as explained in treatises on dynamics, is to be preferred.

12. The following example presents itself in a discussion by M. Liouville*, of a very interesting case of the problem of three bodies.

Ex. Given

$$\frac{d^2u}{dt^2} + n^2\{u - 3x'(ux' + vy')\} = 0,$$

$$\frac{d^2v}{dt^2} + n^2\{v - 3y'(ux' + vy')\} = 0;$$

where, for brevity, x' is put for $\cos(at + b)$, y' for $\sin(at + b)$.

If we transform the above equation by assuming

$$nx' + vy' = U, \quad uy' - vx' = V,$$

* *Sur un cas particulier du Problème des trois corps. Journal de Mathématiques,* Tom. I. 2nd series, p. 248.

we find, after all reductions are effected,

$$\frac{d^2U}{dt^2} + 2a\frac{dV}{dt} + (a^2 + 2n^2)\,U = 0,$$

$$\frac{d^2V}{dt^2} - 2a\frac{dU}{dt} + (n^2 - a^2)\,V = 0.$$

And these equations being linear and with constant coefficients, may be integrated by the process of the previous section.

EXERCISES.

1. $\dfrac{dx}{dt} + 4x + \dfrac{y}{4} = 0,\quad \dfrac{dy}{dt} + 3y - x = 0.$

2. $\dfrac{dx}{dt} + 7x - y = 0,\quad \dfrac{dy}{dt} + 2x + 5y = 0.$

3. $4\dfrac{dx}{dt} + 9\dfrac{dy}{dt} + 44x + 49y = t,\quad 3\dfrac{dx}{dt} + 7\dfrac{dy}{dt} + 34x + 38y = \epsilon^t.$

4. $\dfrac{dx}{dt} + 5x + y = \epsilon^t,\quad \dfrac{dy}{dt} + 3y - x = \epsilon^{2t}.$

5. $\dfrac{dx}{2y - 5x + \epsilon^t} = \dfrac{dy}{x - 6y + \epsilon^{2t}} = dt.$

6. $-dx = \dfrac{dy}{3y + 4z} = \dfrac{dz}{2y + 5z}.$

7. $\dfrac{d^2x}{dt^2} - 3x - 4y + 3 = 0,\quad \dfrac{d^2y}{dt^2} + x - 8y + 5 = 0.$

8. $\dfrac{d^2x}{dt^2} - 3x - 4y + 3 = 0,\quad \dfrac{d^2y}{dt^2} + x + y + 5 = 0.$

9. $\dfrac{d^2x}{dt^2} + m^2x = 0,\quad \dfrac{d^2y}{dt^2} - m^2x = 0.$

10. Given $\dfrac{dx}{X + T_1} = \dfrac{dy}{Y + T_2} = \dfrac{dz}{Z + T_3} = \dfrac{dt}{T}$ where

$X = ax + by + cz$, $Y = a'x + b'y + c'z$, $Z = a''x + b''y + c''z$, and T, T_1, T_2, T_3 are functions of t.

11. What is the general form of the solution of a system of n simultaneous equations of the first order between $n+1$ variables?

12. What number of constants will be involved in the solution of a system of three simultaneous equations of the first, second and fourth order respectively between four variables?

13. Of the system of dynamical equations,

$$\frac{d^2x}{dt^2} + \frac{\mu x}{r^3} = 0, \quad \frac{d^2y}{dt^2} + \frac{\mu y}{r^3} = 0, \quad \frac{d^2z}{dt^2} + \frac{\mu z}{r^3} = 0,$$

where $r = (x^2 + y^2 + z^2)^{\frac{1}{2}}$, seven first integrals are obtained of which it is subsequently found that five only are independent. How many final integrals can hence be deduced without proceeding to another integration?

14. Given $a\dfrac{dx}{dt} = (b - c)\,yz$ .... (1),

$$b\frac{dy}{dt} = (c - a)\,zx \;\ldots\ldots\; (2), \qquad c\frac{dz}{dt} = (a - b)\,xy \;\ldots\ldots\; (3).$$

Putting $\dfrac{a}{b-c} = l$, $\dfrac{b}{c-a} = m$, $\dfrac{c}{a-b} = n$ we find, on eliminating dt,

$$lxdx = mydy = nzdz,$$

from which y and z will be found in terms of x, and their values will reduce (1) to a differential equation of the first order between x and t.

Or multiply the given equations, first by x, y, z, respectively, add the results and integrate; 2ndly by ax, by, cz, respectively, add the results and integrate. Then by means of the integrals obtained eliminate two of the variables from any of the given equations.

15. Shew that in the example of Art. 12, the transformation

$$x = x' \cos(rt + \epsilon) + y' \sin(rt + \epsilon),$$

$$y = -x' \sin(rt + \epsilon) + y' \cos(rt + \epsilon),$$

ϵ being an arbitrary constant, would not lead to a more general solution than the one actually arrived at.

CHAPTER XIV.

OF PARTIAL DIFFERENTIAL EQUATIONS.

1. Partial differential equations are distinguished by the fact that they involve partial differential coefficients in their expression, and therefore indicate the existence of more than one independent variable. Chap. I. Art. 2.

The nature of these equations will be best explained by one or two examples of the mode of their formation.

Ex. 1. The general equation of cylindrical surfaces is

$$x - lz = \phi(y - mz) \quad \text{.....................(1)},$$

ϕ being a functional symbol, and l and m constants determining the direction of the generating line. As this is a relation connecting three variables we are permitted to regard two of them as independent. Choosing x and y as the independent variables, and differentiating with respect to them in succession, z being regarded as dependent on them both, we have

$$1 - l\frac{dz}{dx} = -m\phi'(y - mz)\frac{dz}{dx} \quad \text{.................(2)},$$

$$-l\frac{dz}{dy} = \phi'(y - mz)\left(1 - m\frac{dz}{dy}\right) \quad \text{............(3)}.$$

Eliminating the function $\phi'(y - mz)$, there results

$$l\frac{dz}{dx} + m\frac{dz}{dy} = 1 \quad \text{..............................(4)},$$

the partial differential equation of cylindrical surfaces. Of this equation (1) is termed the general primitive.

In the above example a linear partial differential equation of the first order has been formed by the elimination of a single arbitrary function.

Ex. 2. If we assume as a primitive equation

$$z = ax + by - ab \text{..........................} (5),$$

and, after regarding x and y as independent, differentiate with respect to these variables in succession, we have

$$\frac{dz}{dx} = a, \quad \frac{dz}{dy} = b.$$

Eliminating a and b by substitution in the primitive, there results

$$z = x\frac{dz}{dx} + y\frac{dz}{dy} - \frac{dz}{dx}\frac{dz}{dy} \text{..................} (6),$$

a partial differential equation of the first order, *but not* linear.

Now this equation has been formed by the elimination not of an arbitrary function but of two arbitrary constants. The equation (5) is here, by way of distinction, called the *complete* primitive. The epithets *general* and *complete* seem to have been employed by Lagrange to denote the two kinds of generality which arise from arbitrary functions, and from arbitrary constants, respectively.

Ex. 3. Given $z = \phi(y + ax) + \psi(y - ax)$, where ϕ and ψ are arbitrary symbols of functionality.

Proceeding to differential coefficients of the second order we find

$$\frac{d^2z}{dx^2} = a^2\{\phi''(y + ax) + \psi''(y - ax)\},$$

$$\frac{d^2z}{dy^2} = \phi''(y + ax) + \psi''(y - ax)$$

whence

$$\frac{d^2z}{dx^2} = a^2\frac{d^2z}{dy^2} \text{..} (7),$$

a partial differential equation of the second order and of the first degree.

And this equation has been formed by the elimination of two arbitrary functions from the *general* primitive.

These examples illustrate the usual, and what may perhaps with propriety be termed the *primary*, modes of genesis of partial differential equations, viz. the elimination of arbitrary functions, and the elimination of arbitrary constants. It is to be noted that these modes are perfectly distinct. Thus we might in Ex. 1, by specifying the form of the function ϕ, eliminate the constants l and m from the primitive (1), and the derived equations (2) and (3), instead of eliminating the functional forms from the two latter; but the result would differ in character, as well as in the mode of its origin, from that which has been actually obtained. We must bear in mind that when from a primitive equation of given form different partial differential equations are derived, it is owing to a difference of assumption as to what is to be regarded as arbitrary; so that we are not permitted to say that to the *same* primitive, considered in the same sense of generality, different partial differential equations belong.

In Ex. 1, a partial differential equation of the first order has been formed from a general primitive containing one arbitrary function, and in Ex. 3 a partial differential equation of the second order has been formed from a general primitive containing two arbitrary functions. These examples exhibit a certain analogy with the genesis of ordinary differential equations, the order of the equation being equal to the number of constants in its primitive. But this analogy is not general. For let

$$F\{x, y, z, \phi(u), \psi(v)\} = 0,$$

be an assumed primitive containing two arbitrary functions $\phi(u)$, $\psi(v)$, where u and v are given functions of x, y, z. Then representing the first member by F, regarding x and y as independent variables, and forming all possible derived equations up to the second order, we have

$$\frac{dF}{dx} = 0, \quad \frac{dF}{dy} = 0,$$

$$\frac{d^2F}{dx^2} = 0, \quad \frac{d^2F}{dx\,dy} = 0, \quad \frac{d^2F}{dy^2} = 0,$$

which with the given equation make six equations. But these containing the six functions

$$\phi(u), \quad \psi(v), \quad \phi'(u), \quad \psi'(v), \quad \phi''(u), \quad \psi''(v),$$

do not, in general, suffice to enable us by the elimination of the latter, to form a partial differential equation of the second order free from arbitrary functions.

We see then, 1st that partial differential equations do not arise from the elimination of arbitrary functions only; 2ndly, that even as respects this mode of genesis, no general canons exist similar to those which govern the connexion of ordinary differential equations with their primitives. On both these grounds it will be proper, in considering special classes of equations, to examine their special origin and to seek therein the clue to their solution.

Solution of partial differential equations.

2. The following preliminary theorem is of much importance.

THEOREM. When the partial differential coefficients in the given equation are all taken with respect to one only of the independent variables, we are permitted to integrate as if the other independent variables were constant, adding at the last an arbitrary function of the latter instead of an arbitrary constant.

The reason of this will appear in the following examples.

Ex. 1. Given $x + y\dfrac{dz}{dx} = 0$.

Multiplying by dx, integrating with respect to x, and adding an arbitrary function of y, we have

$$\frac{x^2}{2} + yz = \phi(y),$$

the solution required.

It is *permitted* in the above, and in all similar cases, to complete the solution by adding an arbitrary function of y, because, with reference to the integration effected, y is constant; and it is *necessary* to add such a complementary function in order to obtain the most general solution, because an arbitrary function of one of the variables is more general than an arbitrary constant not involving that variable.

Ex. 2. Given $y\frac{dz}{dy} - 2x - 2z - y = 0.$

This equation may be expressed in the form

$$\frac{dz}{dy} - \frac{2}{y}z = 1 + \frac{2x}{y}.$$

Involving no differential coefficient with respect to x, it may be treated as a linear differential equation of the first order in which y is the independent, and z the dependent variable; only instead of an arbitrary constant we must add an arbitrary function of x. The final solution is

$$x + y + z = y^2\phi(x).$$

It sometimes happens that equations not belonging to the above class are reducible to it by a transformation.

Ex. 3. Given $\frac{d^2z}{dxdy} = x^2 + y^2.$

Let $\frac{dz}{dx} = w$, then we have

$$\frac{dw}{dy} = x^2 + y^2,$$

whence integrating with respect to y, and adding an arbitrary function of x,

$$w = x^2y + \frac{y^3}{3} + \phi(x).$$

Restoring to w its value $\frac{dz}{dx}$, integrating with respect to x, and adding an arbitrary function of y, we have

$$z = \frac{x^3y}{3} + \frac{y^3x}{3} + \int\phi(x)\,dx + \psi(y).$$

Now $\phi(x)$ being arbitrary, $\int\phi(x)dx$ is also arbitrary, and may be represented by $\chi(x)$, whence

$$z = \frac{x^3y + y^3x}{3} + \chi(x) + \psi(y).$$

Linear partial differential equations of the first order.

3. When there are but three variables, z dependent, x and y independent, the equations to be considered assume the form

$$P\frac{dz}{dx} + Q\frac{dz}{dy} = R,$$

P, Q, and R being given functions of x, y, z, or constant. This form we shall first consider.

Usually the differential coefficients $\frac{dz}{dx}$ and $\frac{dz}{dy}$ are represented by p and q respectively. The equation thus becomes

$$Pp + Qq = R \quad\text{.....................}\quad (1).$$

The mode of solution is due to Lagrange, and was first established by the following considerations.

Since z is a function of x and y, we have

$$dz = pdx + qdy.$$

Hence eliminating p between the above and the given equation, we have

$$Pdz - Rdx = q\,(Pdy - Qdx).$$

Suppose in the first place that $Pdz - Rdx$ is the exact differential of a function u, and $Pdy - Qdx$ the exact differential of a function v, then we have

$$du = q\,dv.$$

Now the first member being an exact differential, the second must also be such. This requires that q should be a function of v, but does not limit the form of the function. Represent it by $\phi'(v)$, then we have $du = \phi'(v)\,dv$, whence

$$u = \phi(v) \quad\text{.........................}\quad (2).$$

The functions u and v are determined by integrating the equations

$$Pdz - Rdx = 0, \quad Pdy - Qdx = 0,$$

symmetrically expressible in the form

$$\frac{dx}{P} = \frac{dy}{Q} = \frac{dz}{R} \quad\text{.....................}(3),$$

and of which the solution, Chap. XIII. Art. 5, assumes the form

$$u = a, \quad v = b \quad\text{.......................}(4),$$

a and b being arbitrary constants.

Dismissing the particular hypothesis above employed, Lagrange then proves that if in any case we can obtain two integrals of the system (3) in the forms (4), then $u = \phi(v)$ will satisfy the partial differential equation, in perfect independence of the form of the function ϕ.

We shall adopt a somewhat different course. We shall first establish a general Rule for the formation of a partial differential equation whose primitive is of the form $u = \phi(v)$, u and v being given functions of x, y, and z. Upon the solution of this direct problem we shall ground the solution of the inverse problem of ascending from the partial differential equation to its primitive.

PROPOSITION. *A primitive equation of the form* $u = \phi(v)$, *where* u *and* v *are given functions of* x, y, z, *gives rise to a partial differential equation of the form*

$$Pp + Qq = R \quad\text{....................}(5),$$

where P, Q, R are functions of x, y, z.

Before demonstrating this proposition we stop to observe that the form $u = \phi(v)$ is equivalent to the form

$$F\{x, y, z, \phi(v)\} = 0,$$

ϕ being an arbitrary, but F a definite functional symbol.

For solving the latter equation with respect to $\phi(v)$ we have a result of the form

$$\phi(v) = F_1(x, y, z), \text{ or } \phi(v) = u$$

on representing $F_1(x, y, z)$ by u. Thus the proposition affirmed amounts to this, viz. that any equation between x, y, and z which involves an arbitrary function will give rise to a linear partial differential equation of the first order.

Differentiating the primitive $u = \phi(v)$, first with respect to x, secondly with respect to y, we have

$$\frac{du}{dx} + \frac{du}{dz} p = \phi'(v) \left(\frac{dv}{dx} + \frac{dv}{dz} p\right),$$

$$\frac{du}{dy} + \frac{du}{dz} q = \phi'(v) \left(\frac{dv}{dy} + \frac{dv}{dz} q\right).$$

Eliminating $\phi'(v)$ by dividing the second equation by the first, we have

$$\frac{\dfrac{du}{dy} + \dfrac{du}{dz} q}{\dfrac{du}{dx} + \dfrac{du}{dz} p} = \frac{\dfrac{dv}{dy} + \dfrac{dv}{dz} q}{\dfrac{dv}{dx} + \dfrac{dv}{dz} p},$$

or, on clearing of fractions,

$$\left(\frac{du}{dy}\frac{dv}{dz} - \frac{du}{dz}\frac{dv}{dy}\right) p + \left(\frac{du}{dz}\frac{dv}{dx} - \frac{du}{dx}\frac{dv}{dz}\right) q$$
$$= \frac{du}{dx}\frac{dv}{dy} - \frac{du}{dy}\frac{dv}{dx} \quad \text{.........} \ (6).$$

Now this is a partial differential equation of the form (5). For u and v being given functions of x, y and z, the coefficients of p and q, as well as the second member, are known. The proposition is therefore proved.

As an illustration, we have in Ex. 1. Art. 1, $u = x - lz$, $v = y - mz$, whence

$$\frac{du}{dx} = 1, \quad \frac{du}{dy} = 0, \quad \frac{du}{dz} = -l,$$

$$\frac{dv}{dx} = 0, \quad \frac{dv}{dy} = 1, \quad \frac{dv}{dz} = -m.$$

Substituting these values in (6) there results,

$$lp + mq = 1,$$

which agrees with the result before obtained.

4. The general equation (6), of which the above theorem is a direct consequence, has been established by the direct elimination of the arbitrary function. But the same result may also be established in the following manner, which has the advantage of shewing the real nature of the dependence of the coefficients P, Q, R upon the given functions u and v.

Differentiating the equation $u = f(v)$ with respect to all the variables, we have

$$\frac{du}{dx}dx + \frac{du}{dy}dy + \frac{du}{dz}dz = f'(v)\left(\frac{dv}{dx}dx + \frac{dv}{dy}dy + \frac{dv}{dz}dz\right) \dots (7),$$

and as this equation is to hold true independently of the form of the function $f(v)$, and therefore of the form of the derived function $f'(v)$, we must have

$$\left.\begin{array}{l} \frac{du}{dx}dx + \frac{du}{dy}dy + \frac{du}{dz}dz = 0 \\ \frac{dv}{dx}dx + \frac{dv}{dy}dy + \frac{dv}{dz}dz = 0 \end{array}\right\} \dots\dots\dots\dots\dots (8),$$

whence we find

$$\frac{dx}{\frac{du}{dy}\frac{dv}{dz} - \frac{du}{dz}\frac{dv}{dy}} = \frac{dy}{\frac{du}{dz}\frac{dv}{dx} - \frac{du}{dx}\frac{dv}{dz}} = \frac{dz}{\frac{du}{dx}\frac{dv}{dy} - \frac{du}{dy}\frac{dv}{dx}} \dots (9).$$

Introducing now the condition that z is the dependent, x and y the independent variables, we have

$$pdx + qdy = dz.$$

To eliminate the differentials, let the terms of this equation be divided by the respectively equal members of (9), and we have

$$\left(\frac{du}{dy}\frac{dv}{dz} - \frac{du}{dz}\frac{dv}{dy}\right)p + \left(\frac{du}{dz}\frac{dv}{dx} - \frac{du}{dx}\frac{dv}{dz}\right)q$$

$$= \frac{du}{dx}\frac{dv}{dy} - \frac{du}{dy}\frac{dv}{dx} \dots\dots (10),$$

which agrees with (6).

Now if in the above general form we represent as before the coefficient of p by P, that of q by Q, and the second member by R, we see from (9) that P, Q, R are proportional to dx, dy and dz, in the system (8). But that system is precisely the same as we should obtain by differentiating the equations

$$u = a, \quad v = b,$$

a and b being arbitrary constants. Hence, the partial differential equation whose complete primitive is $u = f(v)$, may be formed by the following simple rule.

RULE. *Forming the equations $u = a$, $v = b$, where a and b are arbitrary constants, differentiate them, and determine the ratios of dx, dy, dz in the form*

$$\frac{dx}{P} = \frac{dy}{Q} = \frac{dz}{R} \quad \text{.......................} (11).$$

Then will $Pp + Qq = R$ be the differential equation required.

Or, the Rule may more briefly be stated thus. *Eliminate dx, dy, dz between the three equations,*

$$du = 0, \quad dv = 0, \quad dz - pdx - qdy = 0 \quad \text{........} (12).$$

It is worth while to notice that the partial differential equation here presents itself, like many other results of analysis, in the form of a *determinant.*

Ex. The functional equation of surfaces of revolution, the axis passing through the origin, is

$$lx + my + nz = \phi\,(x^2 + y^2 + z^2)\,;$$

their partial differential equation is required.

Here, proceeding according to the Rule, we have

$$ldx + mdy + ndz = 0,$$

$$xdx + ydy + zdz = 0,$$

whence

$$\frac{dx}{mz - ny} = \frac{dy}{nx - lz} = \frac{dz}{ly - mx}.$$

The partial differential equation therefore is

$$(mz - ny)\,p + (nx - lz)\,q = ly - mx.$$

5. We proceed in the second place to apply the above results to the inverse problem of solution.

From what has been said of the origin of partial differential equations of the form $Pp + Qq = R$ it is evident that their solution will be effected by the following rule.

RULE. *Form the system of ordinary differential equations*

$$\frac{dx}{P} = \frac{dy}{Q} = \frac{dz}{R} \quad \ldots\ldots\ldots\ldots\ldots\ldots\ldots\ldots \quad (13),$$

and express their integrals in the forms $u = a$, $v = b$; *then will the equation* $u = f(v)$, *where* f *is a symbol of arbitrary functionality, express the solution required.*

For, setting out from the assumed primitive, $u = f(v)$, we should, by the application of the previous and direct Rule, be led to the partial differential equation in question.

The difficulty of the process consisting therefore solely in the integration of the system of *ordinary* differential equations (13), is referred to the methods of the last chapter.

Ex. 1. Given $xp + yq = nz$.

Here, the system of *ordinary* differential equations is

$$\frac{dx}{x} = \frac{dy}{y} = \frac{dz}{nz},$$

and the variables therein are separated. The integrals may obviously be expressed in the forms

$$\frac{y}{x} = c, \quad \frac{z}{x^n} = c'.$$

Hence, the required solution is

$$\frac{z}{x^n} = \phi\left(\frac{y}{x}\right),$$

indicating that z is a homogeneous function of x and y of the nth degree.

Ex. 2. Given $(mz - ny)\,p + (nx - lz)\,q = ly - mx$.

Here the system of ordinary differential equations is

$$\frac{dx}{mz - ny} = \frac{dy}{nx - lz} = \frac{dz}{ly - mx}.$$

From these we readily deduce

$$ldx + mdy + ndz = 0, \quad xdx + ydy + zdz = 0,$$

the integrals of which are

$$lx + my + nz = a, \quad x^2 + y^2 + z^2 = b,$$

the final solution is therefore

$$lx + my + nz = \phi\,(x^2 + y^2 + z^2).$$

Ex. 3. Given $(y^3x - 2x^4)\dfrac{d\mu}{dx} + (2y^4 - x^3y)\dfrac{d\mu}{dy} = 9\,(x^3 - y^3)\,\mu$.

This is the partial differential equation on the solution of which would depend the determination of the general integrating factor of the equation $(x^3y - 2y^4)\,dx + (y^3x - 2x^4)\,dy = 0$. Chap. IV. Art. 6.

The system of ordinary differential equations is

$$\frac{dx}{y^3x - 2x^4} = \frac{dy}{2y^4 - x^3y} = \frac{d\mu}{9\,(x^3 - y^3)\,\mu} \quad\text{.............}\,(a).$$

The first equation of the system is

$$(x^3y - 2y^4)\,dx + (y^3x - 2x^4)\,dy = 0,$$

and of this the complete solution is

$$\frac{x}{y^2} + \frac{y}{x^2} = c.$$

We may also deduce from (a)

$$3\left(\frac{dx}{x} + \frac{dy}{y}\right) + \frac{d\mu}{\mu} = 0.$$

Of which the complete primitive

$$x^3y^3\mu = c'.$$

Hence the solution of the partial differential equation is

$$\mu = \frac{1}{x^3y^3}\,\phi\left(\frac{x}{y^2}+\frac{y}{x^2}\right);$$

and this agrees with the result obtained by other considerations in the chapter referred to.

We may note that in this, as in all similar cases, the differential equation whose integrating factor is sought, presents itself as one of the equations of the system on whose solution the *complete* determination of the factor rests.

6. The above theory may be obviously extended to partial differential equations of the first order and degree involving any number of variables.

Let $x_1, x_2 \dots x_n$ represent the independent variables and z the dependent variable. Let moreover the primitive functional equation be expressed in the form

$$u = \phi\,(v_1, v_2 \dots v_{n-1}) \dots\dots\dots\dots\dots\dots\dots (14),$$

where $u, v_1, v_2 \dots v_n$ are known functions of the variables.

Differentiating with respect to all the variables, and for brevity representing $\phi\,(v_1, v_2 \dots v_n)$ by ϕ, we have

$$du = \frac{d\phi}{dv_1}\,dv_1 + \frac{d\phi}{dv_2}\,dv_2 \dots + \frac{d\phi}{dv_{n-1}}\,dv_{n-1}$$

But ϕ being an arbitrary function of the quantities $v_1, v_2 \dots v_n$, it is evident that the supposition that the above equation is generally true involves the supposition that the system of equations

$$du = 0,\ \ dv_1 = 0,\ \ dv_2 = 0,\ \dots dv_{n-1} = 0,$$

is true, a system of which the developed form is

$$\left.\begin{array}{l}
\dfrac{du}{dx_1}\,dx_1 \dots + \dfrac{du}{dx_n}\,dx_n + \dfrac{du}{dz}\,dz = 0 \\[2ex]
\dfrac{dv_1}{dx_1}\,dx_1 \dots + \dfrac{dv_1}{dx_n}\,dx_n + \dfrac{dv_1}{dz}\,dz = 0 \\[2ex]
\dots\dots\dots\dots\dots\dots\dots\dots\dots\dots \\[2ex]
\dfrac{dv_{n-1}}{dx_1}dx_1 \dots + \dfrac{dv_{n-1}}{dx_n}dx_n + \dfrac{dv_{n-1}}{dz}dz = 0
\end{array}\right\} \dots\dots (15).$$

Now this system may be converted into an equivalent system determining the *ratios* of the differentials $dx_1, dx_2 \dots dx_n, dz$, in the form

$$\frac{dx_1}{P_1} = \frac{dx_2}{P_2} \dots = \frac{dx_n}{P_n} = \frac{dz}{R} \dots\dots\dots\dots\dots\dots (16),$$

where $P_1, P_2 \dots P_n$ and R are functions of the variables or are constants.

Introducing the condition that z is to be regarded as a function of $x_1, x_2, \dots x_n$, we have

$$p_1 dx_1 + p_2 dx_2 \dots + p_n dx_n = dz \dots\dots\dots\dots\dots (17),$$

where $p_1, p_2 \dots p_n$ are the several first differential coefficients of z. And now eliminating the differentials $dx_1, dx_2,, \dots dx_n, dz$ from (16) and (17) by division, we have

$$P_1 p_1 + P_2 p_2 \dots + P_n p_n = R \dots\dots\dots\dots\dots (18),$$

for the partial differential equation sought.

Conversely, to integrate the above equation it is only necessary to form and to integrate the system (16). Representing the integrals of that system in the forms

$$u = a,\ v_1 = b_1,\ v_2 = b_2 \dots v_n = b_n,$$

the final solution will be

$$u = \phi (v_1, v_2, \dots v_n) \dots\dots\dots\dots\dots\dots\dots (19),$$

This solution may also be put in the form

$$\Phi (u, v_1, v_2, \dots v_n) = 0 \dots\dots\dots\dots\dots\dots (20).$$

Ex. $(y+z+u)\frac{du}{dx} + (z+x+u)\frac{du}{dy} + (x+y+u)\frac{du}{dz} = x+y+z.$

(Lacroix, Tom. II. p. 542.)

Here the auxiliary system of equations is

$$\frac{dx}{y+z+u} = \frac{dy}{z+x+u} = \frac{dz}{x+y+u} = \frac{du}{x+y+z},$$

which is reducible to the form

$$\frac{du - dx}{x - u} = \frac{du - dy}{y - u} = \frac{du - dz}{z - u} = \frac{dx + dy + dz + du}{3(x+y+z+u)},$$

each term being now an exact differential. The system of integrals will evidently be

$$\frac{c_1}{x-u} = \frac{c_2}{y-u} = \frac{c_3}{z-u} = (x+y+z+u)^{1}.$$

Or, representing the function $x+y+z+u$ by S,

$$S^{\frac{1}{3}}(x-u) = c_1, \quad S^{\frac{1}{3}}(y-u) = c_2, \quad S^{\frac{1}{3}}(z-u) = c_3.$$

Whence the complete integral symmetrically exhibited will be

$$\Phi\left\{S^{\frac{1}{3}}(x-u),\ S^{\frac{1}{3}}(y-u),\ S^{\frac{1}{3}}(z-u)\right\} = 0.$$

The solution of all partial differential equations of the form

$$X_1\frac{dz}{dx_1} + X_2\frac{dz}{dx_2}\ldots + X_n\frac{dz}{dx_n} = Z,$$

where X_1, X_2, ... X_n and Z are any linear functions of the variables x_1, x_2, ...x_n, z, may be completely effected.

For it depends on the solution of the system of ordinary differential equations

$$\frac{dx_1}{X_1} = \frac{dx_2}{X_2}\ldots = \frac{dx_n}{X_n} = \frac{dz}{Z},$$

which has been fully discussed in Chap. XIII.

Hesse has integrated the still more general equation which, according to the above notation, would present itself in the form

$$X_1\frac{dz}{dx_1} + X_2\frac{dz}{dx_2}\ldots + X_n\frac{dz}{dx_n}$$

$$+ X_{n+1}\left(x_1\frac{dz}{dx_1} + x_2\frac{dz}{dx_2}\ldots + x_n\frac{dz}{dx_n} - z\right) = X_{n+2},$$

where X_1, X_2, ... X_{n+2} are any linear functions of the variables. (Crelle, Tom. XXV. p. 171.)

Non-linear equations of the first order with three variables.

7. Partial differential equations of the first order with two independent variables x, y, and one dependent variable z, have for their typical form

$$F(x, y, z, p, q) = 0 \qquad (1).$$

Those which are linear with respect to p and q, we have considered apart. Those which are non-linear we proceed to consider. The genesis of an equation of this class from a complete primitive involving two arbitrary constants has been illustrated in Ex. 2, Art. 1; and the mode is general. From a given primitive, involving x, y, z with two arbitrary constants, and from its two derived equations of the first order formed by differentiating with respect to x and y respectively, it is possible to eliminate both the constants. The result is a partial differential equation of the first order. Conversely the integration of such an equation consists mainly in the discovery of its complete primitive—not that this is its only form of solution, but because out of it all other forms are developed. From the complete primitive involving arbitrary constants arise, 1st, the general primitive involving arbitrary functions; 2ndly, the singular solution. The terminology of Lagrange is here adopted. (*Calcul des Fonctions, Leçon* XX.)

The process of solution usually employed was originated by Lagrange, and completed by Charpit, whose name it commonly bears.

To deduce the complete primitive of a partial differential equation of the form $F(x, y, z, p, q) = 0$.

The existence of a primitive relation between x, y, z involves the supposition that the equation

$$dz = pdx + qdy \qquad (2),$$

should satisfy the condition of integrability,

$$\left(\frac{dp}{dy}\right) = \left(\frac{dq}{dx}\right) \qquad (3),$$

where $\left(\frac{dp}{dy}\right)$ represents the differential coefficient of p with respect to y on the assumption that p is expressed as a function of x and y, and $\left(\frac{dq}{dx}\right)$ the differential coefficient of q with respect to x, on a similar assumption as to the expression of q. Now regarding p for the sake of greater generality as a function of x, y, z, z being at the same time an unknown function of x and y, we have

$$\left(\frac{dp}{dy}\right) = \frac{dp}{dy} + \frac{dp}{dz}\frac{dz}{dy}$$

$$= \frac{dp}{dy} + q\frac{dp}{dz}.$$

Again, suppose that by means of the given differential equation, q may be expressed as a function of x, y, z, p. Regarding in such expression z as a function of x, y, and p as a function of x, y, and z, we have

$$\left(\frac{dq}{dx}\right) = \frac{dq}{dx} + \frac{dq}{dz}\frac{dz}{dx} + \frac{dq}{dp}\left(\frac{dp}{dx} + \frac{dp}{dz}\frac{dz}{dx}\right),$$

$$= \frac{dq}{dx} + \frac{dq}{dz}p + \frac{dq}{dp}\frac{dp}{dx} + \frac{dq}{dp}\frac{dp}{dz}p.$$

Substituting these values in (3), we have on transposition

$$-\frac{dq}{dp}\frac{dp}{dx} + \frac{dp}{dy} + \left(q - p\frac{dq}{dp}\right)\frac{dp}{dz} = \frac{dq}{dx} + p\frac{dq}{dz} \text{.........} (4).$$

Now the coefficients $-\frac{dq}{dp}$, $q - p\frac{dq}{dp}$, and the second member $\frac{dq}{dx} + p\frac{dq}{dz}$ being known functions of x, y, z, p, since q as determined by the given equation is such, the above presents itself as a linear partial differential equation of the first order in which p is the dependent and x, y, z the independent variables.

Applying therefore Lagrange's process, Art. 6, we have the auxiliary system

$$\frac{dx}{-\frac{dq}{dp}} = dy = \frac{dz}{q - p\frac{dq}{dp}} = \frac{dp}{\frac{dq}{dx} + p\frac{dq}{dz}} \text{..............} (5);$$

and this, it is to be observed, is a system of *ordinary* differential equations between x, y, z, and p. It may further be noted that while it has been formed in order to secure the integrability of the equation $dz = pdx + qdy$, it also includes that equation. For it gives

$$dz = \left(q - p\frac{dq}{dp}\right) dy = pdx + qdy,$$

since by the equation of the first and second members

$$-\frac{dq}{dp}dy = dx.$$

Accordingly if from the system (5) we can deduce a value of p involving an arbitrary constant, that value together with the corresponding value of q drawn from the given equation will render the equation $dz = pdx + qdy$ integrable. Effecting the integration we shall obtain an equation between x, y, z and two arbitrary constants which will constitute a complete primitive.

We say *a* and not *the* complete primitive, because the system (5) may furnish more than one value of p involving an arbitrary constant, and so give occasion to deduce more than one complete primitive. Lagrange had indeed proposed to employ the general value of p involving arbitrary functions, furnished by the solution of the partial differential equation (4). The sufficiency of a value involving only an arbitrary constant was remarked by Charpit and subsequently recognised by Lagrange.

The practical rule for the discovery of a complete primitive of the equation $F(x, y, z, p, q) = 0$ is therefore the following. *Express q in terms of x, y, z, p. Substitute this value in the auxiliary system* (5), *and deduce by integration a value of p involving an arbitrary constant. Substitute that value of p with the corresponding value of q in the equation $dz = pdx + qdy$, also included in the auxiliary system* (5), *and again integrate.*

Ex. 1. Required a complete primitive of the equation $z = pq$.

Substituting $\frac{z}{p}$ for q, the system (5) becomes

$$\frac{p^2 dx}{z} = dy = \frac{p\,dz}{2z} = dp.$$

The equation $dp = dy$ gives $p = y + a$, whence $q = \frac{z}{y+a}$.

Therefore $$dz = (y+a)\,dx + \frac{z}{y+a}\,dy,$$
of which the integral is

$$z = (y+a)(x+b) \text{..................} (6),$$

a and b being arbitrary constants. This then is a complete primitive.

Another will be found by employing the equation

$$\frac{pdz}{2z} = dp;$$

integrating which, we have

$$p = cz^{\frac{1}{2}}, \quad q = \frac{z^{\frac{1}{2}}}{c},$$

whence $$dz = cz^{\frac{1}{2}}dx + \frac{z^{\frac{1}{2}}}{c}\,dy.$$

Integrating, we find

$$2z^{\frac{1}{2}} = cx + \frac{1}{c}y + e,$$

or $$z = \frac{\left(cx + \frac{y}{c} + e\right)^2}{4} \text{......................} (7),$$

e being a new arbitrary constant. It will be found on trial that both (6) and (7) satisfy the equation $z = pq$.

8. Prop. *Given a complete primitive of a partial differential equation of the first order, to deduce the general primitive and the singular solution.*

Expressing the complete primitive in the form

$$z = f(x, y, a, b) \text{.....................} (8),$$

a and b being its arbitrary constants, the partial differential equation is itself obtained by eliminating a and b between the above equation and the derived equations

$$p = \frac{df(x, y, a, b)}{dx}, \quad q = \frac{df(x, y, a, b)}{dy},$$

or, as we may for brevity write,

$$p = \frac{df}{dx}, \quad q = \frac{df}{dy} \ldots\ldots\ldots\ldots\ldots\ldots (9).$$

Now reasoning as in Chap. VIII., the effect of the elimination will be the same if a and b, instead of being constants, are made functions of x and y, so determined as to preserve to the equations (9) their actual form. But a and b being made variable, we have

$$p = \frac{df}{dx} + \frac{df}{da}\frac{da}{dx} + \frac{df}{db}\frac{db}{dx},$$

$$q = \frac{df}{dy} + \frac{df}{da}\frac{da}{dy} + \frac{df}{db}\frac{db}{dy}.$$

Hence the equations for determining a and b are

$$\frac{df}{da}\frac{da}{dx} + \frac{df}{db}\frac{db}{dx} = 0 \ldots\ldots\ldots\ldots (10),$$

$$\frac{df}{da}\frac{da}{dy} + \frac{df}{db}\frac{db}{dy} = 0 \ldots\ldots\ldots\ldots (11).$$

Now this system may be satisfied in two distinct ways,

1st by assuming

$$\frac{df}{da} = 0, \quad \frac{df}{db} = 0 \ldots\ldots\ldots\ldots (12).$$

The values of a and b hence found lead, on substitution in the complete primitive, to that solution which Lagrange terms singular.

2ndly, Supposing $\frac{df}{da}$ and $\frac{df}{db}$ not to vanish, we have, on elimination of them from (10), (11),

$$\frac{da}{dx}\frac{db}{dy} - \frac{da}{dy}\frac{db}{dx} = 0 \ldots\ldots\ldots\ldots (13).$$

Now this supposes either, 1st, that a and b are constant, which leads us back to the complete primitive; or, 2ndly, that b is an arbitrary function of a. Chap. IV. Art. 3. Again, multiplying (10) by dx and (11) by dy, and adding, we have

$$\frac{df}{da}da + \frac{df}{db}db = 0 \ldots\ldots\ldots\ldots (14).$$

Thus the system (10), (11) is now replaced by the system (13), (14).

Making then, in accordance with (13), $b = \phi(a)$, the expression for z in (8) becomes

$$z = f\{x, y, a, \phi(a)\},$$

while (14) becomes

$$\frac{d}{da}f\{x, y, a, \phi(a)\} = 0.$$

And these together constitute what Lagrange terms the *general primitive*. To apply them it is only necessary to give a particular form to $\phi(a)$, and then eliminate a. Hence the following theorem.

THEOREM. *The complete primitive of a partial differential equation of the first order being expressed in the form*

$$z = f(x, y, a, b) \ldots\ldots\ldots\ldots (15),$$

the general primitive will be obtained by eliminating a between the equations

$$\left.\begin{aligned} z &= f\{x, y, a, \phi(a)\} \\ 0 &= \frac{df(x, y, a, \phi(a))}{da} \end{aligned}\right\} \ldots\ldots\ldots\ldots (16),$$

the singular solution, by eliminating a and b between (15) *and the equations*

$$\frac{df(x, y, a, b)}{da} = 0, \qquad \frac{df(x, y, a, b)}{db} = 0 \ldots\ldots (17).$$

It will be observed that the process for obtaining the general primitive is virtually equivalent to that by which we should seek the envelope of the surfaces defined by the corresponding complete primitive, the constants a and b being treated as variable parameters connected by an arbitrary relation, while the process for obtaining the singular solution is that by which we should seek the envelope of (15), supposing a and b to be independent parameters.

Thus, of the system of solutions which consists of a *complete* primitive, a *general* primitive, and a singular solution, the complete primitive must be regarded as forming the basis, and the system itself geometrically interpreted includes the surfaces represented by the complete primitive together with the whole of their possible envelopes.

Ex. To deduce the general primitive and singular solution of the equation $z = pq$.

A complete primitive being

$$z = (y + a)(x + b) \ldots\ldots\ldots\ldots (a),$$

the corresponding general primitive will be expressed by the system

$$\left.\begin{aligned} z &= (y + a)\{x + \phi(a)\} \\ 0 &= x + \phi(a) + (y + a)\phi'(a) \end{aligned}\right\} \ldots\ldots\ldots\ldots (b),$$

from which a must be eliminated when the form of $\phi(a)$ is assigned. Another form of the complete primitive being

$$z = \frac{(cx + \frac{y}{c} + e)^2}{4} \ldots\ldots\ldots\ldots (c),$$

the corresponding form of the general primitive will be

$$\left.\begin{aligned} z &= \tfrac{1}{4}\{cx + \frac{y}{c} + \psi(c)\}^2 \\ 0 &= x - \frac{y}{c^2} + \psi'(c) \end{aligned}\right\} \ldots\ldots\ldots (d),$$

from which c must be eliminated when the form of $\psi(c)$ is assigned.

To deduce the singular solution, we have from (a),

$$\frac{dz}{da} = x + b = 0,$$

$$\frac{dz}{db} = y + a = 0.$$

Hence, $b = -x$, $a = -y$ which, substituted in (a), gives $z = 0$, a singular solution. The same result is deducible from (c).

9. In the last example, two complete primitives, two corresponding forms of general primitive, and one common form of singular solution are presented. Two systems of solution appear, and the question arises: Does either system suffice alone? The answer is given in the following theorem.

THEOREM. *All possible solutions of a partial differential equation of the first order, are virtually contained in the system consisting of a single complete primitive, with the derived general primitive and singular solution.*

As before, we shall represent the proposed differential equation and its given complete primitive in the forms,

$$F(x, y, z, p, q) = 0 \quad\ldots\ldots\ldots\ldots\ldots\ldots\quad (18),$$

$$z = f(x, y, a, b) \quad\ldots\ldots\ldots\ldots\ldots\ldots\ldots\quad (19).$$

We shall also represent in the form,

$$z = \chi(x, y) \quad\ldots\ldots\ldots\ldots\ldots\ldots\ldots\ldots\quad (20),$$

some solution of (18), of which nothing more is known than that it *is* a solution. We are to shew that such solution is included in the system of solutions of which the common primitive (19) constitutes the basis.

If we represent for brevity the values of z in (19) and (20) by f and χ respectively, we shall have, since both are solutions of (18),

$$F\left(x, y, f, \frac{df}{dx}, \frac{df}{dy}\right) = 0 \quad\ldots\ldots\ldots\ldots\quad (21),$$

$$F\left(x, y, \chi, \frac{d\chi}{dx}, \frac{d\chi}{dy}\right) = 0 \text{ .... (22).}$$

From the form of the above equations it appears that if a and b are so determined as to satisfy two of the conditions,

$$f = \chi, \ \frac{df}{dx} = \frac{d\chi}{dx}, \ \frac{df}{dy} = \frac{d\chi}{dy} \text{ .... (23),}$$

they will satisfy the third. For suppose they satisfy the first two, then the system (21), (22) may be expressed in the form

$$F\left(x, y, \chi, \frac{d\chi}{dx}, \frac{df}{dy}\right) = 0, \ F\left(x, y, \chi, \frac{d\chi}{dx}, \frac{d\chi}{dy}\right) = 0 \text{ ... (24),}$$

in which the truth of the third equation of (23) is involved.

Now, as (19) satisfies (18) whatever constant values we assign to a and b, it still will do so if, *after* the differentiations by which $\frac{df}{dx}$ and $\frac{df}{dy}$ are found, we substitute for a and b any functions of x and y.

But a and b can be determined so as to satisfy two conditions. Hence they can be determined so as to satisfy the system (23). Differentiating the equation $f = \chi$ on the hypothesis that a and b are functions so determined, we have

$$\frac{df}{dx} + \frac{df}{da}\frac{da}{dx} + \frac{df}{db}\frac{db}{dx} = \frac{d\chi}{dx},$$

$$\frac{df}{dy} + \frac{df}{da}\frac{da}{dy} + \frac{df}{db}\frac{db}{dy} = \frac{d\chi}{dy}.$$

Here, $\frac{df}{dx}, \frac{df}{dy}$ have the same values as in (23), being obtained by differentiating as if a and b were constant. Hence, reducing by (23), we have

$$\left.\begin{aligned} \frac{df}{da}\frac{da}{dx} + \frac{df}{db}\frac{db}{dx} = 0 \\ \frac{df}{da}\frac{da}{dy} + \frac{df}{db}\frac{db}{dy} = 0 \end{aligned}\right\} \text{ .... (25).}$$

But these are the equations (10) (11), Art. 8, by which the system of solutions founded upon the complete primitive is constructed.

The argument then is briefly this. If $z=\chi(x, y)$ is a solution of the given partial differential equation, it is possible to determine a and b in the given complete primitive so as to satisfy the equations (23); therefore so as to satisfy the equations (25); therefore so as to indicate a necessary inclusion of $z=\chi(x, y)$ in the system which is founded upon the given complete primitive.

COR. 1. Hence the connexion of a given solution with a given complete primitive may be determined in the following manner. Adopting the foregoing notation, determine the values of a and b which satisfy the system (23). If those values are constant, the solution is a particular case of the complete primitive; if they are variable, but so that the one is a function of the other, the solution is a particular case of the general primitive; if they are variable and unconnected it is a singular solution.

COR. 2. Hence also any two systems of solutions founded upon distinct complete primitives are equivalent. For each is virtually composed of all possible particular solutions.

Ex. The equation $z=pq$, has for its complete primitive $z=(x+a)(y+b)$, and for a particular solution $z=\dfrac{(y+x)^2}{4}$. What is the connexion of this solution with the complete primitive?

We have by (23),

$$(x+a)(y+b)=\frac{(y+x)^2}{4},$$

$$y+b=\frac{y+x}{2},\quad x+a=\frac{y+x}{2}.$$

These equations are not independent, the first being the product of the last two. Any two of them give

$$a=\frac{y-x}{2},\quad b=\frac{x-y}{2},$$

whence $b = -a$. Thus, the values of a and b being variable, but such that b is a function of a, the proposed solution is a particular case of the general primitive.

Some general questions, but of minor importance, relating to the functional connexion of different forms of solution, will be noticed in the Exercises at the end of this chapter.

In quitting this part of the subject, we may observe that there are two modes in which the questions it involves may be considered. The first consists in shewing that the gain of generality, which in Charpit's process accrues in the transition from the complete to the general primitive, is equal to that which Lagrange's original but far more difficult process secures by the employment of the general value of p drawn from (4), instead of a particular value drawn from its auxiliary system. The proof of this equivalence, as developed with more or less of completeness, by Lagrange and Poisson, (*Lacroix*, Tom. II. p. 564, III. p. 705), and recently by Prof. De Morgan, (*Cambridge Journal*, Vol. VII. p. 28), is, from its complexity, unsuitable to an elementary work. The other mode is that developed in the foregoing sections.

Derivation of the singular solution from the differential equation.

10. The complete primitive expresses z in terms of x, y, a, b. The differential equation expresses z in terms of x, y, p, q. Either is convertible into the other by means of the two equations derived from the complete primitive by differentiating with respect to x and y respectively. Hence it is not difficult to establish the two following equations,

$$\left.\begin{aligned} \frac{dz}{dp} &= \frac{\dfrac{dz}{da}\dfrac{d^2z}{dbdy} - \dfrac{dz}{db}\dfrac{d^2z}{dady}}{\dfrac{d^2z}{dadx}\dfrac{d^2z}{dbdy} - \dfrac{d^2z}{dady}\dfrac{d^2z}{dbdx}} \\ \frac{dz}{dq} &= -\frac{\dfrac{dz}{da}\dfrac{d^2z}{dbdx} - \dfrac{dz}{db}\dfrac{d^2z}{dadx}}{\dfrac{d^2z}{dadx}\dfrac{d^2z}{dbdy} - \dfrac{d^2z}{dady}\dfrac{d^2z}{dbdx}} \end{aligned}\right\} \quad \ldots\ldots\ldots\ldots (26),$$

in the first members of which z is supposed to be expressed in terms of x, y, p, q by means of the differential equation, in the second members, in terms of x, y, a, b by means of the complete primitive.

Now the singular solution is deduced from the complete primitive by means of the equations

$$\frac{dz}{da}=0, \quad \frac{dz}{db}=0 \qquad (27);$$

and it is evident from the form of (26), that this will generally involve the conditions

$$\frac{dz}{dp}=0, \quad \frac{dz}{dq}=0 \qquad (28).$$

Such then will generally be the conditions for determining the singular solution from the differential equation.

The conditions (28) will not present themselves, should the denominator of the right-hand members of (26) vanish identically. But it may be shewn that in this case the conditions (27) do not lead to a singular solution. And analogy renders it probable that *whenever* the conditions (28) are satisfied the result, if it be a solution at all, will be a singular solution. The complete investigation of this point, however, would involve inquiries similar to those of Chapter VIII.

The Rule indicated is then *to eliminate p and q from the differential equation by means of the equations* (28) *thence derived.*

11. The following geometrical applications are intended to illustrate the preceding sections.

Ex. 1. Required to determine the general equation of the family of surfaces in which the length of that portion of the normal which is intercepted between the surface and the plane x, y, is constant and equal to unity.

As the length of the intercept above described in any surface is $z(1+p^2+q^2)^{\frac{1}{2}}$, we have to solve the equation

$$z^2(1+p^2+q^2)=1 \qquad (a).$$

Hence $q=(z^{-2}-1-p^2)^{\frac{1}{2}}$, and the auxiliary system (5), Art. 7, becomes, on substitution and division by $(z^{-2}-1-p^2)^{\frac{1}{2}}$,

$$\frac{dx}{p}=\frac{dy}{(z^{-2}-1-p^2)^{\frac{1}{2}}}=\frac{dz}{z^{-2}-1}=\frac{-z^3dp}{p}\text{ ....}(b).$$

From the last two members we have on integration

$$\text{..................}\ p=\frac{c\,(1-z^2)^{\frac{1}{2}}}{z}.$$

Substituting this, with the corresponding value of q derived from (a), in the equation $dz=pdx+qdy$ we have

$$dz=\frac{c\,(1-z^2)^{\frac{1}{2}}\,dx}{z}+(1-c^2)^{\frac{1}{2}}\frac{(1-z^2)^{\frac{1}{2}}}{z}\,dy,$$

integrating which in the usual way, we find

$$(1-z^2)^{\frac{1}{2}}=-cx-(1-c^2)^{\frac{1}{2}}y-c',$$

or, changing the signs of c and c',

$$(1-z^2)^{\frac{1}{2}}=cx-(1-c^2)^{\frac{1}{2}}y+c'\text{ ....}(c),$$

which is a complete primitive. The corresponding form of the general primitive will be

$$\left.\begin{aligned}(1-z^2)^{\frac{1}{2}}&=cx-(1-c^2)^{\frac{1}{2}}y+\phi(c)\\ 0&=x-c\,(1-c^2)^{-\frac{1}{2}}y+\phi'(c)\end{aligned}\right\}\text{ .... }(d),$$

from which c must be eliminated.

But another system of solutions exists; for from the first, third, and fourth members of (b) we may deduce

$$pdz+zdp+dx=0,$$

whence $pz+x=a$, from which, and from the given equation determining p and q, we have to integrate

$$dz=\frac{a-x}{z}\,dx+\frac{\{1-(a-x)^2-z^2)\}^{\frac{1}{2}}}{z}\,dy.$$

The result is

$$(x-a)^2+(y-b)^2+z^2=1\text{ ....}(e),$$

a complete primitive. The corresponding general primitive is

$$\left.\begin{array}{r}(x-a^2)+\{y-\psi(a)\}^2+z^2=1\\ x-a+\{y-\psi(a)\}\,\psi'(a)=0\end{array}\right\}\ldots\ldots\ldots\ldots (f).$$

To deduce the singular solution from the differential equation (a) we have

$$\frac{dz}{dp}=p\,(1-p^2-q^2)^{-\frac{1}{2}}=0,\quad \frac{dz}{dq}=q\,(1-p^2-q^2)^{-\frac{1}{2}}=0,$$

whence $p=0$, $q=0$; substituting which in (a) we find

$$z=\pm 1.$$

The above example illustrates the importance of obtaining, if possible, a choice of forms of the complete primitives. The second, of those above obtained, leads to the more interpretable results. It represents a sphere whose radius is unity and whose centre is in the plane x, y, while the derived general primitive represents the tubular surface generated by that sphere moving but not ceasing to obey the same conditions. The singular solution represents the two planes between which the motion would be confined. All these surfaces evidently satisfy the conditions of the problem.

Ex. 2. Required to determine a system of surfaces such that the area of any portion shall be in a constant ratio $(m:1)$ to the area of its projection on the plane xy.

The differential equation is evidently

$$1+p^2+q^2=m^2,$$

and it will readily be found that it has only one complete primitive, viz.

$$z=ax+\sqrt{(m^2-a^2-1)}\,y+b.$$

Thus the general primitive is

$$z=ax+\sqrt{(m^2-a^2-1)}\,y+\phi(a)$$

$$0=x-\frac{a}{\sqrt{(m^2-a^2-1)}}\,y+\phi'(a);$$

and this represents various systems of cones and other developable surfaces.

Similar but more interesting applications may be drawn from the problem of the determination of equally attracting surfaces.

12. Attention has already been directed to the different forms in which the solution of a non-linear equation may sometimes be presented. It may be added that linear equations admit generally of a duplex form of solution. The ordinary method gives directly the equation of the system of surfaces which they represent; Charpit's method leads to a form of solution which exhibits rather the mode of their genesis.

Ex. Lagrange's method presents the solution of the equation

$$(mz - ny)\,p + (nx - lz)\,q = ly - mx \ldots\ldots (a),$$

in the form

$$lx + my + nz = \phi\,(x^2 + y^2 + z^2) \ldots\ldots\ldots (b),$$

the known equation of surfaces of revolution whose axes pass through the origin of coordinates.

Charpit's method presents as the complete primitive of (a)

$$(x - cl)^2 + (y - cm)^2 + (z - cn)^2 = r^2 \ldots\ldots\ldots (c),$$

c and r being arbitrary constants. This is the equation of the generating sphere. The general primitive represents its system of possible envelopes.

These solutions are manifestly equivalent.

Symmetrical and more general solution of partial differential equations of the first order.

13. The method of Charpit labours under two defects, 1st, It supposes that from the given equation q can be expressed as a function of x, y, z, p; 2ndly, It throws little light of analogy on the solution of equations involving *more* than two independent variables—a subject of fundamental importance in connexion with the highest class of researches on Theoretical Dynamics. We propose to supply these defects.

It will have been noted that Charpit's method consists in determining p and q as functions of x, y, z, which render the equation $dz = pdx + qdy$ integrable. This determination presupposes the existence of two algebraic equations between x, y, z, p, q; viz. 1st, the equation given, 2ndly, an equation obtained by integration and involving an arbitrary constant. Let us represent these equations by

$$F(x, y, z, p, q) = 0, \quad \Phi(x, y, z, p, q) = a \ldots\ldots (29),$$

respectively. And let us now endeavour to obtain in a general manner the relation between the functions F and Φ.

Simply differentiating with respect to x, y, z, p, q, and representing $\frac{dF}{dx}$ by X, $\frac{d\Phi}{dx}$ by X', $\frac{dF}{dp}$ by P, $\frac{d\Phi}{dp}$ by P', &c. we have $Xdx + Ydy + Zdz + Pdp + Qdq = 0$,

$$X'dx + Y'dy + Z'dz + P'dp + Q'dq = 0;$$

or, substituting $pdx + qdy$ for dz,

$$(X + pZ)\,dx + (Y + qZ)\,dy + Pdp + Qdq = 0 \ldots (30),$$

$$(X' + pZ')\,dx + (Y' + qZ')\,dy + P'dp + Q'dq = 0 \ldots (31).$$

But, representing for brevity $\frac{d^2z}{dx^2}$, $\frac{d^2z}{dxdy}$ and $\frac{d^2z}{dy^2}$, by r, s, t, respectively, we have

$$\left.\begin{aligned} dp &= rdx + sdy \\ dq &= sdx + tdy \end{aligned}\right\} \ldots\ldots\ldots\ldots\ldots (32).$$

Substituting these values in (31) we have

$$(X' + pZ' + rP' + sQ')\,dx + (Y' + qZ' + sP' + tQ')\,dy = 0,$$

which, since dx and dy are independent, can only be satisfied by separately equating to 0 their coefficients. These furnish then the two equations

$$\left.\begin{aligned} -(X' + pZ') &= rP' + sQ' \\ -(Y' + qZ') &= sP' + tQ' \end{aligned}\right\} \ldots\ldots\ldots\ldots (33).$$

Now these equations are of the same *form* as (32). They establish the same relations between the functions

$$-(X'+pZ'),\ -(Y'+qZ'),\ P',\ Q', \ldots\ldots (34),$$

as (32) does between the differentials dp, dq, dx, dy.

It follows that if we give to dx and dy, which are arbitrary, the ratio of the last two of the functions (34) then will dp and dq have the ratio of the first two, so that the following will be a consistent scheme of relations, viz.

$$\frac{dx}{P'} = \frac{dy}{Q'} = -\frac{dp}{X'+pZ'} = -\frac{dq}{Y'+qZ'} \ldots\ldots (35).$$

Now dividing the successive terms of (30) by the successive members of (35) we have

$$(X+pZ)\,P' + (Y+qZ)\,Q' - P(X'+pZ') - Q\,(Y'+qZ') = 0 \ldots\ldots (36).$$

This is the relation sought. It might be obtained by direct elimination by multiplying the equations of (33) by P and Q respectively, and the corresponding equations derived from (30) by P' and Q' respectively, and subtracting the sum of the former from the sum of the latter.

It is obvious too, and the remark is important, that we might pass directly from (30) to (36) by substituting for dx, dy, dp, dq, the functions of (34), and that this substitution is justified by the identity of relations established in (32) and (33).

If in (36) we substitute for X, Y, &c. their values, and transpose the second and third terms, we have

$$\left(\frac{dF}{dx}+p\frac{dF}{dz}\right)\frac{d\Phi}{dp} - \left(\frac{d\Phi}{dx}+p\frac{d\Phi}{dz}\right)\frac{dF}{dp} + \left(\frac{dF}{dy}+q\frac{dF}{dz}\right)\frac{d\Phi}{dq} - \left(\frac{d\Phi}{dy}+q\frac{d\Phi}{dz}\right)\frac{dF}{dq} = 0 \ldots\ldots\ldots\ldots (37).$$

Such is the relation which connects the functions F and Φ. When F is given it assumes the form of a linear partial differential equation of the first order for determining Φ. If from its auxiliary system we can deduce any integral involving an arbitrary constant, and such that in conjunction with the given equation it enables us to determine p and q as functions of x, y, z, the subsequent integration of $dz = pdx + qdy$ will lead to a form of the complete primitive.

14. Analogy now points out the method to be pursued for the solution of equations involving more than two independent variables.

Prop. To deduce the complete primitive of the partial differential equation

$$F(x_1, x_2 \dots x_n, z, p_1, p_2 \dots p_n) = 0 \dots\dots\dots (38),$$

where $$p_1 = \frac{dz}{dx_1}, \dots p_n = \frac{dz}{dx_n}.$$

In the first place we must seek to determine values of p_1, $p_2, \dots p_n$ in terms of the primitive variables x_1, $x_2 \dots x_n$, z, such as will render integrable the equation

$$dz = p_1 dx_1 + p_2 dx_2 \dots + p_n dx_n \dots\dots\dots (39).$$

Suppose one of the equations requisite in conjunction with (38) for this determination to be

$$\Phi(x_1, x_2, \dots x_n, z, p_1, p_2, \dots p_n) = a_1 \dots\dots (40).$$

Then representing the first members of (38) and (40) by their characteristics F and Φ, differentiating, and substituting for dz its value given in (39), we have results which may be thus expressed,

$$\Sigma_i \left\{\left(\frac{dF}{dx_i} + p_i \frac{dF}{dz}\right) dx_i + \frac{dF}{dp_i} dp_i\right\} = 0 \dots\dots\dots (41),$$

$$\Sigma_i \left\{\left(\frac{d\Phi}{dx_i} + p_i \frac{d\Phi}{dz}\right) dx_i + \frac{d\Phi}{dp_i} dp_i\right\} = 0 \dots\dots\dots (42),$$

where Σ_i represents summation from $i = 1$ to $i = n$.

But since $p_i = \dfrac{dz}{dx_i}$, we have

$$dp_i = \frac{d^2z}{dx_i dx_1} dx_1 + \frac{d^2z}{dx_i dx_2} dx_2 \ldots + \frac{d^2z}{dx_i dx_n} dx_n \ldots\ldots (43).$$

Substituting this value in (42), we shall be permitted, in consequence of the independence of the differentials $dx_1\ dx_2 \ldots dx_n$, to equate their respective coefficients to 0.

It is easy to see that the coefficient of dx_r will be

$$\frac{d\Phi}{dx_r} + p_r \frac{d\Phi}{dz} + \Sigma_i \frac{d\Phi}{dp_i} \frac{d^2z}{dx_i dx_r}.$$

Equating this to 0, we have, on transposition,

$$-\left(\frac{d\Phi}{dx_r} + p_r \frac{d\Phi}{dz}\right) = \Sigma_i \frac{d^2z}{dx_i dx_r} \frac{d\Phi}{dp_i}.$$

Hence, changing i into r and r into i,

$$-\left(\frac{d\Phi}{dx_i} + p_i \frac{d\Phi}{dz}\right) = \Sigma_r \frac{d^2z}{dx_r dx_i} \frac{d\Phi}{dp_r} \ldots\ldots\ldots\ldots (44).$$

Now comparing this with (43), and observing that $\frac{d^2z}{dx_i dx_r} = \frac{d^2z}{dx_r dx_i}$, we see that the systems of differentials represented by dp_i and dx_r respectively are connected by the same relations as the systems of functions represented by

$$-\left(\frac{d\Phi}{dx_i} + p_i \frac{d\Phi}{dz}\right) \text{ and } \frac{d\Phi}{dp_r} \text{ respectively.}$$

Hence, by the reasoning of the previous example, it is permitted to substitute in (41), for the differentials, the corresponding functions, viz. for dp_i, $-\left(\frac{d\Phi}{dx_i} + p_i \frac{d\Phi}{dz}\right)$; and for dx_i, $\frac{d\Phi}{dp_i}$. We thus find

$$\Sigma_i \left\{ \left(\frac{dF}{dx_i} + p_i \frac{dF}{dz}\right) \frac{d\Phi}{dp_i} - \frac{dF}{dp_i} \left(\frac{d\Phi}{dx_i} + p_i \frac{d\Phi}{dz}\right) \right\} = 0 \ldots\ldots (45),$$

the summation extending from $i = 1$ to $i = n$. This is the relation sought, and it is seen to be symmetrical with respect to F and Φ. When F is given, it becomes a linear partial differential equation for determining Φ. From its auxiliary system of ordinary differential equations it suffices to obtain $n-1$ integrals,

$$\Phi_1 = a_1, \quad \Phi_2 = a_2, \ldots \Phi_{n-1} = a_{n-1} \ldots\ldots\ldots\ldots (46),$$

such as, in conjunction with the given equation, will enable us to determine $p_1, p_2, \ldots p_n$ in terms of the original variables; then integrating (39), we shall obtain the complete primitive in the form

$$f(x_1, x_2, \ldots x_n, z, a_1, a_2, \ldots a_n) = 0 \ldots\ldots\ldots (47).$$

All other forms of solution are hence deducible by regarding $a_1, a_2, \ldots a_n$ as parameters varying, independently or in subjection to connecting relations, but so as to leave unaffected the *forms* of $p_1, p_2, \ldots p_n$.

It is proper to observe that the given equation $F = 0$ is itself included among the particular integrals of (45). In fact F is one of the forms of Φ which make $\Phi = a$, a solution, as will be found on trial. The given equation is therefore a particular integral. And therefore the $n-1$ integrals of the system (46) must be independent of it in order to render the determination of $p_1, p_2, \ldots p_n$ possible.

The equation (45) may be expressed as follows:

$$\Sigma_i \left(\frac{dF}{dx_i}\frac{d\Phi}{dp_i} - \frac{dF}{dp_i}\frac{d\Phi}{dx_i}\right) + \frac{dF}{dz}\Sigma_i p_i \frac{d\Phi}{dp_i} - \frac{d\Phi}{dz}\Sigma_i p_i \frac{dF}{dp_i} = 0.$$

And under this elegant form, obtained however by a more complex analysis, the solution is presented by Brioschi (*Tortolini*, Tom. VI. p. 426, *Intorno ad una proprietà delle equazioni alle derivate parziali del primo ordine*).

The problem of the integration of partial differential equations of the first order, irrespectively of the number of the variables, appears to have been first solved by Pfaff, but the most complete discussion of it will be found in a memoir by Cauchy (*Exercices d'Analyse*, Tom. II. p. 238. *Sur l'intégration des équations aux dérivées partielles du premier ordre*), in which the determination of the arbitrary functions of the general primitive so as to satisfy given initial conditions is fully considered, The connexion of the subject with Theoretical Dynamics was first established by the researches of Sir W. Hamilton and Jacobi. The truth, illustrated above, that the solution of a partial differential equation of the first order is reducible to that of a system of ordinary differential equations, and the truth that the solutions of certain systems of differential equations (including that of dynamics) may be reduced to the discovery of a single function defined by a partial differential equation, are correlative. The researches above referred to, together with those of Liouville, Bertrand, and Bour, founded partly upon their results and partly upon the allied discoveries of Lagrange and Poisson concerning the variation of the arbitrary constants in dynamical problems, contain the most important of recent additions to our speculative knowledge of Differential Equations. For this reason we have dwelt upon their history. Fuller information will be found in Mr Cayley's excellent Report on the Recent Progress of Theoretical Dynamics. (*Report of British Association*, 1857).

XIV. EXERCISES.

1. How are equations, in which all the differential coefficients have reference to only one of the variables, solved?

2. $\dfrac{dz}{dx}=\dfrac{y}{\sqrt{(y^2-x^2)}}$. 3. $\dfrac{dz}{dx}=\dfrac{y}{x+z}$.

4. The partial differential equation of the first order which results from a primitive of the form $u=f(v)$, where u and v are determinate functions of x, y, z, is necessarily linear. Prove this.

5. $ap+bq=1$. 6. $p+q=\dfrac{z}{a}$.

7. $yp+xq=z$. 8. $x^2p-xyq+y^2=0$.

9 Integrate the equation of conical surfaces

$$(a-x)p+(b-y)q=c-z.$$

10. $xzp+yzq=xy$.

11. $(y^2+z^2-x^2)p-2xyq+2xz=0$.

12. Required the equation of the surface which cuts at right angles all the spheres which pass through the origin of coordinates and have their centres in the axis of x.

It will be found that this leads to the partial differential equation of the last problem.

13. $z - xp - yq = a(x^2 + y^2 + z^2)^{\frac{1}{2}}$.

14. Find the equation of the surface which cuts at right angles the system of ellipsoids represented by the equation

$$Ax^2 + By^2 + Cz^2 = D^2,$$

where D is the variable parameter. *Lacroix*, Tom. II. p. 678.

15. Find the equation of a surface which belongs at once to surfaces of revolution defined by the equation $py - qx = 0$, and to conical surfaces defined by the equation $px + qy = z$.

In problems like the above we must regard the equations as simultaneous, determine p and q as functions of x, y, z, and substitute their values in the equation $dz = pdx + qdy$, which will become integrable by a single equation if the problem is a possible one, but not otherwise.

16. $x\frac{dz}{dx} + y\frac{dz}{dy} + t\frac{dz}{dt} = az + \frac{xy}{t}$.

17. Explain the distinction between a complete primitive and a general primitive of a partial differential of the first order.

18. Find the complete and the general primitive of

$$z = px + qy = pq.$$

19. Deduce a singular solution of the above.

20. $pq = 1$. 21. $q = xp + p^2$.

22. Shew from the form of its integral that $q = f(p)$ belongs only to developable surfaces.

23. Deduce two complete primitives of

$$pq = px + qy.$$

24. Deduce two complete primitives of

$$\sqrt{p} + \sqrt{q} = 2x.$$

25. Given two general primitives of a partial differential equation of the first order, in the forms,

$$\text{1st.} \quad z = F\{x, y, a, \phi(a)\}, \quad 0 = \frac{dF\{x, y, a, \phi(a)\}}{da},$$

$$\text{2nd.} \quad z = \Phi\{x, y, c, \psi(c)\}, \quad 0 = \frac{d\Phi\{x, y, c, \psi(c)\}}{dc},$$

shew that the dependence of the functions $\psi(c)$ and $\phi(c)$, when the two primitives lead to the same particular integral, may be determined by the following rule. Eliminate x and y from any four independent equations of the system

$$F = \Phi, \quad \frac{dF}{dx} = \frac{d\Phi}{dx}, \quad \frac{dF}{dy} = \frac{d\Phi}{dy}, \quad \frac{dF}{da} = 0, \quad \frac{d\Phi}{dc} = 0.$$

The two resulting equations will involve the relation required, and when the form of $\phi(a)$ is given, the elimination of a from both will give a differential equation for determining the form of $\psi(c)$.

26. The equation $z = pq$ has two general primitives,

$$\text{1st.} \quad z = (y + a)\{x + \phi(a)\}, \quad 0 = \frac{d}{da}[\{y + a\}\{x + \phi(a)\}],$$

$$\text{2nd.} \quad 4z = \{cx + \frac{y}{c} + \psi(c)\}^2, \quad 0 = \frac{d}{dc}\{cx + \frac{y}{c} + \psi(c)\}^2;$$

shew hence that the relation between $\phi(a)$ and $\psi(c)$ is expressed by the equations

$$\phi'(a) + \frac{1}{c^2} = 0, \quad c\psi(c) - c^2\psi'(c) = 2a.$$

CHAPTER XV.

PARTIAL DIFFERENTIAL EQUATIONS OF THE SECOND ORDER.

1. THE general form of a partial differential equation of the second order is

$$F(x, y, z, p, q, r, s, t) = 0 \quad\ldots\ldots\ldots\ldots\ldots(1),$$

where $p = \dfrac{dz}{dx}$, $q = \dfrac{dz}{dy}$, $r = \dfrac{d^2z}{dx^2}$, $s = \dfrac{d^2z}{dx\,dy}$, $t = \dfrac{d^2z}{dy^2}$.

It is only in particular cases that the equation admits of integration, and the most important is that in which the differential coefficients of the second order present themselves only in the first degree; the equation thus assuming the form

$$Rr + Ss + Tt = V \quad\ldots\ldots\ldots\ldots\ldots(2),$$

in which R, S, T and V are functions of x, y, z, p and q. This equation we propose to consider. The most usual method of solution, due to Monge, consists in a certain procedure for discovering either one or two first integrals of the form

$$u = f(v) \quad\ldots\ldots\ldots\ldots\ldots(3),$$

u and v being determinate functions of x, y, z, p, q, and f an arbitrary functional symbol. From these first integrals, singly or in combination, the second integral involving two arbitrary functions is obtained by a subsequent integration.

An important remark must here be made. Monge's method involves the assumption that the equation (2) admits of a first integral of the form (3). Now this is not always the case. There exist primitive equations, involving two arbitrary functions, from which by proceeding to a second differentiation both functions may be eliminated and an equation of the form (2) obtained, but from which it is impossible to eliminate one function only so as to lead to an intermediate equation of the form (3). Especially this happens if the primitive involve an arbitrary function and its derived function together. Thus the primitive

$$z=\phi(y+x)+\psi(y-x)-x\{\phi'(y+x)-\psi'(y-x)\}\ldots(4),$$

leads to the partial differential equation of the second order

$$r-t=\frac{2p}{x}\ldots\ldots\ldots\ldots\ldots\ldots\ldots(5),$$

but not through an intermediate equation of the form (3).

It is necessary therefore not only to explain Monge's method, but also to give some account of methods to be adopted when it fails.

2. It is not only not true that the equation (2) has necessarily a first integral of the form (3), but neither is the converse proposition true. We propose therefore, 1st, to inquire under what conditions an equation of the first order of the form (3) does lead to an equation of the second order of the form (2); 2ndly, to establish upon the results of this direct inquiry the inverse method of solution. And this procedure, though somewhat longer than that usually followed, is more simple, because exact and thorough.

PROP. 1. *A partial differential equation of the first order of the form $u=f(v)$ can only lead to a partial differential equation of the second order of the form*

$$Rr+Ss+Tt=V\ldots\ldots\ldots\ldots\ldots\ldots(6),$$

when u and v are so related as to satisfy identically the condition

$$\frac{du}{dp}\frac{dv}{dq}-\frac{du}{dq}\frac{dv}{dp}=0\ldots\ldots\ldots\ldots\ldots\ldots(7).$$

For, differentiating the equation $u=f(v)$ with respect to x, and observing that $\frac{dz}{dx}=p$, $\frac{dp}{dx}=r$, $\frac{dq}{dx}=0$, we have

$$\frac{du}{dx}+p\frac{du}{dz}+r\frac{du}{dp}+s\frac{du}{dq}=f'(v)\left(\frac{dv}{dx}+p\frac{dv}{dz}+r\frac{dv}{dp}+s\frac{dv}{dq}\right).$$

In like manner differentiating $u=f(v)$ with respect to y, we have

$$\frac{du}{dy}+q\frac{du}{dz}+s\frac{du}{dp}+t\frac{du}{dq}=f'(v)\left(\frac{dv}{dy}+q\frac{dv}{dz}+s\frac{dv}{dp}+t\frac{dv}{dq}\right).$$

Eliminating $f'(v)$ there results

$$\left(\frac{du}{dx}+p\frac{du}{dz}+r\frac{du}{dp}+s\frac{du}{dq}\right)\left(\frac{dv}{dy}+q\frac{dv}{dz}+s\frac{dv}{dp}+t\frac{dv}{dq}\right)$$

$$-\left(\frac{dv}{dx}+p\frac{dv}{dz}+r\frac{dv}{dp}+s\frac{dv}{dq}\right)\left(\frac{du}{dy}+q\frac{du}{dz}+s\frac{du}{dp}+t\frac{du}{dq}\right)=0\ldots(8).$$

On reduction it will be found that the only terms involving r, s, and t in a degree higher than the first will be those which contain rt and s^2. The equation will in fact assume the form

$$Rr+Ss+Tt+U(rt-s^2)=V\ldots\ldots\ldots\ldots(9),$$

in which $U=\frac{du}{dp}\frac{dv}{dq}-\frac{du}{dq}\frac{dv}{dp}$. The forms of the other co-efficients it is unnecessary to examine.

Now this equation assumes the form (6) when the condition (7) is satisfied—and then only.

3. The proposition might also be proved in the following manner. Since $u=f(v)$ we have $du=f'(v)\,dv$, an equation which, since $f(v)$ is arbitrary, involves the two equations $du=0$, $dv=0$. Hence

$$\left.\begin{array}{l}\frac{du}{dx}dx+\frac{du}{dy}dy+\frac{du}{dz}dz+\frac{du}{dp}dp+\frac{du}{dq}dq=0\\ \frac{dv}{dx}dx+\frac{dv}{dy}dy+\frac{dv}{dz}dz+\frac{dv}{dp}dp+\frac{dv}{dq}dq=0\end{array}\right\}\ldots(10).$$

But $dz = pdx + qdy$, $dp = rdx + sdy$, $dq = sdx + tdy$. Whence on substitution

$$\left(\frac{du}{dx}+p\frac{du}{dz}+r\frac{du}{dp}+s\frac{du}{dq}\right)dx+\left(\frac{du}{dy}+q\frac{du}{dz}+s\frac{du}{dp}+t\frac{du}{dq}\right)dy=0.$$

$$\left(\frac{dv}{dx}+p\frac{dv}{dz}+r\frac{dv}{dp}+s\frac{dv}{dq}\right)dx+\left(\frac{dv}{dy}+q\frac{dv}{dz}+s\frac{dv}{dp}+t\frac{dv}{dq}\right)dy=0.$$

Whence eliminating dx and dy, we have the same result as before.

4. A consequence, which, though not affecting the present inquiry, is important, may here be noted. It is that it would be in vain to seek a first integral of the form $u = f(v)$ for any partial differential equation of the second order which is not of the form (9).

PROP. 2. To deduce when possible a first integral, of the form $u = f(v)$, for the partial differential equation (6).

By the last proposition u and v must satisfy the condition (7), which is expressible in the form,

$$\frac{du}{dq} \div \frac{du}{dp} = \frac{dv}{dq} \div \frac{dv}{dp} \text{ ....(11).}$$

Hence, if we represent each member of this equation by m, we have

$$\frac{du}{dq} = m\frac{du}{dp}, \quad \frac{dv}{dq} = m\frac{dv}{dp} \text{ .... (12).}$$

Substituting these values in (10), we have

$$\left.\begin{aligned}\frac{du}{dx}dx+\frac{du}{dy}dy+\frac{du}{dz}dz+\frac{du}{dp}(dp+mdq)=0\\ \frac{dv}{dx}dx+\frac{dv}{dy}dy+\frac{dv}{dz}dz+\frac{dv}{dp}(dp+mdq)=0\end{aligned}\right\} \text{......(13);}$$

and we are to remember that this system, being equivalent to $du = 0$, $dv = 0$ modified by the condition (7), can only have an integral system of the form,

$$u = a, \quad v = b \text{ .... (14),}$$

a and b being arbitrary constants, and u and v connected by the condition (11).

Making $dz = pdx + qdy$ in (13), we have

$$\left.\begin{aligned}\left(\frac{du}{dx}+p\frac{du}{dz}\right)dx+\left(\frac{du}{dy}+q\frac{du}{dz}\right)dy+\frac{du}{dp}(dp+mdq)=0\\ \left(\frac{dv}{dx}+p\frac{dv}{dz}\right)dx+\left(\frac{dv}{dy}+q\frac{dv}{dz}\right)dy+\frac{dv}{dp}(dp+mdq)=0\end{aligned}\right\}\ldots(15).$$

From these and from the equations

$$dp = rdx + sdy, \quad dq = sdx + tdy \;\ldots\ldots\ldots\ldots\; (16),$$

if we eliminate the differentials dx, dy, dp, dq, we shall necessarily obtain a result of the form (6). For in thus doing we only repeat the process of Art. 3, with the added condition (7).

To effect this elimination, we have from (16),

$$dp + mdq = (r + ms)\,dx + (s + mt)\,dy;$$

$$\text{or,}\quad rdx + s\,(dy + mdx) + tmdy = dp + mdq \;\ldots\ldots(17).$$

Now the system (15) enables us to determine the ratios of dy and $dp + mdq$ to dx, and these ratios substituted in (17), reduce it to the form (6).

But in order that it may be, not only of the form (6), but actually equivalent to (6), it is necessary and sufficient that we have

$$\frac{dx}{R} = \frac{dy + mdx}{S} = \frac{mdy}{T} = \frac{dp + mdq}{V} \;\ldots\ldots\ldots\ldots(18).$$

This system of relations among the differentials must thus include the equations (15). The same system (18), together with the equation $dz = pdx + qdy$, must therefore include the system (13). It must therefore in its final integral system include the equations $u = a$, $v = b$ with their implied condition.

We conclude then, that if the equation $Rr + Ss + Tt = V$, result from an equation of the first order of the form $u = f(v)$, the system (18), together with the equation,

$$dz = pdx + qdy \quad \text{.......................} \quad (19),$$

must admit of an integral system determining u and v in equations of the form $u = a$, $v = b$.

To eliminate m from (18) we have, on determining its value from the first and third members, substituting it in the second and fourth, and reducing,

$$Rdy^2 - Sdxdy + Tdx^2 = 0 \quad \text{..............} \quad (20),$$

$$Rdpdy + Tdqdx - Vdxdy = 0 \quad \text{.............} \quad (21),$$

and these, with (19), make three ordinary differential equations among the five variables x, y, z, p, q. But among five variables there ought to exist four ordinary differential equations in order to render the final relations determinate. And this confirms what was said in Art. 1, of the hypothetical character of Monge's method. It is only when the proposed equation originates in an equation of the form $u = f(v)$, that the above system admits of two integrals of the form,

$$u = a, \quad v = b.$$

As (20) is of the second degree it will, unless it is a complete square, be resolvable into two equations of the first degree, and either of these in conjunction with (21) and (19) may lead to a final integral system determining u and v. It follows that when the given equation admits of a first integral at all, it will admit of two such—excepting the case in which (20) is a complete square.

5. As yet no account has been taken of the quantity m. The mode in which it is involved in the equation (18), leads however to a remarkable consequence developed in the following Proposition.

PROP. If by the last proposition we obtain two first integrals of the form

$$u_1 = f(v_1), \quad u_2 = \phi(v_2) \quad \text{.................} \quad (22),$$

and if, regarding these as simultaneous, we determine p and q as functions of x, y, z, those values will be such as to render the equation $dz = pdx + qdy$ integrable, and thus to lead to the second or final integral.

For simplicity, we shall represent $u_1 - f(v_1)$ by F, and $u_2 - \phi(v_2)$ by Φ. Thus the supposed first integrals are simply

$$F = 0, \quad \Phi = 0 \quad \text{.......................} \quad (23).$$

Now reverting to the system (18), and representing the ratio $dy : dx$ by n, its first two equations assume the form,

$$\frac{1}{R} = \frac{n+m}{S} = \frac{nm}{T},$$

and shew that m and n are the two roots of the equation

$$R\mu^2 - S\mu + T = 0.$$

Hence, the value of the ratio $dy : dx$ corresponding to one of the first integrals (23), is the same as the value of m corresponding to the other.

Now for the value of m corresponding to the integral $F = 0$, we have by definition,

$$m = \frac{\frac{du_1}{dq}}{\frac{du_1}{dp}} = \frac{\frac{dv_1}{dq}}{\frac{dv_1}{dp}}$$

$$= \frac{\frac{du_1}{dq} - f'(v)\frac{dv_1}{dq}}{\frac{du_1}{dp} - f'(v)\frac{dv_1}{dp}}$$

$$= \frac{\frac{dF}{dq}}{\frac{dF}{dp}} \quad \text{..............................} \quad (24).$$

Again, seeking the value of the ratio $dy : dx$, corresponding to the integral $\Phi = 0$, we have

$$\left(\frac{d\Phi}{dx}+\frac{d\Phi}{dz}p+\frac{d\Phi}{dp}r+\frac{d\Phi}{dq}s\right)dx$$
$$+\left(\frac{d\Phi}{dy}+\frac{d\Phi}{dz}q+\frac{d\Phi}{dp}s+\frac{d\Phi}{dq}t\right)dy=0.$$

Equating the value of $dy : dx$ hence found to that of m given in (24), we have, on reduction,

$$\frac{dF}{dp}\frac{d\Phi}{dx}+\frac{dF}{dq}\frac{d\Phi}{dy}+\frac{dF}{dp}\frac{d\Phi}{dz}p+\frac{dF}{dq}\frac{d\Phi}{dz}q$$
$$+\frac{dF}{dp}\frac{d\Phi}{dp}r+\left(\frac{dF}{dp}\frac{d\Phi}{dq}+\frac{dF}{dq}\frac{d\Phi}{dp}\right)s+\frac{dF}{dq}\frac{d\Phi}{dq}t=0 \ldots (25).$$

In like manner equating the values of m corresponding to the integral $F=0$, and of $dy : dx$ corresponding to the integral $\Phi=0$, we have

$$\frac{dF}{dx}\frac{d\Phi}{dp}+\frac{dF}{dy}\frac{d\Phi}{dq}+\frac{dF}{dz}\frac{d\Phi}{dp}p+\frac{dF}{dz}\frac{d\Phi}{dq}q$$
$$+\frac{dF}{dp}\frac{d\Phi}{dp}r+\left(\frac{dF}{dq}\frac{d\Phi}{dp}+\frac{dF}{dp}\frac{d\Phi}{dq}\right)s+\frac{dF}{dq}\frac{d\Phi}{dq}t=0 \ldots (26).$$

Subtracting (25) from (26), there results

$$\frac{dF}{dx}\frac{d\Phi}{dp}-\frac{dF}{dp}\frac{d\Phi}{dx}+\frac{dF}{dy}\frac{d\Phi}{dq}-\frac{dF}{dq}\frac{d\Phi}{dy}$$
$$+\left(\frac{dF}{dz}\frac{d\Phi}{dp}-\frac{dF}{dp}\frac{d\Phi}{dz}\right)p+\left(\frac{dF}{dz}\frac{d\Phi}{dq}-\frac{dF}{dq}\frac{d\Phi}{dz}\right)q=0 \ldots (27).$$

Now this is identical with the equation (37), Chap. XIV. Art. 13, expressing the very condition which must be fulfilled in order that the values of p and q given by $F=0$, $\Phi=0$, may render the equation $dz=pdx+qdy$ an exact differential. Hence the proposition is established.

It is interesting to observe that the two first integrals stand in a certain conjugate relation. Each of them satisfies that partial differential equation of the first order and degree which we should have to construct in attempting, by the process of Charpit, to integrate the other. Hence also, although the knowledge of both is desirable, that of either is sufficient to enable us to proceed by integration to the final solution.

6. The statement of Monge's method, as derived from the above investigation, is contained in the following Rule.

RULE. The equation being $Rr + Ss + Tt = V$, form first, the equation

$$Rdy^2 - Sdx\,dy + Tdx^2 = 0 \quad \text{.................. (28)},$$

and resolve it, supposing the first member not a complete square, into two equations of the form

$$dy - m_1 dx = 0, \quad dy - m_2 dx = 0 \quad \text{........... (29)}.$$

From the first of these, and from the equation

$$Rdp\,dy + Tdq\,dx - Vdx\,dy = 0 \quad \text{........... (30)},$$

combined if needful with the equation $dz = pdx + qdy$, seek to obtain two integrals $u_1 = a$, $v_1 = b$. Proceeding in the same way with the second equation of (29), seek two other integrals

$$u_2 = \alpha, \quad v_2 = \beta,$$

then the two first integrals of the proposed equation will be

$$u_1 = f_1(v_1), \quad u_2 = f_2(v_2) \quad \text{.................. (31)}.$$

To deduce the second integral, we must either integrate one of these, or, determining from the two p and q in terms of x, y, and z, substitute those values in the equation

$$dz = pdx + qdy,$$

which will then satisfy the condition of integrability. Its solution will give the second integral sought.

If the values of m_1 and m_2 are equal, only one first integral will be obtained, and the final solution must be sought by its integration.

When it is not possible so to combine the auxiliary equations as to obtain two auxiliary integrals $u = a$, $v = b$, no first integral of the proposed equation exists, and some other process of solution must be sought.

We may observe that the determination of p and q from the two first integrals is facilitated by the fact that u and v satisfy the condition (7). Interpreted by Chap. IV. Art. 3, that condition implies that p and q enter, in some single definite combination, into both u and v.

Ex. 1. Given $\dfrac{d^2z}{dx^2} - a^2 \dfrac{d^2z}{dy^2} = 0.$

Here $R = 1$, $S = 0$, $T = -a^2$, $V = 0$. Hence we have by (28) and (30),

$$dy^2 - a^2dx^2 = 0, \quad dpdy - a^2\,dqdx = 0 \text{..........}(a).$$

The former of these is resolvable into the two equations

$$dy + adx = 0, \quad dy - adx = 0 \text{..............}(b),$$

of which the first gives $y + ax = c$, and at the same time reduces the second equation of (a) to the form $dp + adq = 0$, of which the integral is $p + aq = C$. Thus a first integral of the given equation is

$$p + aq = \phi\,(y + ax) \text{......................}(c).$$

Proceeding in like manner with the second equation of (b), we find as another first integral

$$p - aq = \psi\,(y - ax) \text{...................}(d).$$

From these two equations determining p and q, the equation $dz = pdx + qdy$ becomes

$$dz = \frac{\phi\,(y + ax) + \psi\,(y - ax)}{2}\,dx + \frac{\phi\,(y + ax) - \psi\,(y - ax)}{2a}\,dy.$$

Or

$$dz = \frac{\phi\,(y + ax)\,(dy + adx) - \psi\,(y - ax)\,(dy - adx)}{2a}.$$

Hence if $\dfrac{1}{2a}\displaystyle\int \phi\,(t)\,dt = \phi_{,}(t)$ and $-\dfrac{1}{2a}\displaystyle\int \psi\,(t)\,dt = \psi_1(t)$,

we have

$$z = \phi_{,}(y + ax) + \psi_{,}(y - ax).$$

Here $\phi_{,}$, $\psi_{,}$ are arbitrary functions since ϕ and ψ are such.

It is seen that, in each of the first integrals, the condition (7) is satisfied, and assuming

$$p + aq - \phi(y + ax) = F, \ \ p - aq - \psi(y - ax) = \Phi,$$

it is easy to verify the condition (27).

Ex. 2. Given $r + as + bt = 0$.

Proceeding as before, we find

$$p + nq = \phi(y - mx), \ \ p + mq = \psi(y - nx),$$

as the two first integrals of the proposed, m and n being the roots of the equation $t^2 - at + b = 0$. Hence, determining p and q, substituting in the equation $dz = pdx + qdy$, integrating and reducing we have

$$z = \phi_1(y - mx) + \psi_1(y - nx).$$

But when m and n are equal we have only one first integral, viz.

$$p + mq = \phi(y - mx).$$

Treating this by Lagrange's process, we have the auxiliary system

$$dx = \frac{dy}{m} = \frac{dz}{\phi(y - mx)}.$$

From the first two members we find $y - mx = c$. This enables us to reduce the equation of the first and third to the form

$$dx = \frac{dz}{\phi(c)},$$

whence $$z = x\phi(c) + c'.$$

Therefore, restoring to c its value,

$$z - x\phi(y - mx) = c'.$$

Thus we have for the final integral

$$z - x\phi(y - mx) = \psi(y - mx).$$

Ex. 3. Given

$$(b+cq)^2 r - 2(b+cq)(a+cp)s + (a+cp)^2 t = 0.$$

Here the auxiliary equations are

$$(b+cq)^2 dy^2 + 2(b+cq)(a+cp)\,dy\,dx + (a+cp)^2 dx^2 = 0 \ldots (a).$$

$$(b+cq)^2 dp\,dy + (a+cp)^2 dq\,dx = 0 \ldots\ldots\ldots\ldots (b).$$

The first of these equations gives

$$(b+cq)\,dy + (a+cp)\,dx = 0 \ldots\ldots\ldots\ldots (c),$$

which the equation $dz = pdx + qdy$ reduces to the form

$$adx + bdy + cdz = 0;$$

whence

$$ax + by + cz = \alpha \ldots\ldots\ldots\ldots (d).$$

Again, eliminating dy and dx from (b) and (c), we have

$$(b+cq)\,dp - (a+cp)\,dq = 0,$$

whence, integrating

$$\frac{a+cp}{b+cq} = \beta \ldots\ldots\ldots\ldots (e).$$

Thus a first integral of the proposed equation is

$$\frac{a+cp}{b+cq} = \phi(ax+by+cz),$$

or

$$cp - c\phi(ax+by+cz)\,q = b\phi(ax+by+cz) - a;$$

and this must be integrated by Lagrange's process.

The auxiliary system is, on representing $\phi(ax+by+cz)$ by ϕ,

$$\frac{dx}{c} = -\frac{dy}{c\phi} = \frac{dz}{b\phi - a}.$$

From these we find $adx + bdy + cdz = 0$, whence

$$ax + by + cz = C,$$

and thus $\phi(ax+by+cz) = \phi C.$

Hence substituting $\quad dy = -\phi(C)\,dx,$

whence $$y + \phi(C)\,x = C',$$

or $$y + x\phi(ax + by + cz) = C'.$$

Thus the final integral is

$$y + x\phi(ax + by + cz) = \psi(ax + by + cz).$$

This solution may also be expressed in the form

$$z = x\phi_{,}(ax + by + cz) + y\psi_{,}(ax + by + cz),$$

in which it is in fact presented by Monge, (*Application de l'Analyse à la Géométrie*, Liouville's edition, p. 79). The equation solved is that of surfaces formed by the motion of a straight line which is always parallel to a given plane, and always passes through two given curves.

7. In the above examples V is equal to 0, and this always facilitates the application of Monge's method. The following is an example in which V is not equal to 0.

Ex. 4. Given $r - t = -\dfrac{4p}{x+y}$.

The auxiliary equation being

$$dy^2 - dx^2 = 0,\ dp\,dy - dq\,dx + \frac{4p}{x+y}\,dx\,dy = 0,$$

one of the systems hence derived is

$$dy - dx = 0,\ dp - dq + \frac{4p}{x+y}\,dx = 0.$$

There is also another system, but it is not integrable in the form $u = a,\ v = b$.

From the first of the above equations we get

$$y - x = a,\ dp - dq + \frac{4p\,dx}{2y - a} = 0,$$

the latter of which may, since $dz = pdx + qdy$, be reduced to the form

$$d(2y - a)(p - q) + 2dz = 0,$$

whence $$(2y-a)(p-q)+2z=b,$$

or, replacing a by $y-x$,

$$(x+y)(p-q)+2z=b.$$

Hence a first integral of the proposed equation will be

$$(x+y)(p-q)+2z=f(y-x).$$

Now this being linear, we have, by Lagrange's method, the auxiliary system

$$\frac{dx}{x+y}=\frac{-dy}{x+y}=\frac{dz}{f(y-x)-2z}.$$

The equation of the first two members gives $y+x=a$, and this reduces the equation of the second and third to the form

$$\frac{-dy}{a}=\frac{dz}{f(2y-a)-2z},$$

or $$\frac{dz}{dy}-\frac{2z}{a}=-\frac{f(2y-a)}{a},$$

whence $$z=\frac{-\epsilon^{\frac{2y}{a}}}{a}\int\epsilon^{-\frac{2y}{a}}f(2y-a)\,dy+b.$$

The final integral will therefore be found by substituting in the above, after integration, $y+x$ for a, and $f(y+x)$ for b.

8. Monge's method fails in so many cases, owing to the non-existence of a first integral of the assumed form $u=f(v)$, that it becomes important to inquire how its defects may be supplied. And various methods, all of limited generality, have been discovered. Thus Laplace has developed a method applicable to all equations of the form

$$Rr+Ss+Tt+Pp+Qq+Uz=V;$$

R, S, T, P, Q, U, and V being functions of x and y only,—which consists in a series of transformations, each of which has the effect of reducing the equation to the form

$$S+Pp+Qq+Rr=V,$$

P, Q, R and V being functions of x and y, to which each transformation gives new forms. It may be that among these successive forms, some one will be found which will admit of resolution into two linear equations of the first order. But there are probably no instances in which this method has been applied in which the solution may not be effected with far greater elegance, and with far greater simplicity, by the symbolical methods of the following chapters. And even Laplace's method is better exhibited in a symbolical form. The subject will be resumed.

The following sections contain miscellaneous but important additions.

Miscellaneous Theorems.

9. Poisson has shewn how to deduce a particular integral of any partial differential equation of the form

$$P = (rt - s^2)^n Q \ldots\ldots\ldots\ldots\ldots\ldots\ldots (45),$$

where P is a function of p, q, r, s, t, homogeneous with respect to the three last, n a positive index, and Q any function of x, y, z, and the differential coefficients of z of any order which does not become infinite when $rt - s^2 = 0$.

Assuming $q = \phi(p)$, we have

$$s = \phi'(p)\, r, \quad t = \phi'(p)\, s = \{\phi'(p)\}^2 r \ldots\ldots\ldots (46),$$

values which make $rt - s^2 = 0$. Hence, substituting in (45), the second member vanishes, while in the first, which is homogeneous with respect to r, s, t, some power of r only will remain as a common factor. Dividing by that factor, we shall have an equation involving only p, $\phi(p)$, and $\phi'(p)$, i.e. p, q, and $\frac{dq}{dp}$. Integrating this as an ordinary differential equation we obtain a relation between p, q and an arbitrary constant; and this, integrated as a partial differential equation of the first order, gives the solution in question.

Ex. Given $r^2 - t^2 = rt - s^2$.

Proceeding as above, we find

$$1-\{\phi'(p)\}^2=0,$$

whence $$\frac{dq}{dp}=\pm 1, \quad q=\pm p+c;$$

$$\therefore\ z-cy=\phi(y\pm x),$$

a particular integral.

The above method is applicable to all equations of the second order which are simply homogeneous with respect to r, s, t, for then we have only to suppose $Q=0$.

10. There exists in partial differential equations a remarkable duality, in virtue of which each equation stands connected with some other equation of the same order by relations of a perfectly reciprocal character. As respects equations of the first order the principle may be thus stated.

Suppose that in the given equation

$$\phi\,(x, y, z, p, q)=0 \quad\text{..................}\quad (47)$$

we interchange x and p, y and q, and change z into $px+qy-z$, giving

$$\phi\,(p, q, px+qy-z, x, y)=0 \quad\text{.........}\quad (48);$$

then, if either of these equations can be integrated in the form $z=\psi\,(x, y)$, the solution of the other will be found by eliminating X, Y, and Z between the equations

$$x=\frac{d\psi\,(X, Y)}{dx}, \quad y=\frac{d\psi\,(X, Y)}{dy}, \quad z=Xx+Yy-Z\ldots(49).$$

For, since $dz=pdx+qdy$, we have

$$z=px+qy-\int(xdp+ydq) \quad\text{............}\quad (50).$$

Hence $xdp+ydq$ is an exact differential. Represent it by dZ, and assume z for dependent variable. Assume also two new independent variables X and Y, connected with the former ones by the relations $X=p$, $Y=q$. Then

$$dZ=xdp+ydq=xdX+ydY.$$

Hence $$\frac{dZ}{dX}=x, \quad \frac{dZ}{dY}=y,$$

$$Z=\int(xdp+ydq)=px+qy-z \text{ by (50)};$$

$$\therefore\ z=px+qy-Z=xX+yY-Z.$$

On examining the above equations we see that x, y, z, and X, Y, Z are reciprocally related. Writing, side by side, the equations which are conjugate to each other, we have

$$X=\frac{dz}{dx}, \qquad x=\frac{dZ}{dX};$$

$$Y=\frac{dz}{dy}, \qquad y=\frac{dZ}{dY};$$

$$Z=Xx+Yy-z, \quad z=xX+yY-Z.$$

We see too that the equations (49) which express one set of the relations suffice to convert any relation found by integration between X, Y, Z into a corresponding relation between x, y, z.

The meaning of x, y, z in (48) in the statement of the theorem is the same as that of X, Y, Z in its demonstration, the actual change of expression being only made after the integration in order to shew more distinctly the interchangeable character of x and p, y and q, z and $px+qy-z$.

Ex. Given $z=pq$.

Here the transformed equation is

$$px+qy-z=xy,$$

of which the integral is $z=xy+xf\left(\frac{y}{x}\right)$. Hence

$$\psi(X, Y)=XY+Xf\left(\frac{Y}{X}\right),$$

and we have to eliminate X, Y, between the equations

$$x = Y - \frac{Y}{X} f'\left(\frac{Y}{X}\right) + f\left(\frac{Y}{X}\right), \quad y = X + f'\left(\frac{Y}{X}\right),$$

$$Z = XY.$$

Each particular form assigned to f gives a distinct particular integral. If we assume $f\left(\frac{Y}{X}\right) = a\frac{Y}{X} + b$, we find

$$x = Y + b, \quad y = X + a, \quad Z = XY,$$

from which, eliminating X and Y, we have $z = (x-b)(y-a)$, and this is one form of the complete primitive assigned in Chap. XIV. Art. 7. We may observe that the elimination may be so effected as to lead to general primitives.

11. *In equations of the second order we should have, in addition to the above transformations, to change*

$$r \text{ into } \frac{t}{rt - s^2}, \quad s \text{ into } \frac{-s}{rt - s^2}, \quad t \text{ into } \frac{r}{rt - s^2} \dots (51),$$

in order to form the reciprocal equation. Then the second integral of either being found in the form $z = \psi(x, y)$, *that of the other will be found as before by eliminating* X *and* Y *from* (49). For since

$$x = \frac{dZ}{dX}, \quad y = \frac{dZ}{dY};$$

$$\therefore\ dx = RdX + SdY, \quad dy = SdX + TdY,$$

whence $$dX = \frac{Tdx - Sdy}{RT - S^2}, \quad dY = \frac{-Sdx + Rdy}{RT - S^2}.$$

But $X = p$, $Y = q$, therefore

$$dp = r\,dx + s\,dy = \frac{Tdx - Sdy}{RT - S^2},$$

$$dq = s\,dx + t\,dy = \frac{-Sdx + Rdy}{RT - S^2},$$

whence, equating coefficients,

$$r = \frac{T}{RT - S^2}, \quad s = \frac{-S}{RT - S^2}, \quad t = \frac{R}{RT - S^2}.$$

The extension of the theorem to higher orders involves no difficulty.

12. It is an immediate consequence of the above, that any equation of the form

$$\phi(p, q)\, r + \psi(p, q)\, s + \chi(p, q)\, t = 0 \text{.........(52)}$$

can be reduced to an equation of the form

$$\chi(x, y)\, r - \psi(x, y)\, s + \phi(x, y)\, t = 0 \text{.........(53)},$$

usually more convenient for solution. Legendre's solution of the equation

$$(1 + q^2)\, r - 2pqs + (1 + p^2)\, t = 0,$$

by the aid of the above transformation, will be found in *Lacroix* (Tom. II. p. 623).

The same transformation makes the solution of any equation of the form $Rr + Ss + Tt = V(rt - s^2)$ dependent on that of an equation of the form

$$Rr + Ss + Tt = V,$$

but with different coefficients. The subject of these transformations has been most fully treated by Prof. de Morgan (*Cambridge Philosophical Transactions*, Vol. VIII. p. 606).

13. Legendre also shews how, by a transformation formally resembling the above, to integrate the equation

$$r = f(s, t).$$

Assuming s and t as independent variables, and $\nu = sx + ty - q$ as dependent variable, the equation is reduced to the form

$$\frac{d^2\nu}{dt^2} + S\frac{d^2\nu}{ds\,dt} - T\frac{d^2\nu}{ds^2} = 0 \text{..............(54)},$$

where S and T are the values of $\frac{dr}{ds}$ and $\frac{dr}{dt}$ furnished by the given equation. *Lacroix*, Tom. II. p. 631.

EXERCISES.

1. To what condition must u and v be subject, in order that $u = f(v)$ may be a first integral of an equation of the form $Rr + Ss + Tt = V$?

Integrate by Monge's method the following equations:

2. $x^2r + 2xys + y^2t = 0.$

3. $q^2r - 2pqs + p^2t = 0.$

4. Integrate $ps - qr = 0.$

5. Integrate by Monge's method the equation

$$q(1+q)r - (p+q+2pq)s + p(1+p)t = 0.$$

6. The solution of Ex. 3 may, by the law of reciprocity, be made to depend on that of Ex. 2.

7. Monge's method would not enable us to solve the equation $r - t = \frac{2p}{x}$.

8. Deduce by Poisson's method a particular integral of

$$(1+q^2)r - 2pqs + (1+p^2)t = 0.$$

9. Shew that the equations

$$rt - s^2 = f(p, q), \text{ and } rt - s^2 = \{f(x, y)\}^{-1},$$

are connected by the law of reciprocity.

10. The solution of the equation $r - t = \frac{4x}{p+q}(rt - s^2)$ may be derived from that of the equation $r - t + \frac{4p}{x+y} = 0$. Art. 6. Ex. 4.

CHAPTER XVI.

SYMBOLICAL METHODS.

1. THE term symbolical is, by a restriction of its wider meaning, applied more peculiarly to those methods in Analysis in which operations, separated by a mental abstraction from the subjects upon which they are performed, are expressed by symbols in whose laws the laws of the operations themselves are represented.

Thus $\frac{du}{dx}$ is written symbolically in the form $\frac{d}{dx}u$, the symbol $\frac{d}{dx}$ denoting an operation of which u is the subject. In thus expressing an operation by a symbol, in studying the laws of that symbol, and in founding processes and methods upon those laws, we introduce no strange or novel principle of Language; for it is the very office of Language to express by symbols the procedure of Thought.

Thus also we may write

$$\frac{du}{dx} + au = \left(\frac{d}{dx} + a\right) u \dots\dots\dots\dots\dots\dots (1),$$

$$\frac{d^2u}{dx^2} + a\frac{du}{dx} + bu = \left(\frac{d^2}{dx^2} + a\frac{d}{dx} + b\right) u \dots\dots\dots (2),$$

and so on. It will be observed that the symbol precedes the subject on which it operates.

Operations may be performed in succession. Thus

$$\left(\frac{d}{dx} + a\right)\left(\frac{d}{dx} + b\right) u$$

denotes that we first perform on the subject u the operation denoted by $\frac{d}{dx}+b$, and then on the result effect the operation denoted by $\frac{d}{dx}+a$. Thus a and b being constant, we have

$$\left(\frac{d}{dx}+a\right)\left(\frac{d}{dx}+b\right)u=\left(\frac{d}{dx}+a\right)\left(\frac{du}{dx}+bu\right)$$

$$=\frac{d}{dx}\left(\frac{du}{dx}+bu\right)+a\left(\frac{du}{dx}+bu\right)$$

$$=\frac{d^2u}{dx^2}+(a+b)\ \frac{du}{dx}+abu\ldots\ldots(3).$$

When an operation is repeated, the number of times which it is understood to be performed is expressed by an index attached to the symbol of operation. Thus

$$\left(\frac{d}{dx}+a\right)^2 u=\left(\frac{d}{dx}+a\right)\left(\frac{d}{dx}+a\right)u$$

$$=\frac{d^2u}{dx^2}+2a\frac{du}{dx}+a^2u\ \ldots\ldots\ldots\ldots\ (4).$$

If in the second member of (3), as in the first, we separate the symbols from their subject, we have

$$\left(\frac{d}{dx}+a\right)\left(\frac{d}{dx}+b\right)u=\left\{\frac{d^2}{dx^2}+(a+b)\frac{d}{dx}+ab\right\}u\ldots(5).$$

Now the symbolic expressions for the equivalent operations performed upon u in the two members of this equation are in *formal* analogy with the algebraic equation

$$(m+a)\ (m+b)\ u=\{m^2+(a+b)\ m+ab\}\ u,$$

and this is a particular illustration of a general theorem to the statement and demonstration of which we shall now proceed.

2. If we compare the symbolical expressions

$$\left(\frac{d}{dx}+a\right)\left(\frac{d}{dx}+b\right),\qquad \frac{d^2}{dx^2}+(a+b)\ \frac{d}{dx}+ab\ \ldots\ldots\ (6),$$

whose equivalence is stated in (5), we see that each involves $\frac{d}{dx}$ together with constant quantities. Each might therefore, to borrow the language of analogy, be described as a function of $\frac{d}{dx}$ and constant quantities, or more briefly as a function of $\frac{d}{dx}$, and expressed in the form $f\left(\frac{d}{dx}\right)$. Again, each expresses a system of operations in the performance of which the presence of the symbol $\frac{d}{dx}$ only indicates *differentiation*, not integration. We may with propriety term any function of $\frac{d}{dx}$ possessing this character a *direct* function of $\frac{d}{dx}$. The theorem in question is then the following.

THEOREM. Any direct function of $\frac{d}{dx}$ and constant quantities may be transformed as if $\frac{d}{dx}$ were itself a quantity.

In the first place it is evident that any direct function of the symbol $\frac{d}{dx}$ according to the above definition is, in form, what we should term a rational and integral function of $\frac{d}{dx}$, were that symbol merely algebraic.

Now the laws, according to which algebraic symbols combine with each other in the composition of all rational and integral expressions, are the following, viz. 1st, the distributive, expressed by the equation

$$m(u+v) = mu + mv \quad \text{.............. (7)},$$

2ndly, the commutative, expressed by the equation

$$ma = am \quad \text{........................ (8)},$$

3rdly, the index law, expressed by the equation

$$m^a m^b = m^{a+b} \quad \text{........................(9)}.$$

These determine, and alone determine, the forms, or, to speak more precisely, the permitted variety of forms, of algebraic expressions of the above class.

But the symbol $\frac{d}{dx}$, when employed in combination with constant quantities to operate on subjects which are not constant, is subject to laws formally agreeing with the above. For we have

$$\frac{d}{dx}(u+v) = \frac{d}{dx}u + \frac{d}{dx}v \text{ ....(10)},$$

$$\frac{d}{dx}au = a\frac{d}{dx}u \text{ ....(11)},$$

$$\left(\frac{d}{dx}\right)^a \left(\frac{d}{dx}\right)^b u = \left(\frac{d}{dx}\right)^{a+b} u \text{ ....(12)},$$

the last of these, however, expressing, not any distinctive property of the operation $\frac{d}{dx}$, but only the fact that it is an operation capable of repetition. These laws, in like manner, determine the possible forms of symbolic expressions involving $\frac{d}{dx}$ with constants, and representing direct systems of operations.

Hence the variety of form permitted in the one case is the same as that permitted in the other. In other words, the same transformations are valid.

Among the consequences of the above theorem the following may be noted.

1st, We can reduce any symbolical expression of the form $\frac{d^n}{dx^n} + a_1\frac{d^{n-1}}{dx^{n-1}} + a_2\frac{d^{n-2}}{dx^{n-2}} \dots + a_n$, in which $a_1, a_2, \dots a_n$ are constants, to an equivalent expression of the form

$$\left(\frac{d}{dx} - m_1\right)\left(\frac{d}{dx} - m_2\right) \dots \left(\frac{d}{dx} - m_n\right),$$

where $m_1, m_2, \ldots m_n$ are the roots of the equations

$$m^n + a_1 m^{n-1} + a_2 m^{n-2} \ldots + a_n = 0.$$

2ndly, The order, in which the component operations

$$\frac{d}{dx} - m_1, \quad \frac{d}{dx} - m_2, \ldots \frac{d}{dx} - m_n$$

are written, is indifferent.

Ex. Thus $\frac{d^2u}{dx^2} - a^2u = 0$ may be reduced to either of the forms

$$\left(\frac{d}{dx} + a\right)\left(\frac{d}{dx} - a\right)u = 0, \quad \left(\frac{d}{dx} - a\right)\left(\frac{d}{dx} + a\right)u = 0.$$

3rdly, The complex operation

$$\frac{d^n}{dx^n} + A_1 \frac{d^{n-1}}{dx^{n-1}} + A_2 \frac{d^{n-2}}{dx^{n-2}} \ldots + A_n$$

is itself, like the elementary operation $\frac{d}{dx}$, distributive; i. e. representing that complex operation by $f\left(\frac{d}{dx}\right)$, we have

$$f\left(\frac{d}{dx}\right)(u+v) = f\left(\frac{d}{dx}\right)u + f\left(\frac{d}{dx}\right)v \ldots\ldots\ldots (13).$$

This conclusion may be verified, by substituting for $f\left(\frac{d}{dx}\right)$ the expression for which it stands, and performing the operations.

Inverse Forms.

3. All that is said above relates to the performance of operations, definite in character, upon subjects supposed to be given. But an inverse problem is suggested, in which it is required to determine, not what will be the result of performing a certain operation upon a given subject, but upon what

subject a certain operation must be performed in order to lead to a given result. Thus, in the equation

$$\left(\frac{d}{dx}+a\right)u=v\text{.......................}(14),$$

if u be given, the performance of the operation $\frac{d}{dx}+a$ determines v; but if v be given, then the inquiry arises, what is that unknown subject u, the performance of the operation $\frac{d}{dx}+a$ upon which will lead to the result v?

As any procedure for determining u from v is inverse to the procedure by which v is determined from u, analogy suggests the notation

$$u=\left(\frac{d}{dx}+a\right)^{-1}v\text{.....................}(15),$$

$\left(\frac{d}{dx}+a\right)^{-1}$ representing the inverse procedure in question, but representing that procedure only in its inverse character, i. e. conveying no information as to how it is to be performed, but only telling us that it must be such, that if, having performed it on v, we perform on the result the operation $\frac{d}{dx}+a$ to which it is inverse, we shall reproduce v. For, substituting in (14) the expression for u given in (15), we have

$$\left(\frac{d}{dx}+a\right)\left(\frac{d}{dx}+a\right)^{-1}v=v.$$

The inverse procedure is thus presented as one, *the effect of which the direct operation simply annuls.* This is its *definition.*

Thus in Arithmetic, division is inverse to multiplication. What is meant by dividing a by b is the seeking of a third number c, which when multiplied by b will produce a. And the very procedure by which this is effected consists not in any new and distinct operation for determining the subject c, but in a series of guesses, suggested by our prior *general* knowledge of the results of multiplication, and tested by multiplication.

And generally, if π represent any operation or series of operations possible when their subject is given, and then termed *direct*, and if, in the equation $\pi u = v$, the subject u be not given but only the result v, then we may write

$$u = \pi^{-1} v.$$

And the problem or inquiry contained in the inverse notation of the second member will be answered, when we have, by whatsoever process, so determined the function u as to satisfy the equation $\pi u = v$ or $\pi\pi^{-1} v = v$. By the latter equation the inverse symbol π^{-1} is defined. Thus it is the office of the inverse symbol to propose a question, not to describe an operation. It is, in its primary meaning, interrogative, not directive.

Suppose the given equation to be

$$\left(\frac{d^n}{dx^n} + A_1 \frac{d^{n-1}}{dx^{n-1}} \dots + A_n\right) u = v \text{ .... } (16).$$

Then on the above principle of notation we should have

$$u = \left(\frac{d^n}{dx^n} + A_1 \frac{d^{n-1}}{dx^{n-1}} \dots + A_n\right)^{-1} v,$$

or, with not less propriety of expression,

$$u = \frac{1}{\dfrac{d^n}{dx^n} + A_1 \dfrac{d^{n-1}}{dx^{n-1}} \dots + A_n} v,$$

the last two equations differing in interpretation from (16), not at all as touching the *relation* between u and v, but only as more distinctly presenting u as the object of search.

Of what avail then, it may be asked, is that analogy upon which the expression of the last two equations is founded?

If a convention, it is at least a very natural one, that we should express an operation performed upon a subject, by attaching, in some way, the symbol denoting the operation to the symbol denoting the subject. The order of writing, in that family of languages to which our own belongs, has doubtless determined the mode of connexion actually adopted, and which is the same as if the symbol of operation were a symbol of quantity employed as a coefficient or multiplier. It comes to pass, moreover, that the formal laws of combination in the *direct* cases investigated in Art. 2 prove to be the same for the symbol $\frac{d}{dx}$ as for a coefficient or multiplier. But inverse symbols derive their meaning from the direct operations to which they stand related: they are forms of interrogation, the answers to which are to be tested by the performance of the direct operations. Hence it may be inferred that the laws for the transformation of inverse expressions involving $\frac{d}{dx}$ with constants will be the same as for the corresponding forms of ordinary algebra. The analogy consists, not in the mere adoption of a common notation, but, as all true analogy does, in a similitude of relations.

4. *Solutions of Linear Equations with constant Coefficients.*

If the equation $\left(\frac{d}{dx} - a\right) u = X$ be given, we have

$$u = \left(\frac{d}{dx} - a\right)^{-1} X,$$

but, the known general solution of the given equation being

$$u = \epsilon^{ax} \int \epsilon^{-ax} X dx,$$

we see that

$$\left(\frac{d}{dx} - a\right)^{-1} X = \epsilon^{ax} \int \epsilon^{-ax} X dx \quad \ldots\ldots\ldots\ldots\ldots (17),$$

an arbitrary constant being introduced by the integration in the second member.

If $X = 0$, we have

$$\left(\frac{d}{dx} - a\right)^{-1} X = C\epsilon^{ax} \quad \ldots\ldots\ldots\ldots\ldots (18).$$

These results we shall have occasion to refer to.

Ex. Now suppose the given equation to be

$$\frac{d^2u}{dx^2} - (a+b)\frac{du}{dx} + abu = X,$$

we have, on separating the symbols,

$$\left\{\frac{d^2}{dx^2} - (a+b)\frac{d}{dx} + ab\right\} u = X,$$

or, by Art. 2,

$$\left(\frac{d}{dx} - a\right)\left(\frac{d}{dx} - b\right) u = X \qquad (19).$$

Hence $$\left(\frac{d}{dx} - b\right) u = \left(\frac{d}{dx} - a\right)^{-1} X$$

$$u = \left(\frac{d}{dx} - b\right)^{-1}\left(\frac{d}{dx} - a\right)^{-1} X \qquad (20).$$

On comparing this with (19) we see that, in inverting a system composed of two operations performed in succession, the order of the operations themselves is inverted. This is evidently true whatever may be the number of successive operations, the last to be performed being always the first to be inverted.

From (20) we might deduce the actual value of u by successive applications of (17). Such was the method once employed. But it is better to proceed as follows.

From (19) we have

$$u = \left\{\left(\frac{d}{dx} - a\right)\left(\frac{d}{dx} - b\right)\right\}^{-1} X \qquad (21).$$

Now by the known theory of the decomposition of rational fractions

$$\{(m-a)(m-b)\}^{-1} = N_1 (m-a)^{-1} + N_2 (m-b)^{-1} \ldots (22),$$

N_1, N_2 being functions of a and b, which may be determined in various ways, but most *directly* by multiplying both sides of the equation by $(m-a)(m-b)$, and equating coefficients.

Now the suggested transformation of the expression for u given in (21) is

$$\left\{\left(\frac{d}{dx}-a\right)\left(\frac{d}{dx}-b\right)\right\}^{-1}X=N_1\left(\frac{d}{dx}-a\right)^{-1}X+N_2\left(\frac{d}{dx}-b\right)^{-1}X\ldots(23).$$

And, from the very definition of inverse forms, the proper test of the validity of this transformation is, that the performance of the direct operation $\left(\frac{d}{dx}-a\right)\left(\frac{d}{dx}-b\right)$ on the second member shall reduce it to X.

Effecting this operation, and remembering in so doing that $\frac{d}{dx}-a$ and $\frac{d}{dx}-b$ are commutative, and that by definition $\left(\frac{d}{dx}-a\right)\left(\frac{d}{dx}-a\right)^{-1}X=X$, the second member becomes

$$N_1\left(\frac{d}{dx}-b\right)X+N_2\left(\frac{d}{dx}-a\right)X,$$

or $$(N_1+N_2)\frac{dX}{dx}-(bN_1+aN_2)\,X\ldots\ldots\ldots\ldots\ldots(24),$$

and this reduces to X if

$$N_1+N_2=0,\quad bN_1+aN_2=-1\ \ldots\ldots\ldots\ (25).$$

But these equations for the determination of N_1 and N_2 are the same, and necessarily the same, as we should have found by multiplying, as above indicated, (22), by $(m-a)(m-b)$, and equating coefficients. The two series of operations only differ in that $\frac{d}{dx}$ occupies in the one the place which m occupies in the other. Determining N_1, N_2, we see that u may be expressed in the form

$$u=\frac{1}{a-b}\left\{\left(\frac{d}{dx}-a\right)^{-1}X-\left(\frac{d}{dx}-b\right)^{-1}X\right\}\ldots\ldots\ldots(26).$$

Hence, by (17),

$$u = \frac{1}{a-b}\{\epsilon^{ax}\int\epsilon^{-ax}Xdx - \epsilon^{bx}\int\epsilon^{-bx}Xdx\} \text{........} \quad (27),$$

and as, on effecting the integrations, two arbitrary constants will be introduced, this is the most general value of u.

5. In like manner if there be given the general linear differential equation with constant coefficients

$$\left(\frac{d^n}{dx^n} + A_1\frac{d^{n-1}}{dx^{n-1}} + A_2\frac{d^{n-2}}{dx^{n-2}} \dots + A_n\right)u = X \text{......} \quad (28),$$

and if we represent by a_1, $a_2 \dots a_n$ the roots, supposed all different, of the algebraic equation

$$m^n + A_1m^{n-1} + A_2m^{n-2} \dots + A_n = 0 \text{..........} (29),$$

then the given equation may be expressed in the form

$$\left(\frac{d}{dx} - a_1\right)\left(\frac{d}{dx} - a_2\right)\dots\left(\frac{d}{dx} - a_n\right)u = X,$$

whence

$$u = \left\{\left(\frac{d}{dx} - a_1\right)\left(\frac{d}{dx} - a_2\right)\dots\left(\frac{d}{dx} - a_n\right)\right\}^{-1}X$$

$$= \left\{N_1\left(\frac{d}{dx} - a_1\right)^{-1} + N_2\left(\frac{d}{dx} - a_2\right)^{-1}\dots + N_n\left(\frac{d}{dx} - a_n\right)^{-1}\right\}X^* \dots (30),$$

the decomposition in the second member formally resembling that of the rational fraction.

If the equation (29) have r roots equal to a, there will exist in the resolved expression for u a series of terms of the form

$$\left\{N_1\left(\frac{d}{dx} - a\right)^{-r} + N_2\left(\frac{d}{dx} - a\right)^{-r+1}\dots + N_r\left(\frac{d}{dx} - a\right)^{-1}\right\}X \dots (31),$$

* This theorem was first published in the *Cambridge Mathematical Journal* (1st series, Vol. II. p. 114), in a memoir written by the late D. F. Gregory, then Editor of the Journal, from notes furnished by the author of this work, whose name the memoir bears. The illustrations were supplied by Mr Gregory. In mentioning these circumstances the author recalls to memory a brief but valued friendship.

or, which is preferable, a single term of the form

$$\left(A+B\frac{d}{dx}+C\frac{d^2}{dx^2}\dots+R\frac{d^{r-1}}{dx^{r-1}}\right)\left(\frac{d}{dx}-a\right)^{-r}X \dots\dots (32),$$

$A, B, \dots R$ being determinate constants.

Now since, by (26), $\left(\frac{d}{dx}-a\right)^{-1}X=\epsilon^{ax}\int\epsilon^{-ax}Xdx$;

$$\therefore \left(\frac{d}{dx}-a\right)^{-2}X=\left(\frac{d}{dx}-a\right)^{-1}\epsilon^{ax}\int\epsilon^{-ax}Xdx$$

$$=\epsilon^{ax}\int\epsilon^{-ax}\left(\epsilon^{ax}\int\epsilon^{-ax}Xdx\right)dx$$

$$=\epsilon^{ax}\iint\epsilon^{-ax}Xdx^2.$$

Proceeding thus, we have

$$\left(\frac{d}{dx}-a\right)^{-r}X=\epsilon^{ax}\int\dots\epsilon^{-ax}Xdx \dots\dots\dots\dots (33).$$

Ex. 1. Given $\frac{d^4y}{dx^4}+4\frac{d^3y}{dx^3}+3\frac{d^2y}{dx^2}-4\frac{dy}{dx}-4y=X.$

This equation gives, on decomposing the complex operation performed on y,

$$\left(\frac{d}{dx}+2\right)^2\left(\frac{d}{dx}+1\right)\left(\frac{d}{dx}-1\right)y=X;$$

$$\therefore y=\left\{\left(\frac{d}{dx}+2\right)^2\left(\frac{d}{dx}+1\right)\left(\frac{d}{dx}-1\right)\right\}^{-1}X.$$

Now $\frac{1}{(m+2)^2(m+1)(m-1)}=\frac{4m+11}{9(m+2)^2}-\frac{1}{2(m+1)}+\frac{1}{18(m-1)}.$

Therefore

$$y=\frac{1}{9}\left(4\frac{d}{dx}+11\right)\left(\frac{d}{dx}+2\right)^{-2}X-\frac{1}{2}\left(\frac{d}{dx}+1\right)^{-1}X+\frac{1}{18}\left(\frac{d}{dx}-1\right)^{-1}X$$

$$=\frac{1}{9}\left(4\frac{d}{dx}+11\right)\epsilon^{-2x}\iint\epsilon^{2x}Xdx^2-\frac{1}{2}\epsilon^{-x}\int\epsilon^{x}Xdx+\frac{1}{18}\epsilon^{x}\int\epsilon^{-x}Xdx.$$

Ex. 2. Given $\dfrac{d^2u}{dx^4}+n^2u=X.$

Here
$$u=\left(\frac{d^2}{dx^2}+n^2\right)^{-1}X.$$

Now $(m^2+n^2)^{-1}=\dfrac{1}{2\sqrt{(-1)}}\left[\{m-n\sqrt{(-1)}\}^{-1}-\{m+n\sqrt{(-1)}\}^{-1}\right].$

Hence
$$u=\frac{1}{2n\sqrt{(-1)}}\left[\left\{\frac{d}{dx}-n\sqrt{(-1)}\right\}^{-1}X-\left\{\frac{d}{dx}+n\sqrt{(-1)}\right\}^{-1}X\right]$$
$$=\frac{1}{2n\sqrt{(-1)}}\{\epsilon^{nx\sqrt{(-1)}}\textstyle\int\epsilon^{-nx\sqrt{(-1)}}Xdx-\epsilon^{-nx\sqrt{(-1)}}\int\epsilon^{nx\sqrt{(-1)}}Xdx\}.$$

But $\epsilon^{nx\sqrt{(-1)}}\int\epsilon^{-nx\sqrt{(-1)}}Xdx$
$$=\{\cos nx+\sqrt{(-1)}\sin x\}\{\textstyle\int\cos nx\,Xdx-\sqrt{(-1)}\int\sin nx\,Xdx\}$$
$$\epsilon^{-nx\sqrt{(-1)}}\textstyle\int\epsilon^{nx\sqrt{(-1)}}Xdx$$
$$=\{\cos nx-\sqrt{(-1)}\sin nx\}\{\textstyle\int\cos nx\,Xdx+\sqrt{(-1)}\int\sin nx\,Xdx\},$$

whence, on substitution and reduction,
$$u=\frac{1}{n}\{\sin nx\textstyle\int\cos nx\,Xdx-\cos nx\int\sin nx\,Xdx\}.$$

6. When the second member X is a rational and integral function of x, the final integration may be avoided. For, representing the given equation in the form $f\left(\dfrac{d}{dx}\right)u=X+0$, we have
$$n=\left\{f\left(\frac{d}{dx}\right)\right\}^{-1}X+\left\{f\left(\frac{d}{dx}\right)\right\}^{-1}0\ \ldots\ldots\ldots\ldots\ (34).$$

A particular value of the first term will be obtained by developing $\left\{f\left(\dfrac{d}{dx}\right)\right\}^{-1}$ in ascending powers (so to speak) of $\dfrac{d}{dx}$, and then performing the differentiations on X, while the general value of the second term will introduce the requisite number of arbitrary constants.

Ex. Given $\frac{d^2u}{dx^2}+n^2u=1+x+x^2$.

$$\text{Here}\quad u=\left(\frac{d^2}{dx^2}+n^2\right)^{-1}(1+x+x^2)+\left(\frac{d^2}{dx^2}+n^2\right)^{-1}0$$

$$=\left(n^{-2}-n^{-4}\frac{d^2}{dx^2}+n^{-6}\frac{d^4}{dx^4}-\&c.\right)(1+x+x^2)$$

$$+\,C_1\cos nx+C_2\sin nx$$

$$=n^{-2}(1+x+x^2)-2n^{-4}+C_1\cos nx+C_2\sin nx.$$

The validity of the transformation of the inverse form $\left(\frac{d^2}{dx^2}+n^2\right)^{-1}$ by development, as of its other transformation by decomposition, is tested by performing on the result the direct operation $\frac{d^2}{dx^2}+n^2$. We take occasion to notice that different transformations, while equally valid, do not of necessity conduct us to solutions equally general, nor have we any right to expect that they should. Each solution is an answer to the question contained in the given inverse form, but that question may admit of different answers, and no solution is *general* which does not include them all.

The final integrations may also be avoided when X consists of a series of exponentials of the form ϵ^{mx} with coefficients which are either constants, or rational and integral functions of x.

Since $\left(\frac{d}{dx}\right)^n\epsilon^{mx}=m^n\epsilon^{mx}$, we have, for all interpretable forms of $f\left(\frac{d}{dx}\right)$, the relation

$$f\left(\frac{d}{dx}\right)\epsilon^{mx}=f(m)\,\epsilon^{mx}\;\ldots\ldots\ldots\ldots\ldots\ldots(35),$$

the second number expressing the complete, because the only, value of the first member when $f\left(\frac{d}{dx}\right)$ is rational and integral, but a particular value of the first member when $f\left(\frac{d}{dx}\right)$ is inverse, the test being as before.

Hence, if the given equation be $f\left(\frac{d}{dx}\right) u = \Sigma A_m \epsilon^{mx}$, we have

$$u = \left\{f\left(\frac{d}{dx}\right)\right\}^{-1} \Sigma A_m \epsilon^{mx} + \left\{f\left(\frac{d}{dx}\right)\right\}^{-1} 0$$

$$= \Sigma A_m \{f(m)\}^{-1} \epsilon^{mx} + \left\{f\left(\frac{d}{dx}\right)\right\}^{-1} 0 \;\ldots\ldots\ldots\ldots\; (36),$$

the second term introducing the requisite number of arbitrary constants.

Again, if, in any expression of the form $f\left(\frac{d}{dx}\right) \epsilon^{mx} X$, we convert $\frac{d}{dx}$ into $\frac{d_1}{dx} + \frac{d_2}{dx}$, where $\frac{d_1}{dx}$ operates on x only as contained in ϵ^{mx}, and $\frac{d_2}{dx}$ operates on x only as contained in X, we have

$$f\left(\frac{d}{dx}\right) \epsilon^{mx} X = f\left(\frac{d_1}{dx} + \frac{d_2}{dx}\right) \epsilon^{mx} X$$

$$= f\left(m + \frac{d_2}{dx}\right) \epsilon^{mx} X, \text{ by (35)}$$

$$= \epsilon^{mx} f\left(m + \frac{d_2}{dx}\right) X.$$

Hence, dropping the suffix which is no longer necessary, since X alone follows the operative symbol, we have

$$f\left(\frac{d}{dx}\right) \epsilon^{mx} X = e^{mx} f\left(\frac{d}{dx} + m\right) X \;\ldots\ldots\ldots\ldots\; (37).$$

When therefore X is a rational and integral function of x, a particular value of the first member may be found from the second, by developing the functional symbol and effecting the differentiations. And that particular value may be made general, as in the following example.

Ex. Given $\frac{d^2u}{dx^2} - 3\frac{du}{dx} + 2u = x\epsilon^{mx}$.

Here $u = \left\{\left(\frac{d}{dx} - 1\right)\left(\frac{d}{dx} - 2\right)\right\}^{-1} x\epsilon^{mx} + \left\{\left(\frac{d}{dx} - 1\right)\left(\frac{d}{dx} - 2\right)\right\}^{-1} 0$

$$= \epsilon^{mx}\left\{\left(\frac{d}{dx} + m - 1\right)\left(\frac{d}{dx} + m - 2\right)\right\}^{-1} x + C_1\epsilon^x + C_2\epsilon^{2x}$$

$$= \epsilon^{mx}\left\{\frac{1}{(m-1)(m-2)} - \frac{2m-3}{\{(m-1)(m-2)\}^2}\frac{d}{dx} + \&c.\right\} x + C_1\epsilon^x + C_2\epsilon^{2x}$$

$$= \frac{x\epsilon^{mx}}{(m-1)(m-2)} - \frac{(2m-3)\epsilon^{mx}}{\{(m-1)(m-2)\}^2} + C_1\epsilon^x + C_2\epsilon^{2x}.$$

Again, the theorem (37) relieves us from any difficulty arising from cases of failure referred to in Chap. IX. Art. 9.

Ex. Given $\left(\frac{d}{dx} - a\right)^n u = \epsilon^{ax}$.

Here $$u = \left(\frac{d}{dx} - a\right)^{-n} \epsilon^{ax} = \epsilon^{ax}\left(\frac{d}{dx}\right)^{-n} 1 \text{ by (37)}$$

$$= \epsilon^{ax}\left(c_1 + c_2 x \ldots + c_{n-1}x^{n-1} + \frac{x^n}{1.2\ldots n}\right).$$

When the second member X involves terms of the form $A\cos mx$, $B\sin mx$, &c., we may either substitute for them their exponential values, or we may employ directly the easily demonstrated theorem

$$f\left(\frac{d^2}{dx^2}\right)\begin{matrix}\sin\\ \cos\end{matrix} mx = f(-m^2)\begin{matrix}\sin\\ \cos\end{matrix} mx.$$

Ex. Given $\dfrac{d^2u}{dx^2}+n^2u=\Sigma a_m \cos(mx+n)$.

Here $u=\left(\dfrac{d^2}{dx^2}+n^2\right)^{-1}\Sigma a_m \cos(mx+n)+\left(\dfrac{d^2}{dx^2}+n^2\right)^{-1}0$

$$=\Sigma\frac{a_m \cos(mx+n)}{n^2-m^2}+C_1\cos nx+C_2\sin nx.$$

In this example, however, the failing case which presents itself when $m=n$, is most simply, though not most satisfactorily, treated by the methods of Chap. IX. Art. 11.

The reduction of an integral of the nth order by the foregoing theory is not devoid of elegance.

We have

$$\int\int\ldots X dx^n=\left(\frac{d}{dx}\right)^{-n}X.$$

Now let $x=\epsilon^\theta$, then

$$\frac{dX}{dx}=\epsilon^{-\theta}\frac{dX}{d\theta},$$

$$\frac{d^2X}{dx^2}=\epsilon^{-\theta}\frac{d}{d\theta}\epsilon^{-\theta}\frac{dX}{d\theta}=\epsilon^{-2\theta}\left(\frac{d}{d\theta}-1\right)\frac{d}{d\theta}X, \text{ by (37)}.$$

Proceeding thus, we have

$$\frac{d^n}{dx^n}X=\epsilon^{-n\theta}\left(\frac{d}{d\theta}-n+1\right)\left(\frac{d}{d\theta}-n+2\right)\ldots\frac{d}{d\theta}X\ldots(38),$$

and therefore the operation denoted by $\left(\dfrac{d}{dx}\right)^n$, and the compound operation denoted by

$$\epsilon^{-n\theta}\left(\frac{d}{d\theta}-n+1\right)\left(\frac{d}{d\theta}-n+2\right)\ldots\frac{d}{d\theta},$$

are absolutely equivalent. Hence inverting both, and observing that the inversion of the latter involves the inversion of the order of its component symbols, we have

$$\left(\frac{d}{dx}\right)^{-n} X = \left\{\left(\frac{d}{d\theta}-n+1\right)\left(\frac{d}{d\theta}-n+2\right)\ldots\frac{d}{d\theta}\right\}^{-1}\epsilon^{n\theta}X$$

$$= \frac{1}{1.2\ldots n-1}\left\{\left(\frac{d}{d\theta}-n+1\right)^{-1} - (n-1)\left(\frac{d}{d\theta}-n+2\right)^{-1}\right.$$

$$\left. + \frac{(n-1)(n-2)}{1.2}\left(\frac{d}{d\theta}-n+2\right)^{-1} \text{\&c.}\right\} \epsilon^{n\theta}X$$

$$= \frac{1}{1.2\ldots n-1}\left\{\epsilon^{(n-1)\theta}\left(\frac{d}{d\theta}\right)^{-1}\epsilon^{\theta} - (n-1)\,\epsilon^{(n-2)\theta}\left(\frac{d}{d\theta}\right)^{-1}\epsilon^{2\theta} + \text{\&c.}\right\}X$$

$$= \frac{1}{1.2\ldots n-1}\{x^{n-1}\int X dx - (n-1)x^{n-2}\int X x dx$$

$$+ \frac{(n-1)(n-2)}{1.2}\int X x^2 dx \ldots \mp \int X x^{n-1} dx\},$$

the result in question.

From (38) we have the theorem

$$x^n \frac{d^n}{dx^n} = \frac{d}{d\theta}\left(\frac{d}{d\theta}-1\right)\ldots\left(\frac{d}{d\theta}-n+1\right) \quad\text{.........(39)},$$

which is important in the transformation of differential equations.

Forms purely symbolical.

7. In any system in which thought is expressed by symbols, the laws of combination of the symbols are determined from the study of the corresponding operations in thought. But it may be that the latter are subject to *conditions of possibility* as well as to laws *when possible*. And thus it may be that two systems of symbols, differing in interpretation, may agree as to their formal laws whenever they both express operations possible in thought, while at the same time there may exist combinations which really represent thought in the one but do not in the other. For instance, there exist forms of the functional symbol f, for which we can attach a meaning to the expression $f(m)$, but cannot directly attach a meaning

to the symbol $f\left(\frac{d}{dx}\right)$. And the question arises: Does this difference restrict our freedom in the use of that principle which permits us to treat expressions of the form $f\left(\frac{d}{dx}\right)$ as if $\frac{d}{dx}$ were a symbol of quantity? For instance, we can attach no *direct* meaning to the expression $\epsilon^{h\frac{d}{dx}}f(x)$, but if we develope the exponential as if $\frac{d}{dx}$ were quantitative, we have

$$\epsilon^{h\frac{d}{dx}}f(x) = (1 + h\frac{d}{dx} + \frac{1}{1.2}h^2\frac{d^2}{dx^2} + \&c.)\,f(x)$$

$$= f(x+h) \text{ by Taylor's theorem.}$$

Are we then permitted, on the above principle, to make use of symbolic language; always supposing that we can, by the continued application of the same principle, obtain a *final* result of interpretable form?

Now all special instances point to the conclusion that this is permissible, and seem to indicate, as a general principle, that the mere processes of symbolical reasoning are independent of the conditions of their interpretation. In the few instances we may have occasion to employ, verification will be easy. We take occasion to notice that, whatever view may be taken of this principle, whether it be contemplated as belonging to the realm of *a priori* truth, or whether it be regarded as a generalization from experience, it would be an error to regard it as in any peculiar sense a mathematical principle. It claims a place among the *general* relations of Thought and Language.

On the principle above stated we should have

$$\epsilon^{h\frac{d}{dx}+k\frac{d}{dy}}f(x, y) = \epsilon^{h\frac{d}{dx}}\epsilon^{k\frac{d}{dy}}f(x, y)$$

$$= f(x+h,\ y+k).$$

And here, the expression $\epsilon^{h\frac{d}{dx}+k\frac{d}{dy}}$, which is without meaning in itself, is to be regarded simply as the representative of the expression

$$1+h\frac{d}{dx}+k\frac{d}{dy}+\frac{1}{1.2}\left(h\frac{d}{dx}+k\frac{d}{dy}\right)^2+\frac{1}{1.2.3}\left(h\frac{d}{dx}+k\frac{d}{dy}\right)^3+\&c.,$$

which has meaning. And the proper test of the validity of the symbolic equation

$$\epsilon^{h\frac{d}{dx}+k\frac{d}{dy}}=\epsilon^{h\frac{d}{dx}}\epsilon^{k\frac{d}{dy}}$$

consists in substituting for each exponential form the series it represents, and comparing the finally developed results, just as we should, by developing the exponentials, verify the algebraic equation,

$$\epsilon^{hm+kn}=\epsilon^{hm}\epsilon^{kn}.$$

It must be noted that $\frac{d}{dx}$ and $\frac{d}{dy}$ are commutative, and combine, in all respects, like symbols of quantity. We are not permitted to write $\epsilon^{x+\frac{d}{dx}}=\epsilon^x\epsilon^{\frac{d}{dx}}$, because x and $\frac{d}{dx}$ are not commutative.

8. The above principle is illustrated in the solution of the following partial differential equations.

Ex. Given $\frac{d^2u}{dx^2}-a^2\frac{d^2u}{dy^2}=\phi(x, y)$.

Here $u=\left(\frac{d^2}{dx^2}-a^2\frac{d^2}{dy^2}\right)^{-1}\phi(x, y)$

$$=\left(2a\frac{d}{dy}\right)^{-1}\left\{\left(\frac{d}{dx}-a\frac{d}{dy}\right)^{-1}-\left(\frac{d}{dx}+a\frac{d}{dy}\right)^{-1}\right\}\phi(x, y)$$

$$=\left(2a\frac{d}{dy}\right)^{-1}\{\epsilon^{ax\frac{d}{dy}}\int\epsilon^{-ax\frac{d}{dy}}\phi(x, y)\,dx-\epsilon^{-ax\frac{d}{dy}}\int\epsilon^{ax\frac{d}{dy}}\phi(x,y)\,dx\}$$

$$=\frac{1}{2a}\int\{\Phi_1(x, y+ax)-\Phi_2(x, y-ax)\}\,dy,$$

the forms of Φ and ψ being given by the equations

$$\Phi_1(x, y) = \int \phi(x, y - ax)\, dx + \psi(y),$$

$$\Phi_2(x, y) = \int \phi(x, y + ax)\, dx + \chi(y),$$

$\psi(y)$ and $\chi(y)$ being arbitrary functions of y.

If $\phi(x, y) = 0$, we hence find

$$u = \frac{1}{2a}\int \{\psi(y + ax) - \chi(y - ax)\}\, dy,$$

or, if we represent $\frac{1}{2a}\int\psi(y)\,dy$ by $\psi_1(y)$, and $\frac{1}{2a}\int\chi(y)dy$ by $\chi_1(y)$,

$$u = \psi_1(y + ax) + \chi_1(y - ax).$$

As ψ and χ are arbitrary, ψ_1 and χ_1 are so too. This agrees with the result on p. 360.

Ex. Given $\frac{d^2u}{dx^2} + \frac{d^2u}{dy^2} + \frac{d^2u}{dz^2} = 0.$

We may put this in the form $\frac{d^2u}{dx^2} + au = 0$, where a stands for $\frac{d^2}{dy^2} + \frac{d^2}{dz^2}$, and integrate with respect to x, as if a were a constant quantity. Remembering that the two arbitrary constants of the complete integral must then be replaced by two arbitrary functions of y, z, we get the symbolical solution

$$u = \cos\left\{x\left(\frac{d^2}{dy^2} + \frac{d^2}{dz^2}\right)^{\frac{1}{2}}\right\}\phi(y, z) + \sin\left\{x\left(\frac{d^2}{dy^2} + \frac{d^2}{dz^2}\right)^{\frac{1}{2}}\right\}\psi(y, z).$$

Developing the cosine and the sine, and replacing

$$\left(\frac{d^2}{dy^2} + \frac{d^2}{dz^2}\right)^{\frac{1}{2}}\psi(y, z)$$

by a new arbitrary function $\chi(y, z)$, we have

$$u=\phi(y, z)-\frac{x^2}{1.2}\left(\frac{d^2}{dy^2}+\frac{d^2}{dz^2}\right)\phi(y, z)$$

$$+\frac{x^4}{1.2.3.4}\left(\frac{d^2}{dy^2}+\frac{d^2}{dz^2}\right)^2\phi(y, z)-\&c.$$

$$+x\chi(y, z)-\frac{x^3}{1.2.3}\left(\frac{d^2}{dy^2}+\frac{d^2}{dz^2}\right)\chi(y, z)$$

$$+\frac{x^5}{1.2.3.4.5}\left(\frac{d^2}{dy^2}+\frac{d^2}{dz^2}\right)^2\chi(y, z), \&c.$$

Under this form, the solution is presented by Lagrange in the *Mécanique Analytique*, Tom. II. p. 320.

Generalization of the foregoing theory.

9. All equations, whatsoever their nature or subject, which are expressible in the form

$$(\pi^n + A_1\pi^{n-1} + A_2\pi^{n-2} \ldots + A_n)\, u = X \ldots\ldots\ldots (1),$$

where π is an operative symbol subject to the laws

$$\pi a u = a\pi u, \quad \pi(u+v) = \pi u + \pi v, \quad \pi^m\pi^n u = \pi^{m+n} u,$$

a being a constant and u and v functions of x, admit of transformations analogous to those of Art. 5.

Thus, since $u = (\pi^n + A_1\pi^{n-1} + A_2\pi^{n-2} \ldots + A_n)^{-1} X$,

we shall have, when the roots a_1, a_2, ... a_n of the auxiliary equation

$$m^n + A_1 m^{n-1} + A_2 m^{n-2} \ldots + A_n = 0$$

are real and unequal, the transformation

$$u = N_1(\pi - a_1)^{-1}X + A_2(\pi - a_2)^{-1}X \ldots + N_n(\pi - a_n)^{-1}X \ldots (2),$$

the coefficients N_1, N_2, ... N_n being determined as in Art. 5.

The *legitimacy* of this transformation is proved by operating on both sides of (2) with $\pi^n + A_1\pi^{n-1} \ldots + A_n$, and shewing that (1) is reproduced with the same conditions for determining $N_1, N_2, \ldots N_n$ as if π were a symbol of quantity. But the question of its *completeness*, of its conducting, through the performance of the inverse operations $(\pi - a_1)^{-1}$, &c., to the most *general* solution of (1), is one that we are not called upon to determine *a priori*. In all the cases we shall have to consider, its completeness will be obvious.

Ex. The equation $\frac{d^2u}{dx^2} - (2x+1)\frac{du}{dx} + (x^2 + x - 1)u = 0$

is reducible to the form $\pi(\pi - 1)u = 0$ where $\pi = \frac{d}{dx} - x$. Hence

$$u = (\pi - 1)^{-1} 0 - \pi^{-1} 0.$$

Let $(\pi - 1)^{-1} 0 = y$, then, since $(\pi - 1)y = 0$, we have

$$\frac{dy}{dx} - (x+1)y = 0, \quad y = c_1 \epsilon^{\frac{(x+1)^2}{2}}.$$

In like manner, if $\pi^{-1}0 = z$, we find

$$\frac{dz}{dx} - xz = 0, \quad z = c_2 \epsilon^{\frac{x^2}{2}}.$$

Therefore $u = c_1 \epsilon^{\frac{(x+1)^2}{2}} - c_2 \epsilon^{\frac{x^2}{2}}$.

A very interesting application of the same theory to the solution of partial differential equations is afforded by what Mr Carmichael has termed the index symbol of homogeneous functions. *Cambridge and Dublin Math. Journal*, Vol. VI. p. 277.

Since, if u_a represent a homogeneous function of the ath degree of the variables $x_1, x_2, \ldots x_n$, we have

$$x_1 \frac{du_a}{dx_1} + x_2 \frac{du_a}{dx_2} \ldots + x_n \frac{du_a}{dx_n} = a u_a \ldots\ldots\ldots\ldots (3),$$

it follows that, if we represent the symbol $x_1 \frac{d}{dx_1} \ldots + x_n \frac{d}{dx_n}$ by π, we shall have

$$\pi u_a = a u_a, \quad \pi^2 u_a = a^2 u_a, \text{ \&c.}$$

and therefore, in accordance with the reasoning of Arts. 3 and 4,

$$f(\pi)\, u_a = f(a)\, u_a \dots\dots\dots\dots\dots\dots (A),$$

an equation of which the second member expresses the complete, because the only, value of the first number when $f(\pi)$ is rational and integral, but a particular value when the first member contains inverse factors.

Hence, if we have any equation $f(\pi)\, u = X$, where $f(\pi)$ is of the form $\pi^n + A_1 \pi^{n-1} + A_2 \pi^{n-2} \dots + A_n$, and X is a series of homogeneous functions of the variables, suppose

$$X = X_a + X_b + \dots \text{\&c.},$$

we get

$$\begin{aligned} u &= \{f(\pi)\}^{-1} X + \{f(\pi)\}^{-1} 0 \\ &= \{f(\pi)\}^{-1} X_a + \{f(\pi)\}^{-1} X_b \dots + \{f(\pi)\}^{-1} 0 \\ &= \{f(a)\}^{-1} X_a + \{f(b)\}^{-1} X_b \dots + \{f(\pi)\}^{-1} 0, \text{ by } (A). \end{aligned}$$

To find the value of the last term, we proceed, as in Art. 5, to reduce it to a series of terms of the form $A_i (\pi - a)^{-i} 0$, i being the number of roots equal to a of the equation $f(m) = 0$. Now it may, by an induction founded on successive applications of Lagrange's method for the solution of linear partial differential equations of the first order, be shewn that

$$(\pi - a)^{-i} 0 = u_a (\log x_1)^{i-1} + v_a (\log x_1)^{i-2} \dots + w_a \dots (B),$$

$u_a, v_a, \dots w_a$ being arbitrary homogeneous functions of $x_1, x_2, \dots x_n$ of the ath degree.

To this result we may give the symmetrical form

$$(\pi - a)^{-i} 0 = u_a L^{i-1} + v_a M^{i-2} \dots + w_a,$$

L, M, &c. being logarithms of any homogeneous functions which are not of the degree 0.

It remains to shew how it may be ascertained whether a proposed partial differential equation can be reduced to the form $f(\pi)\, u = X$.

Let us resolve each symbol $\frac{d}{dx_i}$, entering into π, into two, and let $\frac{d'}{dx_i}$ represent $\frac{d}{dx_i}$ as operating on x_i only as entering into u, and $\frac{d''}{dx_i}$ only as entering into π. Also let

$$x_1\frac{d'}{dx_1}+x_2\frac{d'}{dx_2}\ldots+\&c.=\pi', \text{ and } x_1\frac{d''}{dx_1}+x_2\frac{d''}{dx_2}+\&c.=\pi''.$$

It is easily seen then that $\pi=\pi'+\pi''$. We have therefore $\pi'u=(\pi-\pi'')\,u=\pi u$;

$$\therefore\ \pi'^2u=(\pi-\pi'')\,\pi u\ \ldots\ldots\ldots\ldots\ldots\ldots(C).$$

But as π'', in (C), operates on the variables only as entering into π, which is a homogeneous function of those variables of the first degree, we may replace it by unity. We have therefore $\pi'^2u=(\pi-1)\,\pi u$. In the same way it may be shewn that $\pi'^ru=(\pi-r+1)\,(\pi-r+2)\ldots\pi u$. And thus it is seen that any partial differential equation which is expressible in the form $f(\pi)\,u=X$, on the hypothesis that $\frac{d}{dx_1}$, $\frac{d}{dx_2}$, &c. operate on the variables only as entering into u, is reducible to the form $\phi(\pi)\,u=X$, independently of such restriction. This reduction having been effected, the solution can be found by means of (A) and (B), whenever the second member consists of one or more homogeneous functions of $x_1, x_2, \ldots x_n$.

Ex. $$x^2\frac{d^2u}{dx^2}+2xy\frac{d^2u}{dx\,dy}+y^2\frac{d^2u}{dy^2}-n\left(x\frac{du}{dx}+y\frac{du}{dy}\right)+nu$$
$$=x^2+y^2+x^3.$$

Here we have $(\pi'^2-n\pi'+n)\,u=x^2+y^2+x^3$.

Therefore $\{\pi\,(\pi-1)-n\pi+n\}\,u=x^2+y^2+x^3$,

or $(\pi-n)\,(\pi-1)\,u=x^2+y^2+x^3$,

whence

$$u = \{(\pi - n)(\pi - 1)\}^{-1}\{x^2 + y^2 + x^3\} + \{(\pi - n)(\pi - 1)\}^{-1} 0$$

$$= \frac{x^2 + y^2}{(2-n)(2-1)} + \frac{x^3}{(3-n)(3-1)} + u_n + v_1,$$

u_n, v_1 denoting arbitrary homogeneous functions of the degree n and 1 respectively.

10. We may, by simple transformations, reduce to the above case various other classes of equations differing from the above only as to the form of π; e.g. the class in which $\pi = a_1 x_1 \frac{d}{dx_1} + a_2 x_2 \frac{d}{dx_2} \ldots + a_n x_n \frac{d}{dx_n}$; but, passing over such special forms, we shall consider the general equation $f(\pi)u = X$, where

$$\pi = X_1 \frac{d}{dx_1} + X_2 \frac{d}{dx_2} \ldots + X_n \frac{d}{dx_n},$$

and each of the coefficients $X_1, X_2, \ldots X_n$, as well as X, may be any function whatever of the independent variables. And we design to shew, first, how it may be determined whether a given equation admits of reduction to the more general form above proposed; secondly, how, then, to integrate it.

Suppose the given equation of the nth order; then the symbolical form in question, should the proposed reduction be possible, will be

$$(\pi^n + A_1 \pi^{n-1} + A_2 \pi^{n-2} \ldots + A_n)\, u = X \ldots\ldots\ldots (4).$$

Now the highest differential coefficients in the given equation will arise solely from the symbol π^n, and the terms in which they occur will enable us to determine the form of π. Thus, for two variables, we have

$$\left(M\frac{d}{dx} + N\frac{d}{dy}\right)^2 u = M^2 \frac{d^2u}{dx^2} + 2MN \frac{d^2u}{dx\,dy} + N^2 \frac{d^2u}{dy^2}$$

$$+ \left(M\frac{dM}{dx} + N\frac{dM}{dy}\right)\frac{du}{dx} + \left(M\frac{dN}{dx} + N\frac{dN}{dy}\right)\frac{du}{dy},$$

in which the terms containing $\frac{d^2u}{dx^2}$, $\frac{d^2u}{dx\,dy}$, $\frac{d^2u}{dy^2}$ are the same as they would be, if, in the first member, $\frac{d}{dx}$, $\frac{d}{dy}$ were symbols of quantity. And this law is general for the *highest* differential coefficients.

Again, the form of π being determined, the values of A_1, A_2, ... will, whenever the proposed reduction is possible, be found by effecting the operations implied in the first member of (4), and comparing with the first member of the equation given.

Suppose the equation reduced to the form (4). Then, if the auxiliary equation

$$m^n + A_1 m^{n-1} + A_2 m^{n-2} \dots + A_n = 0 \dots\dots\dots (5)$$

have its roots all unequal, we have a series of terms of the form $(\pi - a)^{-1}X$; and each such term involves the solution of a partial differential equation of the first order of the form

$$X_1 \frac{du}{dx_1} + X_2 \frac{du}{dx_2} \dots + X_n \frac{du}{dx_n} - au = X.$$

But, if the auxiliary equation (5) have equal roots, partial differential equations of higher orders will present themselves. We deem it therefore important to shew how this difficulty may be avoided, or, to speak more precisely, how its solution may be made to flow from that of the corresponding case of linear differential equations with constant coefficients.

Introduce a new system of independent variables $y_1, y_2, \dots y_n$, so conditioned as to give $\pi = \frac{d}{dy_1}$. To prove that such a system exists, and to discover it, let us assume y_1, y_2, ... y_n, in succession, as subjects of the above symbolical equation, and examine whether the results are consistent. And first, assuming y_1 as subject, we have

$$X_1 \frac{dy_1}{dx_1} + X_2 \frac{dy_1}{dx_2} \dots\dots + X_n \frac{dy_1}{dx_n} = 1 \dots\dots\dots\dots (6).$$

Secondly, assuming y_i, representative of any of the remaining variables y_2, y_3, ... y_n, as subject, we have the equation

$$X_1 \frac{dy_i}{dx_1} + X_2 \frac{dy_i}{dx_2} \ldots\ldots + X_n \frac{dy_i}{dx_n} = 0 \ldots\ldots\ldots\ldots (7).$$

It follows from the above that, if we integrate the auxiliary system

$$\frac{dx_1}{X_1} = \frac{dx_2}{X_2} \ldots = \frac{dx_n}{X_n} \ldots\ldots\ldots\ldots\ldots\ldots\ldots\ldots (8),$$

the values of $y_2, y_3, \ldots y_n$ will be the first members of the integrals of that system expressed in the form

$$y_2 = a_2, \quad y_3 = y_3 \ldots y_n = a_n \ldots\ldots\ldots\ldots\ldots (9).$$

And it follows from (6) that if, from the system

$$\frac{dx_1}{X_1} = \frac{dx_2}{X_2} \ldots = \frac{dx_n}{X_n} = dy_1 \ldots\ldots\ldots\ldots\ldots (10),$$

differing from (8) only in that it contains one additional member dy_1, we deduce an additional integral equation connecting y_1 with the original variables $x_1, x_2, \ldots x_n$, that equation will give the value of y_1. We see that the number of distinct auxiliary equations is precisely equal to the number of quantities to be determined, so that the scheme is a consistent one.

The solution of the problem is therefore virtually dependent on the partial differential equation (6), from the auxiliary system of which, (10), it suffices to deduce n integrals, one expressing y_1 in terms of $x_1, x_2, \ldots x_n$, the others determining $y_2, y_3, \ldots y_n$, as functions of $x_1, x_2, \ldots x_n$, in the forms (9). To the arbitrary constant in the value of y_1 we may give any value we please.

Introducing the new variables, the equation given now assumes the form

$$f\left(\frac{d}{dy_1}\right) u = \phi\,(y_1, y_2, \ldots y_n),$$

which must be integrated as if u and y_1 were the only variables, an arbitrary function of $y_2, y_3, \ldots y_n$ being introduced in the place of an arbitrary constant. Finally, we must restore to $y_1, y_2, \ldots y_n$ their values in terms of $x_1, x_2, \ldots x_n$.

Ex. Given $(1-x^2)^2\dfrac{d^2u}{dx^2}+2(1-x^2)(1-xy)\dfrac{d^2u}{dxdy}$

$$+(1-xy)^2\frac{d^2u}{dy^2}-2x(1-x^2)\frac{du}{dx}-(x+y-2x^2y)\frac{du}{dy}+n^2u=0.$$

Here, the form of the first three terms shews that we must have $\pi=(1-x^2)\dfrac{d}{dx}+(1-xy)\dfrac{d}{dy}$, and the equation assumes the form

$$(\pi^2+n^2)\,u=0.$$

To avoid the difficulty arising from the imaginary factors of π^2+n^2, let us assume two new variables, x' and y', such that we may have $\pi=\dfrac{d}{dx'}$. Then by (10)

$$\frac{dx}{1-x^2}=\frac{dy}{1-xy}=dx',$$

corresponding to which we have the integral systems

$$\frac{y-x}{\sqrt{(1-x^2)}}=c, \qquad x'=\log\sqrt{\left(\frac{1+x}{1-x}\right)}+c'.$$

Hence, if we assume

$$x'=\log\sqrt{\left(\frac{1+x}{1-x}\right)}, \qquad y'=\frac{y-x}{\sqrt{(1-x^2)}},$$

we get the transformed equation

$$\left(\frac{d^2}{dx'^2}+n^2\right)u=0;$$

$$\therefore\ u=\cos nx'\,\phi(y')+\sin nx'\,\psi(y'),$$

or, restoring to x' and y' their values,

$$u=\cos\left\{n\log\sqrt{\left(\frac{1+x}{1-x}\right)}\right\}\phi\frac{y-x}{\sqrt{(1-x^2)}}$$

$$+\sin\left\{n\log\sqrt{\left(\frac{1+x}{1-x}\right)}\right\}\psi\frac{y-x}{\sqrt{(1-x^2)}}.$$

EXERCISES.

1. $\frac{d^3y}{dx^3} - 2\frac{d^2y}{dx^2} + \frac{dy}{dx} = \epsilon^x.$

2. $\frac{d^2u}{dx^2} - 5\frac{du}{dx} + 6u = \epsilon^{mx}.$

3. Determine the solution of the above equation when $m = 2$.

4. $\frac{d^2u}{dx^2} - 9\frac{du}{dx} + 20u = x^2\epsilon^{3x}.$

5. $\frac{d^2u}{dx^2} + 3\frac{du}{dx} + 2u = \cos mx.$

6. Solve the equation $\left(\frac{d}{dx} - a\right)^n u = \cos mx.$

In the above example it will be most convenient to proceed thus:

$$u = \left(\frac{d}{dx} - a\right)^{-n} \cos mx + \left(\frac{d}{dx} - a\right)^{-n} 0$$

$$= \frac{\left(\frac{d}{dx} + a\right)^n}{\left(\frac{d^2}{dx^2} - a^2\right)^n} \cos mx + \epsilon^{ax}\left(\frac{d}{dx}\right)^{-n} 0$$

$$= \frac{1}{(-m^2 - a^2)^n}\left(\frac{d}{dx} + a\right)^n \cos mx + \epsilon^{ax}(c_1 + c_2 x \ldots + c_n x^{n-1}).$$

7. Solve the equation $\left(\frac{d}{dx} - a\right)^n u = \epsilon^x \cos mx.$

8. $x^2\frac{d^2u}{dx^2} + 2xy\frac{d^2u}{dx\,dy} + y^2\frac{d^2u}{dy^2} - n\left(x\frac{du}{dx} + y\frac{du}{dy} - u\right) = 0.$

9. $x^2\frac{d^2u}{dx^2} + 2xy\frac{d^2}{dx\,dy} + y^2\frac{d^2u}{dy^2} = (x^2 + y^2)^{\frac{n}{2}}.$

10. Solve, by the method of Art. 10, the equation

$$\left(x\frac{d}{dx}+y\frac{d}{dy}+z\frac{d}{dz}\right)^2 u+n^2u=0.$$

11. The solution of any equation of the form

$$\frac{d^2u}{dx^2}+(2X+a)\frac{du}{dx}+\left(\frac{dX}{dx}+X^2+aX+b\right)u=0$$

may be reduced to that of two linear equations of the first order.

CHAPTER XVII.

SYMBOLICAL METHODS, CONTINUED.

1. THE classes of equations considered in the last chapter might all be gathered up into the one larger class represented by

$$f(\pi)\,u = x,$$

π being a symbol combining with constant quantities as if it were itself a symbol of quantity. But linear differential equations do not, except under particular conditions, admit of expression in this form. Those which are of the ordinary species involve in their general expression two symbols, x and $\frac{d}{dx}$, operating in combination on the sought and dependent variable y; and no substituted form of such equations is *general* which introduces fewer than two symbols in the place of x and $\frac{d}{dx}$. We propose in this chapter to employ a transformation which is general, and which is adapted in a very remarkable degree to the development of general methods of solution. A somewhat fuller account of it will be found in a memoir on a General Method in Analysis, (*Philosophical Transactions*, for 1844, Part II.) Other principles and other methods will also be noticed.

The following theorems, demonstrated in Chap. XVI. will frequently recur.

If $x = \epsilon^\theta$, and if $\frac{d}{d\theta}$ be represented by D, then

$$x^n \frac{d^n u}{dx^n} = D(D-1)\ldots(D-n+1)\,u \text{ .... (1)},$$

while the relations connecting $\frac{d}{d\theta}$ and ϵ^θ, become

$$f(D)\,\epsilon^{m\theta} = f(m)\,\epsilon^{m\theta} \quad\text{.......................... (2)},$$

$$f(D)\,\epsilon^{m\theta}u = \epsilon^{m\theta}f(D+m)\,u \quad\text{................ (3)}.$$

The latter of these relations enables us to transfer the exponential $\epsilon^{m\theta}$ from one side of the expression $f(D)$ to the other, by changing D into $D \pm m$, according as the transference is from right to left or from left to right. Thus, as another form of (3), we should have

$$\epsilon^{m\theta}f(D)\,u = f(D-m)\,\epsilon^{m\theta}u \quad\text{................. (4)}.$$

It is an immediate consequence of the above theorem that *every linear differential equation which can be expressed in the form,*

$$(a+bx+cx^2\ldots)\frac{d^n u}{dx^n} + (a'+b'x+c'x^2\ldots)\frac{d^{n-1}u}{dx^{n-1}} + \&\text{c.} = X \ldots (5),$$

can be reduced to the symbolical form,

$$f_0(D)\,u + f_1(D)\,\epsilon^{\theta}u + f_2(D)\,\epsilon^{2\theta}u + \&\text{c.} = T \quad\text{......... (6)},$$

where T is a function of θ.

For, multiplying the given equation by x^n, and assuming $x = \epsilon^{\theta}$, the first term of the left-hand member becomes, by (1),

$$(a + b\epsilon^{\theta} + c\epsilon^{2\theta} + \&\text{c.})\,D(D-1)\ldots(D-n+1)\,u,$$

and this is reducible, by (4), to the form

$$aD(D-1)\ldots(D-n+1)\,u + b(D-1)(D-2)\ldots(D-n)\,\epsilon^{\theta}u$$
$$+\,c(D-2)(D-3)\ldots(D-n-1)\,\epsilon^{2\theta}u + \&\text{c.},$$

each term of which is of the general form $\phi(D)\,\epsilon^{i\theta}u$. The other terms of the first member of (5) admitting of a similar reduction, while the second member becomes a function of θ, the equation itself assumes the symbolical form (6).

Ex. 1. Given $\dfrac{d^2u}{dx^2} - n^2u = 0$.

Multiplying by x^2, and transforming as above, we get

$$D(D-1)\,u - n^2\epsilon^{2\theta}u = 0.$$

Ex. 2. Given $(1+ax^2)\dfrac{d^2u}{dx^2}+ax\dfrac{du}{dx}\pm n^2u=\phi(x)$.

Multiplying by x^2, we have, by (1),

$$(1+a\epsilon^{2\theta})\,D(D-1)\,u+a\epsilon^{2\theta}\,Du\pm n^2\epsilon^{2\theta}u=\epsilon^{2\theta}\phi(\epsilon^{\theta}).$$

But

$$\epsilon^{2\theta}\,D(D-1)\,u=(D-2)(D-3)\,\epsilon^{2\theta}u\,;\ \text{and}\ \epsilon^{2\theta}Du=(D-2)\,\epsilon^{2\theta}u,$$

whence, substituting, and collecting together terms like with respect to the exponentials, we have

$$D(D-1)\,u+\{a(D-2)^2\pm n^2\}\,\epsilon^{2\theta}u=\epsilon^{2\theta}\phi(\epsilon^{\theta})$$

as the symbolical form.

To return from the symbolical to the ordinary form of a differential equation, we must, by (3), transfer the exponentials to the *left* of each symbolic function $f(D)$, convert the latter into a series of factorials of the from $D(D-1)\ldots(D-n+1)$, and then apply the transformation (1).

Ex. 3. Given $D(D-1)\,u+D(D+1)\,\epsilon^{\theta}u=0$.

We have in succession,

$$D(D-1)\,u+\epsilon^{\theta}(D+1)(D+2)\,u=0,$$

$$D(D-1)\,u+\epsilon^{\theta}\{D(D-1)+4D+2\}\,u=0,$$

$$x^2\frac{d^2u}{dx^2}+x\left(x^2\frac{d^2u}{dx^2}+4x\frac{du}{dx}+2u\right)=0.$$

Therefore, dividing by x,

$$(x+x^2)\frac{d^2u}{dx^2}+4x\frac{du}{dx}+2u=0.$$

A symbolical equation which has only two terms in its first member may be termed a binomial equation; one which has three terms a trinomial equation, and so on. We may determine by inspection to which of these classes an ordinary differential equation is reducible. For multiplying it by such a power of x as to permit its expression in the form

$$Ax^n \frac{d^n y}{dx^n} + Bx^{n-1} \frac{d^{n-1}y}{dx^{n-1}} + \&c. = X,$$

where A, B, &c. are algebraic polynomials with respect to x; the number of *distinct* powers of x involved in those polynomials will determine the number of terms in the reduced symbolical equation.

Ex. 4. Thus the equation

$$(a + bx) \frac{d^2u}{dx^2} + (c + ex) \frac{du}{dx} + qu = 0,$$

being expressed in the form

$$(a + bx) x^2 \frac{d^2u}{dx^2} + (cx + ex^2) x \frac{du}{dx} + (qx^2) u,$$

it is seen that its symbolical form will be trinomial, since the terms within the brackets involve x in the degrees 0, 1, and 2.

Finite solution of differential equations expressed in the symbolical form.

2. If we affect both sides of the symbolical equation (6) with $\{f_0(D)\}^{-1}$, then for $f_0(D)^{-1}f_1(D)$ write $\phi_1(D)$ &c., and for $\{f_0(D)\}^{-1}\, T$ write U, we shall have

$$u + \phi_1(D)\, \epsilon^{\theta}u + \phi_2(D)\, \epsilon^{2\theta}u \ldots + \phi_n(D)\, \epsilon^{n\theta}u = U \ldots\ldots\ldots (7);$$

and under this form the equation will be discussed in the following section.

PROP. 1. *The equation*

$$u + a_1\phi(D)\, \epsilon^{\theta}u + a_2\phi(D)\, \phi(D-1)\, \epsilon^{2\theta}u \ldots$$
$$+ a_n\phi(D)\, \phi(D-1) \ldots \phi(D-n+1)\, \epsilon^{n\theta}u = U \ldots (8),$$

may be resolved into a system of equations of the form

$$u - q\phi(D)\, \epsilon^{\theta}u = U,$$

the values of q being determined by the equation

$$q^n + a_1q^{n-1} + a_2q^{n-2} \ldots + a_n = 0.$$

For

$$\phi(D)\,\phi(D-1)\,\epsilon^{2\theta}u = \phi(D)\,\epsilon^{\theta}\phi(D)\,\epsilon^{\theta}u = \{\phi(D)\,\epsilon^{\theta}\}^2 u,$$

and in general

$$\phi(D)\,\phi(D-1)\ldots\phi(D-n+1)\,\epsilon^{n\theta}u = \{\phi(D)\,\epsilon^{\theta}\}^n u.$$

So that, if we represent the symbol $\phi(D)\,\epsilon^{\theta}$ by ρ, the equation in question becomes

$$(1 + a_1\rho + a_2\rho^2 \ldots + a_n\rho^n)\,u = U;$$

$$\therefore\ u = (1 + a_1\rho + a_2\rho^2 \ldots + a_n\rho^n)^{-1}\,U$$

$$= \{N_1(1-q_1\rho)^{-1} + N_2(1-q_2\rho)^{-1} \ldots + N_n(1-q_n\rho)^{-1}\}\,U,$$

provided that $q_1, q_2 \ldots q_n$ are roots of the equation

$$q^n + a_1q^{n-1} + a_2q^{n-2} \ldots + a_n = 0,$$

and that $N_1, N_2 \ldots N_n$ are of the forms

$$N_1 = \frac{q_1^{\,n-1}}{(q_1-q_2)(q_1-q_3)\ldots(q_1-q_n)}\ldots$$

$$N_n = \frac{q_n^{\,n-1}}{(q_n-q_1)(q_n-q_2)\ldots(q_n-q_{n-1})}.$$

Let $(1-q_1\rho)^{-1}U = u_1$, $(1-q_2\rho)^{-1}U = u_2$, and so on, then

$$u = N_1u_1 + N_2u_2 \ldots + N_nu_n,$$

where, in general, u_i is given by the solution of the equation

$$u_i - q_i\phi(D)\,\epsilon^{\theta}u_i = U \ldots\ldots\ldots\ldots\ldots\ldots(9).$$

The solution of the general equation (8) is therefore dependent on that of the binomial equation (9).

When $\phi(D)$ is of the form D^{-1} the equation (8) corresponds to the ordinary linear differential equation with constant coefficients.

Thus the equation $u - \frac{q^2}{D(D-1)} \epsilon^{2\theta} u = 0$, which may be integrated by the above process, is only the symbolical form of the equation $\frac{d^2u}{dx^2} - q^2 u = 0$ (see Ex. 1); and its solution, expressed in terms of x, is

$$u = C\epsilon^{qx} + C'\epsilon^{-qx}.$$

In like manner the equation $u + \frac{q^2}{D(D-1)} \epsilon^{2\theta} u = 0$ has for its solution, expressed in terms of x,

$$u = C \cos qx + C' \sin qx.$$

But, when $\phi(D)$ is not of the form D^{-1}, the equation (8) will represent an ordinary equation with variable coefficients.

Ex. 5. Given

$$(x^2 - 3x^3 + 2x^4) \frac{d^2u}{dx^2} + (4x - 6x^2) \frac{du}{dx} + (2 + 6x) u = ax^n.$$

The symbolical form of this equation is

$$(D+1)(D+2)u - 3(D+1)(D-2)\epsilon^{\theta} u$$
$$+ 2(D-2)(D-3)\epsilon^{2\theta} u = a\epsilon^{n\theta},$$

whence

$$u - 3\frac{D-2}{D+2}\epsilon^{\theta} u + 2\frac{(D-2)(D-3)}{(D+2)(D+1)}\epsilon^{2\theta} u = \frac{a\epsilon^{n\theta}}{(n+2)(n+1)},$$

or, putting $\frac{D-2}{D+2}\epsilon^{\theta} = \rho$, $\frac{a\epsilon^{n\theta}}{(n+2)(n+1)} = T$,

$$(1 - 3\rho + 2\rho^2) u = T.$$

Hence $u = \frac{1}{1 - 3\rho + 2\rho^2} T = \left(\frac{2}{1-2\rho} - \frac{1}{1-\rho}\right) T$

$$= 2u_1 - u_2 \quad \ldots\ldots\ldots\ldots\ldots\ldots\ldots\ldots \quad (a),$$

where $u_1 = (1-2\rho)^{-1} T$, $u_2 = (1-\rho)^{-1} T$.

From the former we have

$$(1-2\rho)\,u_1 = T, \text{ or } u_1 - 2\frac{D-2}{D+2}\,\epsilon^{\theta}u_1 = \frac{a\epsilon^{n\theta}}{(n+2)(n+1)},$$

whence
$$(D+2)\,u_1 - 2\,(D-2)\,\epsilon^{\theta}u_1 = \frac{\epsilon^{n\theta}}{n+1};$$

and this gives

$$(x-2x^2)\frac{du_1}{dx} + (2+2x)\,u_1 = \frac{ax^n}{n+1} \text{ ....} (b).$$

In like manner we find, for u_2,

$$(x-x^2)\frac{du_2}{dx} + (2+x)\,u_2 = \frac{ax^n}{n+1} \text{ ....} (c).$$

The values of u_1 and u_2, determined from (b) and (c), and substituted in (a), will give the complete solution.

If $a=0$, we find

$$u = \frac{C_1(1-2x)^3 + C_2(1-x)^3}{x^2}.$$

3. We proceed to consider more fully the theory of the binomial equation

$$u + \phi(D)\,\epsilon^{r\theta}u = U \text{ ....} (10).$$

PROP. 2. *The equation* $u+\phi(D)\,\epsilon^{r\theta}u = U$ *will be converted into* $v+\phi(D+n)\,\epsilon^{r\theta}v = V$, *by the relations*

$$u = \epsilon^{n\theta}v, \quad U = \epsilon^{n\theta}V.$$

For assume $u=\epsilon^{n\theta}v$, and, substituting in the original equation, we have

$$\epsilon^{n\theta}v + \phi(D)\,\epsilon^{(n+r)\theta}v = U;$$

$$\therefore\ \epsilon^{n\theta}v + \epsilon^{n\theta}\phi(D+n)\,\epsilon^{r\theta}v = U, \text{ by (3)}$$

$$v+\phi(D+n)\,\epsilon^{r\theta}v = \epsilon^{-n\theta}U.$$

Let $U=\epsilon^{n\theta}V$; then the above becomes

$$v+\phi(D+n)\,\epsilon^{r\theta}v = V,$$

as was to be shewn.

Thus in any binomial equation we can convert $\phi(D)$ *into* $\phi(D+n)$, *n being any constant.*

PROP. 3. *The equation* $u+\phi(D)\,\epsilon^{r\theta}u=U$ *will be converted into* $v+\psi(D)\,\epsilon^{r\theta}v=V$, *by the relations,*

$$u=P_r\frac{\phi(D)}{\psi(D)}v,\quad U=P_r\frac{\phi(D)}{\psi(D)}V,$$

where $P_r\dfrac{\phi(D)}{\psi(D)}$ *denotes the symbolical product*

$$\frac{\phi(D)\,\phi(D-r)\,\phi(D-2r)\ldots}{\psi(D)\,\psi(D-r)\,\psi(D-2r)\ldots}.$$

For, assume $u=f(D)v$, and, substituting in the original equation, we have

$$f(D)\,v+\phi(D)\,\epsilon^{r\theta}f(D)\,v=U;$$

$$\therefore f(D)\,v+\phi(D)f(D-r)\,\epsilon^{r\theta}v=U,\text{ by (4).}$$

$$v+\frac{\phi(D)f(D-r)}{f(D)}\epsilon^{r\theta}v=\{f(D)\}^{-1}U\ldots\ldots\ldots(11).$$

Comparing this with the equation $v+\psi(D)\,\epsilon^{r\theta}v=V$, we have

$$\frac{\phi(D)f(D-r)}{f(D)}=\psi(D);$$

$$\therefore f(D)=\frac{\phi(D)}{\psi(D)}f(D-r).$$

Hence
$$f(D-r)=\frac{\phi(D-r)}{\psi(D-r)}f(D-2r),$$

and so on; wherefore the value of $f(D)$ will be represented by the infinite product $\dfrac{\phi(D)\,\phi(D-r)\,\phi(D-2r)\ldots}{\psi(D)\,\psi(D-r)\,\psi(D-2r)\ldots}$. Hence (11) becomes

$$v+\psi(D)\,\epsilon^{r\theta}v=V$$

with the relations

$$u = P_r \frac{\phi(D)}{\psi(D)} v, \quad U = P_r \frac{\phi(D)}{\psi(D)} V \text{.........} (12).$$

As this proposition is of great importance in the solution of differential equations, it will be proper to examine the conditions which its application involves. Evidently they consist in such a choice of the form of $\psi(D)$ as will render the symbolical product $P_r \frac{\phi(D)}{\psi(D)}$ finite, and the transformed equation (11) integrable.

That the expression of $P_r \frac{\phi(D)}{\psi(D)}$ may be finite, it is sufficient that for every elementary factor $\chi(D)$ occurring in the numerator there should correspond a similar factor $\chi(D \pm ir)$ in the denominator, i being any integer or 0; and *vice versa;* for

$$P_r \frac{\chi(D)}{\chi(D+ir)} = \frac{\chi(D)\,\chi(D-r)\,\chi(D-2r)\ldots}{\chi(D+ir)\,\chi\{D+(i-1)\,r\}\ldots}$$

$$= \frac{1}{\chi(D+ir)\,\chi\{D+(i-1)\,r\}\ldots\chi(D+r)},$$

which is a finite expression. Again

$$P_r \frac{\chi(D)}{\chi(D-ir)} = \frac{\chi(D)\,\chi(D-r)\ldots}{\chi(D-ir)\,\chi\{D-(i+1)\,r\}\ldots}$$

$$= \chi(D)\,\chi(D-r)\ldots\chi\{D-(i-1)\,r\},$$

which is also finite; the product of any number of such expressions is finite also.

Hence, if $\chi(D)$ be any elementary factor of $\phi(D)$, it may be converted into $\chi(D \pm ir)$; for let $\phi(D) = \chi(D)\,\chi_1(D)$, and let $\psi(D) = \chi(D \pm ir)\,\chi_1(D)$, wherein $\chi_1(D)$ denotes the product of the remaining factors, then

$$P_r \frac{\phi(D)}{\psi(D)} = P_r \frac{\chi(D)}{\chi(D \pm ir)},$$

which is finite.

Hence also, if $\phi(D)$ involve any factor of the form $\frac{\chi(D)}{\chi(D\pm ir)}$, it may be made to disappear; for let $\phi(D) = \frac{\chi(D)}{\chi(D\pm ir)}\chi_1(D)$, and let $\psi(D) = \chi_1(D)$, then

$$P_r \frac{\phi(D)}{\psi(D)} = P_r \frac{\chi(D)}{\chi(D \pm ir)},$$

which is finite.

4. We see, then, that there are two distinct kinds of transformation to which the Proposition may be applied. In the first kind $\phi(D)$ is converted into another symbolic function $\psi(D)$ without any loss of component factors, whether of numerator or of denominator, but only with such change as consists in the conversion of D into $D \pm ir$. And here the order of the transformed equation is the same as that of the equation given, and, its solution introducing a sufficient number of arbitrary constants, no others need to be introduced, either in the prior determination of V or in the subsequent derivation of u. But in the second species of transformation some component factor of $\phi(D)$ (usually of the form $\frac{D+a}{D+b}$ where $a-b$ is a multiple of r) is lost, and the transformed equation being of an order lower than that of the equation given, the deficient constants of its solution must be supplied, either beforehand in the determination of V, or subsequently in the derivation of u. If in the former, any constants, sufficient in number, introduced by the performance of $\left\{P_r \frac{\phi(D)}{\psi(D)}\right\}^{-1} U$ will serve the purpose. If in the latter, all the constants introduced by the performance of $P_r \frac{\phi(D)}{\psi(D)} v$ must be retained, but their subsequent relation must be determined by means of the differential equation.

The reason why the constants connected with the disappearing factors are arbitrary in V alone, is, that V enters into no other equation than the one in whose solution those constants are found. If, however, the entire series of constants in V be retained, they will be reduced to one by the subsequent differentiations in passing to the value of u.

All that may seem obscure in the above statement will be made clear by the following examples.

Ex. 6. Given $\frac{d^2u}{dx^2}+q^2u-\frac{6u}{x^2}=0$, an equation occurring in the theory of the earth's figure.

The symbolical form is

$$u+\frac{q^2}{(D+2)(D-3)}\epsilon^{2\theta}u=0 \text{............} (a).$$

Now we may, by Prop. 3, directly reduce this equation to the form

$$v+\frac{q^2}{(D+2)(D+1)}\epsilon^{2\theta}v=0,$$

which, by Prop. 1, is resolvable into two equations of the first order. But it is better to assume as the transformed equation

$$v+\frac{q^2}{D(D-1)}\epsilon^{2\theta}v=0 \text{................} (b),$$

the solution of which is known already. Art. 2.

By Prop. 2, assuming $u=\epsilon^{-2\theta}w$, we have

$$w+\frac{q^2}{D(D-5)}\epsilon^{2\theta}w=0 \text{..............} (c).$$

Again, by Prop. 3, we can pass from (*c*) to (*b*) by assuming

$$w=P_2\frac{D-1}{D-5}v=(D-1)(D-3)v.$$

Hence $\quad u=\epsilon^{-2\theta}(D-1)(D-3)v$

$$=\epsilon^{-2\theta}\{D(D-1)-3D+3\}v,$$

$$=\frac{1}{x^2}\left(x^2\frac{d^2}{dx^2}-3x\frac{d}{dx}+3\right)c\sin(qx+c')$$

on restoring x and putting for v its value in terms of x.

Effecting the differentiations, we find

$$u = c\left\{\left(\frac{3}{x^2} - q^2\right)\sin(qx + c') - \frac{3q}{x}\cos(qx + c')\right\} \ldots\ldots (d).$$

We might have proceeded directly from (a) to (b) by Prop. 3; but, had we done so, the final reductions would not have depended on differentiations alone. Thus we should have had

$$u = P_2 \frac{D(D-1)}{(D+2)(D-3)} v = \frac{D-1}{D+2} v$$
$$= \{1 - 3(D+2)^{-1}\} v = (1 - 3\epsilon^{-2\theta} D^{-1} \epsilon^{2\theta}) v$$
$$= v - 3\epsilon^{-2\theta} \int \epsilon^{2\theta} v d\theta,$$

whence, restoring x and giving to v its previous value, we should be led to the same solution as before.

5. The two forms of solution above presented illustrate an important observation, viz. that when in the transition from $\phi(D)$ to $\psi(D)$, by Prop. 3, the reductions consist in augmenting, if we may be allowed the expression, D in factors of the denominator of $\phi(D)$, or in diminishing D in factors of the numerator, they will be effected by differentiations; while those reductions which consist in augmenting D in factors of the numerator of $\phi(D)$, or diminishing it in factors of the denominator, involve integrations. And it is one use of Prop. 2, that it enables us, in many cases, so to prepare the given symbolical equation that the final reductions shall depend on differentiations.

Ex. 7. It is required to determine the symbolical form and character of those differential equations of the nth order, the solution of which depends on that of the equation

$$\frac{d^n v}{dx^n} \pm q^n v = X.$$

The symbolical form of this equation is

$$v \pm \frac{q^n}{D(D-1)\ldots(D-n+1)} \epsilon^{n\theta} v = V \ldots\ldots\ldots (a),$$

where V is the symbolical form of $\left(\frac{d}{dx}\right)^{-n} X$, i.e. the result obtained by writing ϵ^θ for x in the nth integral of Xdx^n, no constants being added in the integration. From inspection of (a), it is evident that the class of equations sought must, on assuming $x = \epsilon^\theta$, be expressible in the form

$$u \pm \frac{q^n}{(D+a_1)(D+a_2)\dots(D+a_n)} \epsilon^{n\theta} u = U \dots\dots (b),$$

in which we shall suppose the quantities $a_1, a_2 \dots a_n$ to be ranged in the order of decreasing magnitude. Put $u = \epsilon^{-a_1\theta} u_1$, then by Prop. 2,

$$u_1 \pm \frac{q^n}{D(D+a_2-a_1)\dots(D+a_n-a_1)} \epsilon^{n\theta} u_1 = \epsilon^{a_1\theta} U \dots. (c).$$

The first factor of the denominator of $\phi(D)$ in (c) now agrees with the first factor in that of $\psi(D)$ in (a). In any of the remaining factors we may, by Prop. 2, convert D into $D \pm in$, i being any integer,—hence, that they may all correspond with the factors of $\psi(D)$, it is necessary that each of the quantities

$$\frac{a_2 - a_1 + 1}{n},\ \frac{a_3 - a_1 + 2}{n},\ \frac{a_4 - a_1 + 3}{n} \dots \frac{a_n - a_1 + n - 1}{n} \dots (d),$$

should be equal to a negative integer or to 0. And in this statement the conditions of finite solution are involved.

The value of u will be deduced from that of v by *differentiation*, for since $a_2 - a_1 < -1$,

$$P_n \frac{D-1}{(D+a_2-a_1)} = (D-1)(D-n+1)\dots(D+a_2-a_1+n),$$

and so on for the remaining factors to which P_n is to be applied.

Ex. 8. Given $\dfrac{d^2u}{dx^2} - \dfrac{i(i+1)}{x^2} u \pm q^2 u = 0$, where i is an integer.

This equation, which includes that of Ex. 6, presents itself in various physical problems (Poisson, *Théorie Mathématique de la Chaleur*, p. 158. Mossotti, on *Molecular Action*, &c.)

Its symbolical form is

$$u \pm \frac{q^2}{(D+i)(D-i-1)} \epsilon^{2\theta} u = 0 \quad \text{.........}(a).$$

Hence, by the last example,

$$u = \epsilon^{-i\theta} P_2 \frac{D-1}{D-2i-1} v$$

$$= \epsilon^{-i\theta} (D-1)(D-3)\dots(D-2i+1) v \quad \text{.........}(b),$$

where v is given by $\dfrac{d^2v}{dx^2} \pm q^2 v = 0.$

The expression (b) may be reduced to a more convenient form, as follows.

Since $f(D-a) = \epsilon^{a\theta} f(D) \epsilon^{-a\theta}$, we have

$$u = \epsilon^{-i\theta} . \epsilon^{\theta} D \epsilon^{-\theta} . \epsilon^{3\theta} D \epsilon^{-3\theta} \dots \epsilon^{(2i-1)\theta} D \epsilon^{-(2i-1)\theta} v$$

$$= \epsilon^{-(i+1)\theta} \left(\epsilon^{2\theta} \frac{d}{d\theta}\right)^i \epsilon^{-(2i-1)\theta} v$$

$$= x^{-(i+1)} \left(x^3 \frac{d}{dx}\right)^i x^{-(2i-1)} v.$$

Hence, according as the upper or lower sign is taken in the original equation, we have

$$u = \frac{1}{x^{i+1}} \left(x^3 \frac{d}{dx}\right)^i \frac{c_1 \cos qx + c_2 \sin qx}{x^{2i-1}} \quad \text{..........}(c).$$

or

$$u = \frac{1}{x^{i+1}} \left(x^3 \frac{d}{dx}\right)^i \frac{c_1 \epsilon^{qx} + c_2 \epsilon^{-qx}}{x^{2i-1}} \quad \text{.....................}(d).$$

Ex. 9. Given $\dfrac{d^2u}{dx^2} - a^2 \dfrac{d^2u}{dy^2} - \dfrac{i(i+1)}{x^2} = 0.$

Comparing this equation with the last, we see that its solution may be derived from (d) by changing therein q into $a\dfrac{d}{dy}$, c_1, c_2 into arbitrary functions of y following the exponentials. Hence we shall have

$$u = \frac{1}{x^{i+1}}\left(x^3\frac{d}{dx}\right)^i \frac{\epsilon^{ax\frac{d}{dy}}\phi(y) + \epsilon^{-ax\frac{d}{dy}}\psi(y)}{x^{2i-1}}$$

$$= \frac{1}{x^{i+1}}\left(x^3\frac{d}{dx}\right)^i \frac{\phi(y+ax) + \psi(y-ax)}{x^{2i-1}}.$$

The reason why the arbitrary function $\phi(y)$ must be placed after $\epsilon^{ax\frac{d}{dy}}$ and not before it, is that, in the derivation of the exemplar form, the arbitrary constant takes its place after, and not before ϵ^{qx}.

For $$\left(\frac{d}{dx} - q\right)^{-1} 0 = \epsilon^{qx}\left(\frac{d}{dx}\right)^{-1} 0 = \epsilon^{qx} c.$$

Here indeed we may transpose the constant, but when q is converted into $a\dfrac{d}{dy}$ we have

$$\left(\frac{d}{dx} - a\frac{d}{dy}\right)^{-1} 0 = \epsilon^{ax\frac{d}{dy}}\left(\frac{d}{dx}\right)^{-1} 0 = \epsilon^{ax\frac{d}{dy}}\phi(y),$$

and here the arbitrary function cannot be transposed, since y and $\dfrac{d}{dy}$ are not commutative.

The principle here illustrated, and which is a very important one, is that all conclusions founded on community of formal laws should stop short of interpretation. The *form* should be kept distinct from the *matter*. There is perfect analogy between the theorems

$$\left(\frac{d}{dx} - q\right)^{-1} 0 = \epsilon^{qx}\left(\frac{d}{dx}\right)^{-1} \epsilon^{qx} 0,$$

and $$\left(\frac{d}{dx} - a\frac{d}{dy}\right)^{-1} 0 = \epsilon^{ax\frac{d}{dy}}\left(\frac{d}{dx}\right)^{-1} \epsilon^{-ax\frac{d}{dy}} 0,$$

but not between the theorems,

$$\left(\frac{d}{dx} - q\right)^{-1} 0 = c\epsilon^{qx}, \text{ and } \left(\frac{d}{dx} - a\frac{d}{dy}\right)^{-1} 0 = \epsilon^{ax\frac{d}{dy}}\phi(y),$$

because in the formation of the latter interpretation has been employed.

The above example is one in which Monge's method of solution would fail, except for the particular case of $i=0$. And this gives occasion to the remark that symbolical methods are not, as they have sometimes been supposed to be, valuable only as abbreviating the processes of analysis. There are innumerable cases in which they afford the only proper mode of procedure.

Ex. 10. Given

$$(1-x^2)\frac{d^2u}{dx^2}+\left\{\frac{p+1}{x}-(4n-p+1)\,x\right\}\frac{du}{dx}-2n\,(2n-p)\,u=0.$$

This equation occurs in some researches of Poisson on definite integrals. The symbolical form is

$$u-\frac{(D+2n-2)\,(D+2n-2-p)}{D\,(D+p)}\,\epsilon^{2\theta}u=0 \;\ldots\ldots\ldots\; (a).$$

This equation is integrable in several distinct cases, but we shall examine here the particular case in which n is an integer.

Assuming as the transformed equation,

$$v-\frac{D+2n-2-p}{D+p}\,\epsilon^{2\theta}v=V \;\ldots\ldots\ldots\ldots\ldots\; (b),$$

it being necessary to introduce V because the transformed equation is of an order lower than that of the equation given, we have,

$$u=P_2\,\frac{D+2n-2}{D}\,v$$

$$=(D+2n-2)\,(D+2n-4)\ldots(D+2)\,v \;\ldots\ldots\; (c),$$

$$0=(D+2n-2)\,(D+2n-4)\ldots(D+2)\,V.$$

The latter equation gives for V the general value,

$$V=c_1\epsilon^{-2\theta}+c_2\epsilon^{-4\theta}\ldots+c_{n-1}\epsilon^{-(2n-2)\theta},$$

of which it suffices to retain one term. Retaining the first, substituting in (*b*), and operating on both sides with $D+p$, we get

$$(D+p)\,v-(D+2n-2-p)\,\epsilon^{2\theta}v=c_1\,(p-2)\,\epsilon^{-2\theta}.$$

Restoring x, and integrating, a value of v is found, involving two arbitrary constants, whence u will be given by

$$u=\left(x\frac{d}{dx}+2n-2\right)\left(x\frac{d}{dx}+2n-4\right)\ldots\left(x\frac{d}{dx}+2\right)v \ldots. (d).$$

The proposed equation is also integrable when p is an odd integer, and when $2n-p$ is an even integer. In the former case we may assume as the transformed equation,

$$v-\frac{(D+2n-1-p)\,(D+2n-1-p-1)}{(D+p)\,(D+p-1)}\,\epsilon^{2\theta}v=0,$$

which must be integrated by Prop. 1. In the latter case we must assume

$$v-\epsilon^{2\theta}v=V;$$

but in this case two constants must be retained in V; viz. one from each set of the reducing operations by which the factors of $\phi\,(D)$ are made to disappear.

6. It will be observed that, in the foregoing examples, we reduce the proposed symbolical equation by Propositions 2 and 3, either directly to an equation of the first order, or to a form which by Prop. 1 is resolvable into a system of equations of the first order. But there exists yet another class of equations admitting of finite solution; viz. such as by Props. 2 and 3 are reducible to either of the primary forms,

$$v+\frac{a\,(D-2)^2\pm n^2}{D\,(D-1)}\,\epsilon^{2\theta}v=0 \ldots\ldots\ldots\ldots\ldots (13),$$

$$v+\frac{(D-1)\,(D-2)}{aD^2\pm n^2}\,\epsilon^{2\theta}v=0 \ldots\ldots\ldots\ldots\ldots (14).$$

The former of these is the symbolical form of the equation

$$(1+ax^2)\frac{d^2u}{dx^2}+ax\frac{du}{dx}\pm n^2u=0,$$

which is reducible to $\frac{d^2u}{dt^2} \pm n^2u = 0$, by the assumption

$$t = \int \frac{dx}{\sqrt{(1 + a x^2)}}.$$

The latter is the symbolical form of the equation

$$(x^2 + a)\, x^2 \frac{d^2u}{dx^2} + (2x^2 + a)\, x \frac{du}{dx} \pm n^2u = 0,$$

which is reducible to the form $\frac{d^2u}{dt^2} \pm n^2u$, by the assumption

$$t = \int \frac{dx}{x\sqrt{(x^2 + a)}}.$$

Hence, the ordinary solutions of (13) and (14) will be obtained by substituting

$$t = \int \frac{dx}{\sqrt{(1 + ax^2)}}, \quad t = \int \frac{dx}{x\sqrt{(x^2 + a^2)}},$$

in the solution of the equation $\frac{d^2u}{dt^2} \pm n^2u = 0$.

It may be added that the forms (13) and (14) are allied, the one being convertible into the other by changing θ to $-\theta$.

Ex. 11. Given

$$(1 - x^2) \frac{d^2u}{dx^2} - (2m + 1)\, x \frac{du}{dx} + (q^2 - m^2)\, u = 0.$$

The symbolical form is

$$u - \frac{(D + m - 2)^2 - q^2}{D\,(D - 1)}\, \epsilon^{2\theta} u = 0.$$

If we apply Prop. 2, so as to convert D into $D - m$, and then by Prop. 3 reduce the equation to the general form (13), we shall obtain the final solution in the form

$$u = \left(\frac{d}{dx}\right)^m \{c_1 \cos (q \sin^{-1} x) + c_2 \sin (q \sin^{-1} x)\}.$$

7. *Pfaff's Equation.* The differential equation,

$$(a+bx^n)\,x^2\frac{d^2u}{dx^2}+(c+ex^n)\,x\frac{du}{dx}+(f+gx^n)\,u=X\ldots(a),$$

which includes all binomial equations of the second order, has been discussed by Euler, and, with greater generality, by Pfaff (*Disquisitiones Analyticæ*). We propose to investigate the conditions under which it admits of finite solution.

It suffices for this purpose to consider the case in which $X=0$.

The symbolical form is then

$$u+\frac{b\,(D-n)\,(D-n-1)+e\,(D-n)+g}{aD\,(D-1)+cD+f}\,\epsilon^{n\theta}u=0\ldots(b).$$

If n is not equal to q, it is convenient to change the independent variable by assuming $n\theta=2\theta'$, whence

$$\frac{d}{d\theta}=\frac{n}{2}\cdot\frac{d}{d\theta'}.$$

So that changing $n\theta$ into $2\theta'$, we must change D into $\frac{n}{2}D$. The result may be expressed in the form,

$$u+\frac{b}{a}\,\frac{(D-\alpha_1)\,(D-\alpha_2)}{(D-\beta_1)\,(D-\beta_2)}\,\epsilon^{2\theta'}u=0\ \ldots\ldots\ldots\ldots\ldots\ (c),$$

where α_1 and α_2 are roots of the equation,

$$b\left(\frac{n\alpha}{2}-n\right)\left(\frac{n\alpha}{2}-n-1\right)+e\left(\frac{n\alpha}{2}-n\right)+g=0\ \ldots\ldots\ (d),$$

and β_1, β_2 are roots of the equation,

$$a\frac{n\beta}{2}\left(\frac{n\beta}{2}-1\right)+c\frac{n\beta}{2}+f=0\ \ldots\ldots\ldots\ldots\ldots\ (e).$$

1st, By Prop. 3, (c) can be immediately reduced to the form

$$v+\frac{b}{a}\,\frac{(D-\alpha_1)\,(D-\alpha_1-1)}{(D-\beta_1)\,(D-\beta_1-1)}\,\epsilon^{2\theta'}v=0,$$

and then resolved into two equations of the first order, if we have at the same time $\alpha_1 - \alpha_2$, and $\beta_1 - \beta_2$ odd integers.

2ndly, The equation can, by Prop. 3, be reduced to an equation of the first order if any one of the four quantities

$$\alpha_1 - \beta_1,\ \alpha_1 - \beta_2,\quad \alpha_2 - \beta_1,\ \alpha_2 - \beta_2$$

is an even integer.

3rdly, It is easily shewn that by Props. 2 and 3 (c) is reducible to the integrable form (13) if the quantities

$$\beta_1 - \beta_2 \text{ and } \alpha_1 + \alpha_2 - \beta_1 - \beta_2$$

are both odd integers.

4thly, It is in like manner reducible to (14) if the quantities

$$\alpha_1 - \alpha_2 \text{ and } \alpha_1 + \alpha_2 - \beta_1 - \beta_2$$

are both odd integers.

These results may be collected into the following theorem. The equation (c) is integrable in finite terms, 1st, if any one of the four quantities represented by $\alpha - \beta$ is an even integer; 2ndly, if any two of the quantities

$$\alpha_1 - \alpha_2,\quad \beta_1 - \beta_2,\quad \alpha_1 + \alpha_2 - \beta_1 - \beta_2$$

is an odd integer.

In this theorem the integral values are supposed to be either positive or negative, and the even ones to include the value 0.

The above results are equivalent to those of Pfaff, as presented with some slight increase of generality in a memoir by Sauer (*Crelle*, Vol. II. p. 93). Pfaff's conditions are however exhibited in so complex a form as to render the comparison difficult. His method, it is needless to say, is wholly different from the above.

Symbolical equations which are not binomial.

8. Although processes of greater or less generality may be established for the treatment of equations which, when symbolically expressed, involve more than two terms in the first member, yet their reduction if possible by some preliminary transformation to the binomial form should always be our first object. We purpose here to illustrate this observation.

Ex. 12. Given $\frac{d^2y}{dx^2}=a-\frac{by}{(2cx-x^2)^2}$.

Writing this equation in the form

$$(2c-x)^2 x^2 \frac{d^2y}{dx^2}+by=a\,(2cx-x^2)^2,$$

we see at once that its symbolical form will not be binomial. Assuming $y=(2c-x)^m u$, we have on reduction

$$(2c-x)\,x^2\frac{d^2u}{dx^2}-2mx^2\frac{du}{dx}+\frac{m\,(m-1)\,x^2+b}{2c-x}u=\frac{ax^2}{(2c-x)^{m-1}}.$$

Now let m be so determined as to make the numerator of the third term divisible by its denominator. This involves the condition

$$m\,(m-1)+\frac{b}{4c^2}=0 \ldots\ldots\ldots\ldots\ldots\ldots (a),$$

while the differential equation becomes

$$(2c-x)\,x^2\frac{d^2u}{dx^2}-2mx^2\frac{du}{dx}-m\,(m-1)\,(2c+x)\,u=\frac{ax^2}{(2c-x)^{m-1}},$$

of which the symbolical form is

$$(D-m)(D+m-1)\,u-\frac{1}{2c}(D+m-1)(D+m-2)\,\epsilon^{\theta}u=\frac{a}{2c}\frac{\epsilon^{2\theta}}{(2c-\epsilon^{\theta})^{m-1}},$$

whence, operating on both sides with $(D+m-1)^{-1}$,

$$(D-m)\,u-\frac{1}{2c}(D+m-2)\,\epsilon^{\theta}u=\frac{a}{2c}\epsilon^{(i-m)\theta}D^{-1}\epsilon^{(m+1)\theta}(2c-\epsilon^{\theta})^{1-m}.$$

Restoring x, and solving the equation, we have, on representing $2c - x$ by X,

$$u = \frac{a}{2c} x^m X^{1-2m} \int x^{-2m} X^{2m-2} \int x^m X^{1-m} dx^2,$$

which integration by parts reduces to the form

$$u = \frac{a \{x^m X^{1-2m} \int X^m x^{1-m} dx - x^{1-m} \int x^m X^{1-m} dx\}}{4c^2 (2m-1)}.$$

Therefore $$y = \frac{a \{x^m X^{1-m} \int x^{1-m} X^m dx - x^{1-m} X^m \int x^m X^{1-m} dx\}}{4c^2 (2m-1)},$$

the integral required, It is to be noted that each integration introduces an arbitrary constant. It is also seen that each value of m derived from (a) leads to the same result.

The above equation occurs in the problem of determining the tendency of an elastic bridge to break, when a heavy body, e.g. a railway train, passes rapidly over it. The equation between y and x is, on a certain hypothesis, that of the trajectory described. See an interesting paper by Prof. Stokes (*Cambridge Phil. Transactions*, Vol. VIII. p. 708).

Ex. 13. Given $$(1-\mu^2) \frac{d}{d\mu} (1-\mu^2) \frac{du}{d\mu} + n(n+1)(1-\mu^2) u + \frac{d^2u}{d\phi^2} = 0,$$

the well-known equation of Laplace's functions.

Representing $\frac{d}{d\phi} \sqrt{(-1)}$ by a, the equation may be expressed in the form

$$(1-\mu^2)^2 \frac{d^2u}{d\mu^2} - 2\mu(1-\mu^2) \frac{du}{d\mu} + \{n(n+1)(1-\mu^2) - a^2\} u = 0,$$

and it is evident that it would not, on assuming $\mu = \epsilon^\theta$, take the binomial form.

Let then $u = (1-\mu^2)^r v$. We find, on substitution, and division of the result by $(1-\mu^2)^r$,

$$(1-\mu^2)\frac{d^2v}{d\mu^2}-(4r+2)\mu\frac{dv}{d\mu}+\{n(n+1)-4r^2-2r\}v+\frac{4r^2-a^2}{1-\mu^2}v=0 \ldots (a).$$

Let $4r^2-a^2=0$. Then $r=\pm\frac{a}{2}$. Either sign may be taken. Choosing the lower, we have

$$(1-\mu^2)\frac{d^2v}{d\mu^2}+2(a-1)\mu\frac{dv}{d\mu}+\{n(n+1)-a(a-1)\}v=0,$$

an equation which, on making $\mu=\epsilon^\theta$, assumes the symbolical form

$$v-\frac{(D-a+n-1)(D-a-n-2)}{D(D-1)}\epsilon^{2\theta}v=0 \ldots\ldots (b).$$

To integrate this, assume

$$w-\frac{(D-a-n-1)(D-a-n-2)}{D(D-1)}\epsilon^{2\theta}w=0 \ldots\ldots (c).$$

Then by Prop. 3,

$$v=(D-a+n-1)(D-a+n-3)\ldots(D-a-n+1)w$$

$$=\mu^{n+a}\left(\frac{d}{d\mu}\frac{1}{\mu}\right)^n\mu^{n-a}w \ldots\ldots (d),$$

while (c), resolved by Prop. 1 and integrated, gives the solution

$$w=(1+\mu)^{n+a}\psi(\phi)+(1-\mu)^{n+a}\chi(\phi) \ldots\ldots (e),$$

ψ and χ being arbitrary functional signs. This expression for w having been substituted in (d), we must write $\frac{d}{d\phi}\sqrt{(-1)}$ for a, and interpret the result.

Now if, instead of $\psi(\phi)$ and $\chi(\phi)$, we write $\psi\{\epsilon^{\phi\sqrt{(-1)}}\}$ and $\chi\{\epsilon^{\phi\sqrt{(-1)}}\}$, as we are evidently permitted to do, and if we observe that generally

$$t^{\frac{d}{d\phi}\sqrt{(-1)}}f\{\epsilon^{\phi\sqrt{(-1)}}\}=\epsilon^{\log t\frac{d}{d\phi}\sqrt{(-1)}}f\{\epsilon^{\phi\sqrt{(-1)}}\}$$

$$=f[\epsilon^{\{\phi+\log t\sqrt{(-1)}\}\sqrt{(-1)}}]=f\{\epsilon^{\phi\sqrt{(-1)}-\log t}\}$$

$$=f\left\{\frac{\epsilon^{\phi\sqrt{(-1)}}}{t}\right\} \;\ldots\ldots\ldots\ldots\ldots\ldots\; (f),$$

we shall ultimately find

$$u=F\left\{\mu,\ \frac{\sqrt{(1-\mu^2)}}{\mu}\,\epsilon^{\phi\sqrt{(-1)}}\right\},$$

where $F\{\mu,\ \epsilon^{\phi\sqrt{(-1)}}\}=\mu^n\left(\frac{d}{d\mu}\frac{1}{\mu}\right)^n\left\{(\mu+\mu^2)^n\,\psi\left(\frac{\mu\epsilon^{\phi\sqrt{(-1)}}}{1+\mu}\right)\right.$

$$\left.+(\mu-\mu^2)^n\,\chi\left(\frac{\mu\epsilon^{\phi\sqrt{(-1)}}}{1-\mu}\right)\right\} \;\ldots\ldots\ldots\; (15),$$

which is the complete integral.

For a discussion of this result, and for the finite expression for Laplace's functions to which it leads, the reader is referred to a paper on the Equation of Laplace's functions in the *Cambridge Mathematical Journal.* (New Series, Vol. I. p. 10.)

If in the equation (*a*) we make the third instead of the fourth term to vanish, which gives for r the values $\frac{n}{2}$ and $-\frac{n+1}{2}$, and then assume $\frac{\mu}{\sqrt{(1-\mu^2)}}=t$, we shall obtain, taking the second value of r, the symbolical equation

$$v+\frac{(D+n-1)^2-a^2}{D(D-1)}\,\epsilon^{2\theta}\,v=0.$$

Now by Propositions 2 and 3 this is reducible to the integrable form

$$w+\frac{(D-2)^2-a^2}{D(D-1)}\,\epsilon^{2\theta}w=0,$$

by the relation

$$v = \epsilon^{-(n+1)\theta} D(D-1) \ldots (D-n)\, w$$

$$= \left(\frac{d}{dt}\right)^{n+1} w.$$

Hence we find

$$v = \left(\frac{d}{dt}\right)^{n+1} [c_1 \{t + \sqrt{(1+t^2)}\}^a + c_2 \{t + \sqrt{(1+t^2)}\}^{-a}],$$

whence u is known.

Let us examine the form of the solution, when, as is common in the expression of Laplace's equation, we replace μ by $\cos\theta$. We find

$$t = \cot\theta, \quad \frac{d}{dt} = -\sin^2\theta \frac{d}{d\theta},$$

whence $$t + \sqrt{(1+t^2)} = \cot\tfrac{1}{2}\theta.$$

Substituting, and observing that $u = (\sin\theta)^{-n-1} v$, we have

$$u = (\sin\theta)^{-n-1} \left(\sin^2\theta \frac{d}{d\theta}\right)^{n+1} \left\{c_1 \left(\cot\frac{\theta}{2}\right)^a + c_2 \left(\tan\frac{\theta}{2}\right)^a\right\}.$$

And hence, restoring to a its meaning, introducing arbitrary functions for constants, and effecting one of the differentiations, we may deduce the following solution of Laplace's equation, viz.:

$$u = (\sin\theta)^{-n} \left(\sin\theta \frac{d}{d\theta} \sin\theta\right)^n \left[F_1 \left\{\epsilon^{\phi\sqrt{(-1)}} \tan\frac{\theta}{2}\right\}\right.$$

$$\left. + F_2 \left\{\epsilon^{-\phi\sqrt{(-1)}} \tan\frac{\theta}{2}\right\}\right] \ldots (16).$$

Under this singularly elegant form the solution, obtained by a different method, was given by Professor Donkin. (*Philosophical Transactions*, for 1857.)

Solution of linear equations by series.

9. PROP. 4. *If a linear differential equation whose second member is* 0 *be reduced to the symbolical form*

$$f_0(D)\,u + f_1(D)\,\epsilon^{\theta}u + f_2(D)\,\epsilon^{2\theta}u \ldots + f_n(D)\,\epsilon^{n\theta}u = 0 \ldots (17)$$

(*Art.* 1), *then a particular solution will be*

$$u = \Sigma u_m \epsilon^{m\theta} \ldots\ldots\ldots\ldots\ldots\ldots\ldots\ldots (18),$$

the value of the index m in the first term being any root of the equation $f_0(m) = 0$, the corresponding value of u_m an arbitrary constant, and the law of the succeeding constants being expressed by the equation,

$$f_0(m)\,u_m + f_1(m)u_{m-1} + f_2(m)u_{m-2} \ldots + f_n(m)u_{m-n} = 0 \ldots (19).$$

For the form of u assigned in (18) will constitute a solution of (17) if, on substituting that form for u in the first member of (17) and arranging the result in ascending powers of ϵ^{θ}, each coefficient should vanish. And this, as we shall see, will take place if the coefficients are subject to the relation expressed by (19).

Assuming then $u = \Sigma u_m \epsilon^{m\theta}$, we find,

$$f_0(D)\,u = \Sigma f_0(D)\,u_m \epsilon^{m\theta} = \Sigma f_0(m)\,u_m \epsilon^{m\theta}, \text{ by (2)},$$

$$f_1(D)\,\epsilon^{\theta}u = \Sigma f_1(m+1)\,u_m \epsilon^{(m+1)\theta},$$

$$f_2(D)\,\epsilon^{2\theta}u = \Sigma f_2(m+2)\,u_m \epsilon^{(m+2)\theta},$$

and so on. In the first of these, we see that the coefficient of any particular term $\epsilon^{m\theta}$ is $f_0(m)\,u_m$. In the second, the coefficient of $\epsilon^{(m+1)\theta}$ is $f_1(m+1)\,u_m$, and therefore the coefficient of $\epsilon^{m\theta}$ is $f_1(m)\,u_{m-1}$. In the third, the coefficient of $\epsilon^{m\theta}$ is $f_2(m)\,u_{m-2}$; and so on. Thus the aggregate coefficient of $\epsilon^{m\theta}$ is

$$f_0(m)\,u_m + f_1(m)\,u_{m-1} + f_2(m)\,u_{m-2} \ldots + f_n(m)\,u_{m-n},$$

and this, equated to 0, expresses the law (19).

Let $u_r\epsilon^{r\theta}$ be the first term in the developed value of u; then must we suppose $u_{r-1}=0$, $u_{r-2}=0$, &c. and (19) becomes

$$f_0(r)\,u_r=0.$$

As, by hypothesis, u_r is not equal to 0, this gives $f_0(r)=0$, for the determination of r, and leaves u_r arbitrary. Hence the proposition is established.

Thus there will, except in particular cases of failure hereafter to be considered, be as many distinct solutions of the form (18), each involving an arbitrary constant, as there are units in the degree of $f_0(m)$.

Ex. 14. Given $\dfrac{d^2u}{dx^2}-\dfrac{a-1}{x}\dfrac{du}{dx}-n^2u=0.$

The symbolical form is

$$D(D-a)\,u-n^2\epsilon^{2\theta}u=0.$$

Hence, we have $u=\Sigma u_m x^m$, the law of formation of the coefficients being

$$m(m-a)\,u_m-n^2u_{m-2}=0, \text{ or } u_m=\frac{n^2}{m(m-a)}\,u_{m-2},$$

while the initial exponent is 0 or a. There are therefore two ascending series, one beginning with C, the other with $C'x^a$. Thus we have

$$u=C+\frac{Cn^2}{2(2-a)}x^2+\frac{Cn^4}{2.4(2-a)(4-a)}x^4+\&c.$$

$$+\,C'x^a+\frac{C'n^2x^{a+2}}{2(a+2)}+\frac{C'n^4x^{a+4}}{2.4.(a+4)(a+2)}+\&c.$$

10. When the equation $f_0(m)=0$, has equal or imaginary roots, the following procedure must be adopted. Let the solution of the equation $f_0(D)\,u=0$, be .

$$u=AP+BQ+CR+\&c. \quad\ldots\ldots\ldots\ldots\ldots (20),$$

A, B, C, &c. being the arbitrary constants. Substitute this value in the given differential equation, regarding A, B, C, &c. as variable, and the result will assume the form

$$A'P + B'Q + C'R + \&c. = 0 \text{ .... } (21),$$

and will be satisfied if we have

$$A' = 0, \quad B' = 0, \quad C' = 0, \&c. \text{............ } (22).$$

This will indeed become a system of linear simultaneous equations for determining A, B, C. And the solution of this system in a series will be of the form

$$A = \Sigma a_m \epsilon^{m\theta}, \quad B = \Sigma b_m \epsilon^{m\theta}, \quad C = \Sigma c_m \epsilon^{m\theta}, \&c.$$

the law of formation of the coefficients a_m, b_m, c_m, &c. being expressed by a system of simultaneous equations formed from (22), by changing therein every term of the form $\phi(D)\,\epsilon^{i\theta} A$ into $\phi(m)\, a_{m-i}$, &c. (*Philosophical Transactions.*)

There is a particular case of exception to the above rule. When two of the roots of $f_0(m) = 0$ differ by a multiple of the common difference of the indices of the ascending development, the equation $f_0(D) = 0$, must be replaced by what that equation would become were the roots in question equal.

Ex. 15. Given $\dfrac{d^2u}{dx^2} - \dfrac{1}{x}\dfrac{du}{dx} + q^2 u = 0$.

The symbolical form is

$$D^2 u + q^2 \epsilon^{2\theta} u = 0 \text{ .... } (a).$$

Now $D^2 u = 0$ gives $u = A + B\theta$. Substituting this value in (a), regarding A and B as variable, we have

$$D^2 A + q^2 \epsilon^{2\theta} A + 2DB + \theta\,(D^2 B + q^2 \epsilon^{2\theta} B) = 0,$$

which furnishes the two equations,

$$D^2 A + q^2 \epsilon^{2\theta} A + 2DB = 0, \quad D^2 B + q^2 \epsilon^{2\theta} B = 0,$$

whence $A = \Sigma a_m \epsilon^{m\theta}$, $B = \Sigma b_m \epsilon^{m\theta}$, with the relations

$$m^2 a_m + q^2 a_{m-2} + 2m b_m = 0, \quad m^2 b_m + q^2 b_{m-2} = 0,$$

from which we have

$$b_m = \frac{-q^2}{m^2} b_{m-2}, \quad a_m = \frac{-q^2}{m^2} a_{m-2} + \frac{2q^2}{m^3} b_{m-2} \dots\dots\dots (b).$$

Thus we find, on substitution, and restoration of x,

$$u = a_0 + a_2 x^2 + a_4 x^4 + \&c.$$
$$+ \log x (b_0 + b_2 x^2 + b_4 x^4 + \&c.),$$

where a_0, b_0 are arbitrary, and the succeeding values determined by (b).

Were the symbolical equation of the form

$$D(D \pm 2i) u + q^2 \epsilon^{2\theta} u = 0,$$

it would still be necessary to determine the form of the primary assumption by solving the equation $D^2 u = 0$, not by $D(D \pm 2i) u = 0$. We should therefore still have $u = A + B\theta$, in which A and B are series to be determined as before.

Ex. 16. Given $x^2 \dfrac{d^2 u}{dx^2} + x \dfrac{du}{dx} + (n^2 + x^2) u = 0.$

The symbolical equation is

$$(D^2 + n^2) u + \epsilon^{2\theta} u = 0 \dots\dots\dots\dots\dots\dots\dots (a).$$

Now the equation $(D^2 + n^2) u = 0$ gives

$$u = A \cos n\theta + B \sin n\theta \dots\dots\dots\dots\dots\dots\dots (b),$$

substituting which in (a), and equating to 0 the coefficients of $\cos n\theta$ and $\sin n\theta$ in the result, we have

$$D^2 A + 2nDB + \epsilon^{2\theta} A = 0,$$

$$D^2 B - 2nDA + \epsilon^{2\theta} B = 0,$$

whence $A = \Sigma a_m \epsilon^{m\theta}$, $B = \Sigma b_m \epsilon^{m\theta}$, with the relations,

$$m^2 a_m + 2mnb_m + a_{m-2} = 0,$$
$$m^2 b_m - 2mna_m + b_{m-2} = 0,$$

and therefore,

$$a_m = -\frac{ma_{m-2} - 2nb_{m-2}}{m(m^2+4n^2)}, \quad b_m = -\frac{mb_{m-2} + 2na_{m-2}}{m(m^2+4n^2)} \text{ ...... } (c).$$

Thus the solution assumes the form,

$$u = \cos(n \log x)(a_0 + a_2x^2 + a_4x^4 + \&c.)$$
$$+ \sin(n \log x)(b_0 + b_2x^2 + b_4x^4 + \&c.),$$

wherein a_0 and b_0 are arbitrary, and the succeeding coefficients determined by c.

The fundamental equation (19), written in a reversed order, determines the law of the formation of the coefficients in those solutions of (17) which are expressible in descending powers of x. The number of such solutions will be equal to the degree of the equation $f_n(m) = 0$, but their respective first exponents will be its roots severally diminished by n.

For the extension of the above theory to the case in which the given differential equation has a second member X, the reader is referred to the original memoir.

Theory of Series.

11. The relations which enable us to express the integrals of differential equations in series, enable us also to reduce the summation of series to the solution of differential equations. Thus, from Proposition 4, it appears that if $u = \Sigma u_m x^m$, where the law of formation of the successive coefficients, is

$$f_0(m)\,u_m + f_1(m)\,u_{m-1} \dots f_n(m)\,u_{m-n} = 0 \text{ .... } (23),$$

the value of u will be obtained by the solution of the differential equation,

$$f_0(D)\,u + f_1(D)\,\epsilon^{\theta}u \dots + f_n(D)\,\epsilon^{n\theta}u = 0 \text{ .... } (24).$$

We suppose here $f_0(m), f_1(m) \dots f_n(m)$, &c. to be polynomials, and that the series is complete; i.e. contains all the terms which can be formed in subjection to its law expressed by (23), the first exponent being therefore a root of $f_0(m) = 0$.

When the series is incomplete, the first member of the differential equation will be the same as for the complete series, while the second member will be formed by substituting in the first member, in the place of u, the series which it represents. It is obvious that all the terms will disappear, except a few derived from that end of the series where the defect of completeness exists, so that the second member of the differential equation will be finite.

Ex. 17. Let

$$u=1-\frac{n^2}{1.2}x^2+\frac{n^2(n^2-2^2)}{1.2.3.4}x^4-\frac{n^2(n^2-2^2)(n^2-4^2)}{1.2.3.4.5.6}x^6, \text{\&c.}$$

Here $u=\Sigma u_m x^m$, with the relation,

$$u_m=-\frac{n^2-(m-2)^2}{m(m-1)}u_{m-2}.$$

Or,

$$m(m-1)u_m-\{(m-2)^2-n^2\}u_{m-2}=0,$$

and we observe that the series is complete, the first index 0 being a root of $m(m-1)=0$.

Hence, the differential equation will be

$$D(D-1)u-\{(D-2)^2-n^2\}\epsilon^{2\theta}u=0,$$

of which the solution, expressed in terms of x, is

$$u=c_1\cos(n\sin^{-1}x)+c_2\sin(n\sin^{-1}x).$$

The constants must be determined by comparison with the original series. We thus find $c_1=1$, $c_2=0$.

The following is a species of application which is of frequent use in the theory of probabilities.

Ex. 18. The series

$$p^a\{1+aq+\frac{a(a+1)}{1.2}q^2\ldots+\frac{a(a+1)\ldots(a+b-1)}{1.2\ldots b}q^b\},$$

occurs as the expression of the probability that an event whose probability of occurrence in a single trial is p, and of failure q, will occur at least a times in $a+b$ trials.

Representing the series within the brackets by u, and assuming $q=\epsilon^{\theta}$, we have $u=\Sigma u_m \epsilon^{m\theta}$, where

$$mu_m-(m+a-1)\,u_{m-1}=0.$$

Hence, we shall have

$$Du-(D+a-1)\,\epsilon^{\theta}u=-\frac{a\,(a+1)\dots(a+b)}{1\,.\,2\dots b}\,\epsilon^{(b+1)\theta},$$

or, restoring q,

$$\frac{du}{dq}-\frac{a}{1-q}u=-\frac{a\,(a+1)\dots(a+b)}{1\,.\,2\dots b}\,\frac{q^b}{1-q}.$$

Integrating which, we have

$$u=(1-q)^{-a}\left\{C-\frac{a\,(a+1)\dots(a+b)}{1\,.\,2\dots b}\int_0^q q^b\,(1-q)^{a-1}dq\right\}.$$

Now the first term of the development of this expression in ascending powers of q will be C; whence, comparing with the bracketed series, we have $C=1$. Substituting, and observing that $p=1-q$, the expression for the probability in question becomes

$$1-\frac{a\,(a+1)\dots(a+b)}{1\,.\,2\dots b}\int_0^q q^b\,(1-q)^{a-1}\,dq \quad\dots\dots\dots (a).$$

To this we may however give a more symmetrical form. For

$$\int_0^q q^b\,(1-q)^{a-1}dq=\left(\int_0^1-\int_q^1\right)q^b\,(1-q)^{a-1}dq$$

$$=\frac{\Gamma\,(b+1)\,\Gamma\,(a)}{\Gamma\,(a+b+1)}-\int_q^1 q^b\,(1-q)^{a-1}dq,$$

by a known theorem of definite integration.

Substituting in (a), and observing that

$$\frac{a\,(a+1)\dots(a+b)}{1\,.\,2\dots b}=\frac{\Gamma\,(a+b+1)}{\Gamma\,(b+1)\,\Gamma\,(a)},$$

we find.

$$\text{Probability} = \frac{\Gamma(a+b+1)}{\Gamma(a)\,\Gamma(b+1)} \int_q^1 q^b (1-q)^{a-1}\, dq,$$

or, as it may be otherwise expressed,

$$\text{Probability} = \frac{\int_q^1 q^b (1-q)^{a-1}\, dq}{\int_0^1 q^b (1-q)^{a-1}\, dq} \ldots\ldots\ldots\ldots\ldots (b).$$

The peculiar advantage of this form of expression is that, precisely in those cases in which the series becomes unmanageable from the largeness of a and b, the integrals admit, as Laplace has shewn, of a rapid approximation, (*Théorie Analytique des Probabilités*).

Ex. 19. The function $(1 - 2\nu\cos\omega + \nu^2)^{-n}$ being expanded in a series of the form $A_0 + 2(A_1 \cos\omega + A_2 \text{ con } 2\omega \ldots + \&c.)$, it is required to determine A_r.

We have

$$(1 - 2\nu\cos\omega + \nu^2)^{-n} = \{1 - \nu\epsilon^{\omega\sqrt{(-1)}}\}^{-n} \times \{1 - \nu\epsilon^{-\omega\sqrt{(-1)}}\}^{-n}.$$

Expanding each factor, and seeking the common coefficient of $\epsilon^{r\omega\sqrt{(-1)}}$ and $\epsilon^{-r\omega\sqrt{(-1)}}$ in the product, we find, putting $t = \nu^{\frac{1}{2}}$,

$$A_r = t^{\frac{r}{2}} \Sigma_{m=0}^{m=\infty} u_m t^m,$$

where generally,

$$m(m+r)\, u_m - (m+n-1)(m+n+r-1)\, u_{m-1} = 0,$$

$$\text{while } u_0 = \frac{n(n+1) \ldots (n+r-1)}{1 \,.\, 2 \ldots r}.$$

Hence the differential equation will be,

$$D(D+r)\,u - (D+n-1)(D+n+r-1)\,\epsilon^\theta u = 0,$$

or,

$$u - \frac{(D+n-1)(D+n+r-1)}{D(D+r)}\,\epsilon^\theta u = 0.$$

Now this can, by Prop. 3, be reduced to the form,

$$v - \epsilon^\theta v = V,$$

by the relations,

$$u = (D+n-1) \ldots (D+1)(D+n+r-1) \ldots (D+r+1)\,v,$$

$$V = \{(D+n-1) \ldots (D+1)(D+n+r-1) \ldots (D+r+1)\}^{-1}U.$$

In determining V from the latter equation, it suffices to introduce two arbitrary constants, one from each of the two sets of inverse operations. The final solution, in the obtaining of which the only difficulty consists in the reductions, is

$$A_r = \frac{1}{\Gamma(n)\,t^{\frac{r}{2}}}\left(\frac{d}{dt}\right)^{n-1}\frac{t^{r+n-1}}{(1-t)^n}.$$

12. When, in the series $\Sigma u_m x^m$, the coefficient u_m is a rational function of m invariable in form, the summation is most readily effected in the following manner.

Let the series be $\Sigma\phi(m)\,x^m$; then putting $x=\epsilon^\theta$,

$$u = \Sigma\phi(m)\,\epsilon^{m\theta} = \Sigma\phi(D)\,\epsilon^{m\theta}$$

$$= \phi(D)\,\Sigma\epsilon^{m\theta} \ldots\ldots\ldots\ldots\ldots\ldots \text{(25)}.$$

Hence, if the summation is from $m=0$ to $m=$ infinity, we have

$$u = \phi(D)\,\frac{1}{1-\epsilon^\theta};$$

but if the summation is from $m=a$ to $m=b$ inclusive,

$$u = \phi(D)\,\frac{\epsilon^{a\theta}-\epsilon^{(b+1)\theta}}{1-\epsilon^\theta}.$$

Ex. 20. Let $u = \dfrac{4x^3}{1.2.3} + \dfrac{5x^4}{2.3.4} + \dfrac{6x^5}{3.4.5} + \&c.$

$$\text{Here } \phi(m) = \frac{m+1}{m(m-1)(m-2)};$$

$$\therefore\ u = \frac{D+1}{D(D-1)(D-2)}\,(\epsilon^{3\theta}+\epsilon^{4\theta}+\&c.)$$

$$= \left\{\frac{1}{2}D^{-1} - 2(D-1)^{-1} + \frac{3}{2}(D-2)^{-1}\right\}\frac{\epsilon^{3\theta}}{1-\epsilon^\theta}.$$

The final result is

$$u = \frac{7x^2}{4} - \frac{x}{2} - \left(\frac{1}{2} - 2x + \frac{3}{2}x^2\right) \log(1-x).$$

Generalization of the foregoing theory.

13. As Propositions (1), (2), (3) are founded solely on the particular law of combination of the symbols D and ϵ^θ, expressed by the equation

$$f(D)\,\epsilon^{m\theta}u = \epsilon^{m\theta} f(D+m)\,u,$$

they remain true for any symbols π and ρ, whatever their interpretation, which combine according to the same formal law; viz.

$$f(\pi)\,\rho^m u = \rho^m f(\pi + m)\,u \text{ .... } (26).$$

Thus, supposing the law obeyed, the symbolical equation,

$$u + \phi(\pi)\,\rho^n u = U \text{ .... } (27),$$

can, by Prop. 3 considered in its purely formal character, be transformed into

$$v + \psi(\pi)\,\rho^n v = V \text{ .... } (28),$$

by the assumption,

$$u = P_n \frac{\phi(\pi)}{\psi(\pi)}\,v, \quad U = P_n \frac{\phi(\pi)}{\psi(\pi)}\,V.$$

The corresponding transformations flowing from Propositions (1) and (2), it is unnecessary to state.

Now the law (26) is obeyed, not alone by the pure symbols D and ϵ^θ, but by certain combinations of those symbols. Thus, if we assume

$$\pi = D - n\phi(D)\,\epsilon^\theta, \quad \rho = \phi(D)\,\epsilon^\theta \text{ .... } (29),$$

the law will still be obeyed. And the importance of the remark consists in this, that an equation which, when expressed by means of the symbols D and ϵ^θ, is not a binomial, may assume the binomial form for some other determination of π and ρ.

If in (26), we make $m=1$, we have $f(\pi)\,\rho u = \rho f(\pi+1)\,u$, which shews that ρ may be transferred from the right to the left of $f(\pi)$, if we, so to speak, add to π the constant increment 1. This then suggests the more general law,

$$f(\pi)\,\rho u = \rho f(\pi+\Delta\pi)\,u \;\ldots\ldots\ldots\ldots\ldots\ldots\ldots\; (29),$$

where $\Delta\pi$ represents any constant quantity regarded as an increment of π. In connexion with this theory, the following proposition is important.

PROP. *Supposing $f(x)$ to represent a function which admits of expansion in ascending positive and integral powers of x, it is required to develope $f(\pi+\rho)$ in ascending powers of ρ, π and ρ being symbols which combine in subjection to the law* (29).

By successive applications of (29) we have, m being a positive integer,

$$f(\pi)\,\rho^m u = \rho^m f(\pi+m\Delta\pi)\,u \;\ldots\ldots\ldots\ldots\ldots\ldots\; (30),$$

of which another form is $\rho^m f(\pi)\,u = f(\pi-m\Delta\pi)\,\rho^m u$. Again, since $f(\pi+\rho)$ is, by hypothesis, expressible in a series of the form

$$A_0 + A_1(\pi+\rho) + A_2(\pi+\rho)^2 + \&c.$$

we shall have

$$(\pi+\rho)\,f(\pi+\rho) = f(\pi+\rho)\,(\pi+\rho) \;\ldots\ldots\ldots\; (31),$$

for either member becomes, on substituting for $f(\pi+\rho)$ the above form,

$$A_0(\pi+\rho) + A_1(\pi+\rho)^2 + \&c.$$

Now, let the form of the unknown and sought expansion of $f(\pi+\rho)$ in ascending powers of ρ, be

$$f(\pi+\rho) = f_0(\pi) + f_1(\pi)\,\rho + f_2(\pi)\,\rho^2 + \&c. \;\ldots\ldots\ldots\; (32),$$

$$= \Sigma f_m(\pi)\,\rho^m,$$

the subject u being understood though not expressed.

Then, by (31),

$$(\pi+\rho)\,\Sigma f_m(\pi)\,\rho^m = \Sigma f_m(\pi)\,\rho^m\,(\pi+\rho).$$

But

$$\begin{aligned}(\pi+\rho)\,\Sigma f_m(\pi)\,\rho^m &= \Sigma\pi f_m(\pi)\,\rho^m + \Sigma\rho f_m(\pi)\,\rho^m \\ &= \Sigma\pi f_m(\pi)\,\rho^m + \Sigma f_m(\pi-\Delta\pi)\,\rho^{m+1},\end{aligned}$$

in which the coefficient of ρ^m is

$$\pi f_m(\pi) + f_{m-1}(\pi-\Delta\pi) \;\ldots\ldots\ldots\ldots\ldots\ldots\ldots\; (33).$$

Again,

$$\begin{aligned}\Sigma f_m(\pi)\,\rho^m(\pi+\rho) &= \Sigma f_m(\pi)\,\rho^m\pi + \Sigma f_m(\pi)\,\rho^{m+1} \\ &= \Sigma f_m(\pi)\,(\pi - m\Delta\pi)\,\rho^m + \Sigma f_m(\pi)\,\rho^{m+1},\end{aligned}$$

in which the aggregate coefficient of ρ^m is

$$f_m(\pi)\,(\pi - m\Delta\pi) + f_{m-1}(\pi).$$

Equating this with (33), we have

$$\pi f_m(\pi) + f_{m-1}(\pi-\Delta\pi) = (\pi - m\Delta\pi)\,f_m(\pi) + f_{m-1}(\pi),$$

whence

$$\begin{aligned}f_m(\pi) &= \frac{1}{m}\,\frac{f_{m-1}(\pi) - f_{m-1}(\pi - m\Delta\pi)}{\Delta\pi} \\ &= \frac{1}{m}\,\frac{\Delta f_{m-1}(\pi)}{\Delta\pi} \;\ldots\ldots\ldots\ldots\ldots\ldots\ldots\; (34),\end{aligned}$$

if we define $\Delta f(\pi)$, not, as is usual, by $f(\pi+\Delta\pi) - f(\pi)$, but by $f(\pi) - f(\pi-\Delta\pi)$. The above equation determines the law of derivation of the coefficients $f_1(\pi)$, $f_2(\pi)$, &c. It only remains to determine $f_0(\pi)$.

That $f_0(\pi) = f(\pi)$ may be shewn by induction from the particular cases in which

$$f(\pi+\rho) = \pi+\rho,\quad (\pi+\rho)^2, \text{ \&c.}$$

or, with more formal propriety, thus:

Let $\rho_1 = n\rho$, where n is a constant,

$$\begin{aligned}f(\pi)\,\rho_1 = f(\pi)\,n\rho &= nf(\pi)\,\rho \\ &= n\rho f(\pi-\Delta\pi) \\ &= \rho_1 f(\pi-\Delta\pi).\end{aligned}$$

Comparing the first and last members, we see that π and ρ_1 combine according to the same law as π and ρ.

Thus, we have,

$$f(\pi + \rho_1) = f_0(\pi) + f_1(\pi)\rho_1 + f_2(\pi)\rho_1^2 + \&c.$$

$f_0(\pi)$, $f_1(\pi)$, &c. being the same as in (32).

Or,

$$f(\pi + n\rho) = f_0(\pi) + f_1(\pi)n\rho + f_2(\pi)n\rho^2 + \&c.;$$

so that, making $n = 0$, we have $f_0(\pi) = f(\pi)$.

Determining then the successive coefficients by (34), we have finally,

$$f(\pi+\rho) = f(\pi) + \frac{\Delta f(\pi)}{\Delta\pi}\rho + \frac{1}{1.2}\frac{\Delta^2 f(\pi)}{(\Delta\pi)^2}\rho^2$$

$$+ \frac{1}{1.2.3}\frac{\Delta^3 f(\pi)}{(\Delta\pi)^3}\rho^3 + \&c. \ldots\ldots\ldots\ldots\ldots (35),$$

wherein it is to be remembered, that

$$\frac{\Delta f(\pi)}{\Delta\pi} = \frac{f(\pi) - f(\pi - \Delta\pi)}{\Delta\pi}.$$

When $\Delta\pi = 0$, the symbols π and ρ become commutative, and (35) assumes the form of Taylor's theorem.

As a particular application of the above, suppose that we have given the trinomial equation

$$(D^2 + aD + b)u + (cD + e)\epsilon^\theta u + f\epsilon^{2\theta}u = 0 \ldots\ldots (a),$$

and that we desire to ascertain whether this can be transformed into a binomial equation by assuming

$$\pi = D - m\epsilon^\theta, \quad \rho = \epsilon^\theta,$$

assumptions which satisfy the law

$$f(\pi)\rho = \rho f(\pi + 1).$$

Here we have $\qquad D = \pi + m\rho,$

whence $\quad f(D) = f(\pi) + \dfrac{\Delta f(\pi)}{\Delta\pi} m\rho + \dfrac{1}{2}\dfrac{\Delta^2 f(\pi)}{\Delta\pi^2} m^2\rho^2 + \&c.,$

where $\Delta\pi = 1$, and

$$\frac{\Delta f(\pi)}{\Delta\pi} = f(\pi) - f(\pi - 1).$$

Hence $D^2 + aD + b = \pi^2 + a\pi + b + (2\pi - 1 + a)\, m\rho + m^2\rho^2,$

$$cD + e = c\pi + e + cm\rho.$$

Thus (a) becomes

$$\{\pi^2 + a\pi + b + (2\pi - 1 + a)\, m\rho + m^2\rho^2\}\, u$$
$$+ (c\pi + e + cm\rho)\, \rho u + f\rho^2 u = 0,$$

or $\pi^2 + a\pi + b + \{(2m + c)\,\pi + m\,(a-1) + e\}\,\rho + (m^2 + cm + f)\,\rho^2 = 0,$

and this reduces to a binomial equation, 1st, if m be a root of the quadratic equation

$$m^2 + cm + f = 0;$$

2ndly, if it be possible to satisfy simultaneously the equations

$$2m + c = 0, \quad m\,(a - 1) + e = 0,$$

equations which imply the condition

$$2e - c\,(a - 1) = 0.$$

The discussion of the binomial equation when obtained involves no difficulty.

For a discussion of the general trinomial equation of the second degree, the reader is referred to the original Memoir.

Laplace's transformation of partial differential equations.

14. Laplace has developed a method for the reduction of the partial differential equation

$$Rr + Ss + Tt + Pp + Qq + Rz = U \ldots\ldots\ldots (36),$$

$R, S, T, \ldots U$ being functions of x and y, which is deserving of attention from its great generality.

One of the auxiliary equations in Monge's method is

$$Rdy^2 - Sdx\,dy + Tdx^2 = 0.$$

Let two integrals of this equation be

$$\phi(x, y) = c, \qquad \psi(x, y) = c',$$

and assume two new variables, ξ and η, connected with x and y by the equations

$$\xi = \phi(x, y), \qquad \eta = \psi(x, y).$$

The student will have no difficulty in proving that the given equation will assume the form

$$\frac{d^2 z}{d\xi d\eta} + L\frac{dz}{d\xi} + M\frac{dz}{d\eta} + Nz = V \ldots\ldots\ldots\ldots (37),$$

L, M, N, V being functions of ξ and η. The theory of the reduction of this equation is then contained in the following propositions:

1st, The equation (37) may be presented in the form

$$\left(\frac{d}{d\xi} + M\right)\left(\frac{d}{d\eta} + L\right)z + \left(N - LM - \frac{dL}{d\xi}\right)z = V \ldots (38).$$

Hence, if the condition

$$N - LM - \frac{dL}{d\xi} = 0 \ldots\ldots\ldots\ldots\ldots\ldots\ldots (39)$$

be satisfied, and we assume $\left(\frac{d}{d\eta} + L\right)z = z'$, we shall have

$$\left(\frac{d}{d\xi} + M\right)z' = V.$$

The solution of the given equation is then dependent on that of two partial differential equations of the first order.

2ndly, Inverting the order of the symbolic factors, the equation is also solvable if we have

$$N - LM - \frac{dM}{d\eta} = 0 \ldots\ldots\ldots\ldots\ldots\ldots\ldots\ldots (40).$$

3rdly, The equation (37) can be transformed into a series of other equations of the same form, and therefore integrated, if, for any of those equations, the condition (39) or (40) is satisfied.

For, expressing it in the form (38), let, as before,

$$\left(\frac{d}{d\eta}+L\right)z=z' \qquad\qquad (41).$$

Then $$\left(\frac{d}{d\xi}+M\right)z'+\left(N-LM-\frac{dL}{d\xi}\right)z=V,$$

whence $$z=\frac{-\dfrac{dz'}{d\xi}-Mz'+V}{N-LM-\dfrac{dL}{d\xi}},$$

which is of the form

$$z=A\frac{dz'}{d\xi}+Bz'+C,$$

A, B, C being functions of ξ and η. Substituting this expression for z in (41), we have a result of the form

$$\frac{d^2z'}{d\xi d\eta}+L'\frac{dz'}{d\xi}+M'\frac{dz'}{d\eta}+N'z'=V' \qquad (42).$$

Thus the from (37) is reproduced, but with changed coefficients. Hence the equation is integrable if either of the following conditions is satisfied, viz.

$$N'-L'M'-\frac{dL'}{d\xi}=0,\quad N'-L'M'-\frac{dM'}{d\eta}=0 \qquad (43).$$

If neither be satisfied, the process of transformation may be indefinitely repeated, and should an equation be obtained in which either of the relations (43) is satisfied, the solution may be found. It has indeed been asserted that "if the given equation be integrable, we shall finally get an equation in which this essential condition is satisfied" (Peacock's *Examples*, p. 464). The state of our knowledge of the conditions of finite integration does not however warrant this confidence.

A discussion of the equation

$$a\frac{d^2z}{dx^2}+b\frac{d^2z}{dxdy}+c\frac{d^2z}{dy^2}+\frac{e\dfrac{dz}{dx}+f\dfrac{dz}{dy}}{hx+ky}+\frac{gz}{(hx+ky)^2}=M \qquad (a)$$

by Laplace's method is given in Lacroix (Tom. II. pp. 611—614), but it is far too long and too complex to find a place here. The best mode of treating the equation is probably the following. Let s and t be two new variables connected with x and y by the *linear* relations

$$hx + ky = s, \quad y + mx = t,$$

of which one is suggested by the form of the given equation, while the other is adopted in order to put us in possession of a disposable constant m. Transforming, and making in the result $s = \epsilon^\theta$, we obtain the symbolical equation

$$\{AD(D-1)+ED+g\}z+\frac{d}{dt}\{B(D-1)+F\}\epsilon^\theta z+C\frac{d^2}{dt^2}\epsilon^{2\theta}z=\epsilon^{2\theta}M\ldots(b),$$

in which

$$A = ah^2 + bhk + ck^2, \quad B = 2ahm + b\,(h + km) + 2ck,$$

$$C = am^2 + bm + c, \quad E = eh + fk, \quad F = em + f.$$

The equation will be a binomial one, if m be determined so as to make $C = 0$. We have then

$$am^2 + bm + c = 0,$$

while the symbolical equation (b) becomes

$$z + \frac{d}{dt}\,\frac{B\,(D-1) + F}{AD\,(D-1) + ED + g}\,\epsilon^\theta z = \{AD(D-1) + ED + g\}^{-1}\epsilon^{2\theta}M,$$

and is integrable if the following condition is satisfied, viz.

$$\frac{F-B}{B} - \frac{A-E \pm \sqrt{\{(A-E)^2 - 4g\}}}{2A} = \text{an integer or } 0.$$

This condition will be found to include the one to which Laplace's method leads.

At the same time it is seen that the equation (b) assumes the binomial form under other conditions than the above; e.g. if we have simultaneously

$$B = 0, \quad F = 0,$$

from which, by elimination of m, we find

$$f\,(2ah + bk) - e\,(bh + 2ck) = 0.$$

This condition being satisfied, and m determined, the symbolical equation becomes

$$z + C\frac{d^2}{dt^2}\,\frac{1}{AD(D-1)+ED+g}\,\epsilon^{2\theta}z = \{AD(D-1)+ED+g\}^{-1}\epsilon^{2\theta}M,$$

and is integrable if the two roots of the equation

$$Am\,(m-1)+Em+g=0$$

differ by an odd integer. There are probably other cases dependent on the reduction of Art. (13).

In one respect Laplace's transformation possesses a generality superior to that of all others. For its tentative application fewer restrictions on the coefficients of the given equation are necessary. But, that the application may succeed, other conditions seem to be demanded which render the estimation of the true measure of its generality difficult. And, in particular instances, it is seen that it is less general than the method of the foregoing sections.

Miscellaneous Notices.

15. Of special additions to the theory of the solution of differential equations by symbolical methods, the following may be noticed.

1st, Professor Donkin has shewn that, if $f(x)$ be any function capable of development in powers of x, then whatever may be the interpretations of the symbols π and ρ, we have

$$f(\rho^{-1}\pi\rho)\,u = \rho^{-1}f(\pi)\,\rho u \;\ldots\ldots\ldots\ldots\ldots\ldots\; (44).$$

This is evident from the consideration of such cases as the following:

$$(\rho^{-1}\pi\rho)^2 = \rho^{-1}\pi\rho\rho^{-1}\pi\rho = \rho^{-1}\pi^2\rho,$$
$$(\rho^{-1}\pi\rho)^{-1} = \rho^{-1}\pi^{-1}(\rho^{-1})^{-1} = \rho^{-1}\pi^{-1}\rho.$$

We are thus enabled to generalize many important theorems.

Thus, since $\left\{\frac{d}{dx}+\phi'(x)\right\}u = \epsilon^{-\phi(x)}\frac{d}{dx}\,\epsilon^{\phi(x)}u$, we have

$$f\left\{\frac{d}{dx}+\phi'(x)\right\}u = \epsilon^{-\phi(x)}f\left(\frac{d}{dx}\right)\epsilon^{\phi(x)}u \;\ldots\;\ldots\ldots\; (45),$$

(*Cambridge Mathematical Journal*, 2nd Series, Vol. v. p. 10.)

2ndly, Mr Hargreave, observing that the symbols $\frac{d}{dx}$ and $-x$ are connected by the same laws as x and $\frac{d}{dx}$, (the proof of this will afford an exercise for the student), has remarked that if in any differential equation and its *symbolic* solution we change x into $\frac{d}{dx}$, and $\frac{d}{dx}$ into $-x$, we shall obtain another form accompanied by its symbolic solution. (*Philosophical Transactions for* 1848, Part I.)

Applying this law of duality to the known solution of the linear differential equation of the first order, it is easy to shew that the equation

$$x\phi(D)\,u + \psi(D)\,u = X$$

has for its symbolic solution,

$$u = \{\phi(D)\}^{-1}\,\epsilon^{\chi(D)}\,x^{-1}\,\epsilon^{-\chi(D)}\,X \;\ldots\ldots\ldots\ldots\ldots\; (46),$$

where $$\chi(D) = \int \frac{\psi(D)}{\phi(D)}\,dD,$$

a form which had before been established on other grounds, (*Philosophical Magazine*, Feb. 1847). Many other illustrations of the same law will be found in the memoir of Mr Hargreave referred to.

3rdly, The method by which the development of $f(\pi+\rho)$ is obtained in Art. 13, leads to other and similar results, of which the following is among the most interesting, viz.

$$f\left(x+\frac{d}{dx}\right)u = \left\{F(x) + F'(x)\frac{d}{dx} + \frac{1}{1.2}F''(x)\frac{d^2}{dx^2} + \&\text{c.}\right\}u \,..\, (47),$$

the coefficients of the expansion in the second member following the law of Taylor's theorem, and the function $F(x)$ being equal to $\epsilon^{\frac{1}{2}\left(\frac{d}{dx}\right)^2} f(x)$. (*Cambridge Mathematical Journal*, 1st Series, Vol. IV. p. 214.)

The last theorem enables us to integrate at once any equation of the form,

$$F(x)\,u + F'(x)\frac{du}{dx} + \frac{F''(x)}{1.2}\frac{d^2u}{dx^2} + \&\text{c.} = X.$$

where $F(x)$ is a rational and integral function of x. For let

$$f(x) = \epsilon^{-\frac{1}{2}\frac{d^2}{dx^2}} F(x) = (1 - \frac{1}{2}\frac{d^2}{dx^2} + \frac{1}{1.2}\frac{1}{4}\frac{d^4}{dx^4} - \&c.)\, F(x),$$

an expression always finite under the conditions supposed. Then the given equation assumes the form

$$f(\pi)\, u = X,$$

where $\pi = x + \dfrac{d}{dx}$, and may be treated by the method of the last section.

Other examples of the expansion of functions whose symbols are non-commutative—some of them admitting of a similar application—will be found in the memoir of Professor Donkin above referred to, and in an interesting memoir by Mr Bronwin, (*Cambridge Mathematical Journal*, Vol. III. p. 36).

4thly, Many important partial differential equations of the second order admit of reduction to the form

$$\frac{du}{dx}\frac{dv}{dy} - \frac{du}{dy}\frac{dv}{dx} = 0,$$

whence an integral $u = f(v)$ may be deduced. Thus the equation

$$\left(\frac{d\phi}{dp}\frac{d\psi}{dq} - \frac{d\phi}{dq}\frac{d\psi}{dp}\right)(rt - s^2) - \frac{d\phi}{dp}r - \left(\frac{d\phi}{dq} + \frac{d\psi}{dp}\right)s - \frac{d\psi}{dq}t + 1 = 0,$$

where ϕ and ψ represent any given functions of p and q, may be expressed in the form

$$\frac{d(\phi - x)}{dx}\frac{d(\psi - y)}{dy} - \frac{d(\phi - x)}{dy}\frac{d(\psi - y)}{dx} = 0;$$

whence $\phi - x = F(\psi - y)$ is a first integral. Mainardi has shewn that nearly all the equations which occur in Monge's *Application de l'Analyse à la Géométrie*, admit either of the above reduction, or of a purely symbolical mode of solution. (*Tortolini*, Vol. v. p. 161).

5thly, The Author is indebted to Mr Spottiswoode of Oxford for an interesting communication on the laws of combination of symbols which are at the same time linear with respect to $\frac{d}{dx}$, $\frac{d}{dy}$, &c. and linear with respect to x, y, &c. The following is one of the results. If, assuming

$$\pi_1 = x\frac{d}{dx} + y\frac{d}{dy}, \quad \pi_2 = y\frac{d}{dx} + x\frac{d}{dy},$$

a partial differential equation can be presented in the form

$$f(\pi_1, \pi_2)\, u = 0$$

on the assumption that $\frac{d}{dx}$ and $\frac{d}{dy}$ operate only on the subject u, then it can be expressed in the form $F(\pi_1, \pi_2)\, u = 0$, independently of such restrictive hypothesis. It might be added, that all such equations are reducible to equations with constant coefficients, by assuming

$$\log (x^2 + y^2)^{\frac{1}{2}} = x', \ \log \left(\frac{x+y}{x-y}\right)^{\frac{1}{2}} = y'.$$

To the above might be added many other special deductions, isolated now, but destined perhaps, at some future time, to be embraced in the unity of a larger theory.

EXERCISES.

1. Integrate $x^2\frac{d^2u}{dx^2} + 4x\frac{du}{dx} - -q^2x^2u = 0.$

2. Integrate $(x^2 - x^3)\frac{d^2u}{dx^2} - (x + 3x^2)\frac{du}{dx} + (1 - x)\, u = 0.$

3. Riccati's equation is reducible to the form

$$\frac{d^2w}{dx^2} + bcx^m w = 0.$$

Hence investigate the conditions of integrability.

The symbolical form is $w + \frac{bc}{D(D-1)}\epsilon^{(m+2)\theta} w = 0$; and this may either be reduced directly by Prop. 3 to a form integrable by Prop. 1, or, by assuming $(m+2)\,\theta = 2\theta'$, converted into a particular case of Ex. 8 in the Chapter.

4. The equation $\frac{d^2u}{dx^2}+\frac{a}{x}\frac{du}{dx}+bu=0$ is integrable in finite terms if a is an even number.

5. The equation $\frac{d^2u}{dx^2}+\frac{2r}{a}\frac{du}{dx}=bx^m u$ is integrable in finite terms if $m=-\frac{4(i\pm r)}{2i\pm 1}$, where i is a positive whole number or 0.

6. The more general equation

$$\frac{d^2u}{dx^2}+\frac{r}{a}\frac{du}{dx}=\left(bx^m+\frac{c}{x^2}\right)u,$$

which includes the above, is integrable in finite terms if

$$m+2=\frac{2\sqrt{\{(1-r)^2+4c\}}}{2i+1},$$

i being a positive whole number or 0. (Malmsten, *Cambridge Mathematical Journal*, 2nd Series, Vol. v. p. 180). Verify this.

7. As an illustration of the theory of disappearing factors, integrate the equation

$$(x^2+qx^3)\frac{d^2u}{dx^2}+\{(a+3)qx^2+(b-i+1)x\}\frac{du}{dx}$$
$$+\{(a+1)qx-bi\}u=0.$$

8. The equation $(1-ax^2)\frac{d^2y}{dx^2}-bx\frac{dy}{dx}-cy=0$, is integrable in finite terms in the following three cases; viz.

1st, If $\frac{b}{a}$ is an odd integer;

2ndly, If $\sqrt{\left\{\left(1-\frac{b}{a}\right)^2+\frac{4c}{a}\right\}}$ is an odd integer;

3rdly, If $\frac{b}{a}+\sqrt{\left\{\left(1-\frac{b}{a}\right)^2+\frac{4c}{a}\right\}}$, or $\frac{b}{a}-\sqrt{\left\{\left(1-\frac{b}{a}\right)^2+\frac{4c}{a}\right\}}$, is an even integer.

9. Integrate the partial differential equation,

$$\frac{d^2z}{dx^2}-\frac{d^2z}{dy^2}-\frac{2}{x}\frac{dz}{dx}=0.$$

10. The partial differential equation,

$$\frac{d^2z}{dx^2}-a^2y^{2b}\frac{d^2z}{dy^2}=0,$$

is integrable in finite terms if $b=\dfrac{2i}{2i\pm 1}$. (*Legendre.* See *Lacroix*, Tom. II. p. 618.) Verify this.

11. Shew that the sum of the series $1.2\ldots nx+2.3\ldots(n+1)x^2\ldots+p(p+1)\ldots(p+n-1)x^p$ may be expressed in the form

$$x\left(\frac{d}{dx}\right)^n\frac{x^n-x^{p+n}}{1-x}.$$

12. Sum the series

$$1+\frac{1^nx}{1}+\frac{2^nx^2}{1.2}+\frac{3^nx^3}{1.2.3}+\&c.$$

13. The equation $(a+bx)\dfrac{d^2u}{dx^2}+(f+gx)\dfrac{du}{dx}+ngu=0$, is integrable in finite terms if n is an integer.

Apply the method of Art. 13 to reduce the symbolical equation to a binomial form. Or assume $a+bx=t$.

14. The differential equation,

$$\frac{d^2u}{dx^2}+2Q\frac{du}{dx}+\left\{Q^2+\frac{dQ}{dx}\pm c^2-\frac{m(m+1)}{x^2}\right\}u=0,$$

can be integrated in finite terms, whatever function of x is represented by Q. (Curtis, *Cambridge Mathematical Journal*, Vol. IX. p. 280).

The equation may be expressed in the form

$$\left(\frac{d}{dx}+Q\right)^2 u \pm c^2 u - \frac{m(m+1)}{x^2}u = 0,$$

$$\text{or } \epsilon^{-\int Q dx}\left(\frac{d}{dx}\right)^2 \epsilon^{\int Q dx} u \pm \left\{c^2 \mp \frac{m(m+1)}{x^2}\right\} u = 0,$$

$$\text{or } \left(\frac{d}{dx}\right)^2 \epsilon^{\int Q dx} u \pm \left\{c^2 \mp \frac{m(m+1)}{x^2}\right\} \epsilon^{\int Q dx} u = 0.$$

Let $\epsilon^{\int Q dx} u = v$; then compare the resulting form with Ex. 8 of the Chapter.

15. Shew generally that, if we can integrate the equation

$$f\left(\frac{d}{dx}\right) u + \phi(x) u = X,$$

we can integrate $f\left(\frac{d}{dx}+Q\right) u + \phi(x) u = X.$

16. We meet the equation

$$\frac{d^2y}{dc^2} + \frac{1-3c^2}{c-c^3}\frac{dy}{dc} - \frac{1}{1-c^2}y = 0,$$

in the theory of the elliptic functions, (Legendre's modular equation). Shew that it is not integrable in finite terms, but is integrable in the form $y = A + B \log c$, where A and B are series expressed in ascending even powers of c.

17. Prove the following generalization of Prop. 3,

$$F\{\phi(D)\epsilon^{r\theta}\} = P_r \frac{\phi(D)}{\psi(D)} F\{\psi(D)\epsilon^{r\theta}\} P_r \frac{\psi(D)}{\phi(D)}.$$

18. Prove the following still more general theorem,

$$F\{D, \phi(D)\epsilon^{r\theta}\} = P_r \frac{\phi(D)}{\psi(D)} F\{D, \psi(D)\epsilon^{r\theta}\} P_r \frac{\psi(D)}{\phi(D)}.$$

CHAPTER XVIII.

SOLUTION OF LINEAR DIFFERENTIAL EQUATIONS BY DEFINITE INTEGRALS.

1. THE solution of linear differential equations by definite integrals was first made a direct object of inquiry by Euler. His method consisted in assuming the form of the definite integral, and then, from its properties, determining the class of equations whose solution it was fitted to express. Laplace first devised a method of ascending from the differential equation to the definite integral. And Laplace's is still the most general method of procedure known. Its application is however not wholly free from difficulties, due partly to the present imperfection of the theory of definite integrals, partly to an occasional failure of correspondence in the conditions upon which continuity of form in the differential equation and continuity of form in its solution depend. Indeed it ought never to be employed without some means of testing the result *a posteriori*, e.g. by comparison with the solution of the proposed differential equation in series. Frequently indeed it is possible to deduce the solution in definite integrals from the solution in series without employing Laplace's method at all.

Laplace's method is applied with peculiar advantage to equations in the coefficients of which x enters only in the first degree, and of which the second member is 0. Expressing any such equation in the form,

$$x\phi\left(\frac{d}{dx}\right)u+\psi\left(\frac{d}{dx}\right)u=0 \quad\text{.................. (1)},$$

we must assume

$$u=\int\epsilon^{xt}\,T\,dt,$$

T being a function of t, the form of which, together with the limits of integration, must be determined by substituting the expression for u in the proposed differential equation. Effecting this substitution, we have a result which may be thus expressed,

$$\int x\phi\left(\frac{d}{dx}\right)\epsilon^{xt}Tdt + \int\psi\left(\frac{d}{dx}\right)\epsilon^{xt}Tdt = 0,$$

or, since

$$\phi\left(\frac{d}{dx}\right)\epsilon^{xt} = \phi(t)\,\epsilon^{xt},$$

$$\int x\epsilon^{xt}\phi(t)\,Tdt + \int\epsilon^{xt}\psi(t)\,Tdt = 0 \;\ldots\ldots\ldots\ldots\ldots\; (2).$$

Of this however, the first term is, by integration by parts, reducible to the form

$$\epsilon^{xt}\phi(t)\,T - \int\epsilon^{xt}\frac{d}{dt}\{\phi(t)\,T\}\,dt.$$

Thus, (2) assumes the form

$$\epsilon^{xt}\phi(t)\,T - \int\epsilon^{xt}\left\{\frac{d}{dt}[\phi(t)\,T] - \psi(t)\,T\right\}dt = 0 \;\ldots\ldots\; (3),$$

and will therefore be satisfied, if we make

$$\epsilon^{xt}\,\phi(t)\,T = 0,$$

$$\frac{d}{dt}\{\phi(t)\,T\} - \psi(t)\,T = 0.$$

The former of these equations has reference only to the limits; the latter, expressed in the form

$$\frac{d}{dt}\{\phi(t)\,T\} - \frac{\psi(t)}{\phi(t)}\{\phi(t)\,T\} = 0,$$

gives on integration,

$$\phi(t)\,T = C\epsilon^{\int\frac{\psi(t)}{\phi(t)}dt},$$

and determines T in the form,

$$T = \frac{C\epsilon^{\int\frac{\psi(t)}{\phi(t)}dt}}{\phi(t)}.$$

Thus, we have

$$u = C\int \frac{\epsilon^{xt+\int\frac{\psi(t)}{\phi(t)}dt}}{\phi(t)}\,dt \;\ldots\ldots\ldots\ldots\ldots\ldots\ldots (4),$$

the limits of integration being determined by the equation

$$\epsilon^{xt+\int\frac{\psi(t)}{\phi(t)}dt} = 0 \;\ldots\ldots\ldots\ldots\ldots\ldots\ldots\ldots (5).$$

Should this equation have n distinct roots, these may evidently be so disposed as to give $n-1$ distinct particular integrals.

Such is the general statement of Laplace's method. Applied to an equation in the coefficients of which the highest power of x involved is the nth, it would make the determination of T depend on the solution of a differential equation of the nth order. Other practical limitations may be noted. For instance, the method is only directly applicable to the expression of integrals which produce on development series of a certain form. Thus, if we develope the exponential in the assumed expression for u, we have

$$u = \int T dt + x\int Tt dt + \frac{x^2}{1\,.\,2}\int Tt^2 dt + \&c.$$

an expansion in which positive and integral powers of x alone present themselves. Integrals of different forms may, however, by preparation of the differential equation, be brought under the dominion of the method. These and other points we propose to illustrate by the detailed examination of a special but very important example, particular forms of which are of very frequent occurrence in physical inquiries. We shall first, in accordance with what has above been said, determine the different kinds of solution in series of which the equation admits. This part of the investigation is intended to be supplementary to Art. 9 of the last Chapter.

Ex. Given $x\dfrac{d^2u}{dx^2} + a\dfrac{du}{dx} - q^2xu = 0.$

Solutions expressed in Series.

2. The symbolical form of the above equation is

$$u - \frac{q^2}{D(D+a-1)} \epsilon^{2\theta} u = 0 \text{ .... (6).}$$

Hence, if an integral be expressible in the form $\Sigma u_m x^m$, the law of formation of the coefficients u_m will be

$$u_m = \frac{q^2 u_{m-2}}{m(m+a-1)} \text{ .... (7),}$$

while the lowest value of m will be 0, or $1-a$. Thus, except in a particular case to be noticed hereafter, the complete integral will be

$$u = A\{1 + \frac{q^2x^2}{2(a+1)} + \frac{q^4x^4}{2.4(a+1)(a+3)} + \&c.\}$$

$$+ Bx^{1-a}\{1 + \frac{q^2x^2}{2.(3-a)} + \frac{q^4x^4}{2.4(3-a)(5-a)} + \&c.\} \text{ ... (8),}$$

The two series in the general value of u are evidently convergent for all values of x. As this question of the convergency of series is sometimes important in connexion with the solution of differential equations, the reader is reminded that according as, in the series of terms or groups of terms

$$u_0 + u_1 + u_2 + \&c.,$$

the ratio $\frac{u_n}{u_{n-1}}$ tends, when n is indefinitely increased, to a limit less or greater than unity, the series is convergent or divergent; but that, when the ratio tends to unity, we must apply a system of criteria developed by Professor De Morgan, (*Differential and Integral Calculus*, p. 325*).

* That this system virtually includes previous special results has been proved by Bertrand (*Liouville*, Tom. VII. p. 35); that it is a legitimate development of the fundamental principles of Cauchy has been established by Paucker (*Crelle*, *Band.* XLII. p. 138).

When a is an odd integer, the general integral will involve a logarithm. In particular if $a = 1$, we shall have

$$u = a_0 + a_2 x^2 + a_4 x^4 + \&c. + \log x\,(b_0 + b_2 x^2 + b_4 x^4 + \&c.) \ldots (9),$$

a_0 and b_0 being arbitrary constants, and the succeeding coefficients determined by

$$m^2 a_m + 2mb_m - q^2 a_{m-2} = 0, \quad m^2 b_m - q^2 b_{m-2} = 0 \ldots\ldots\ldots (10).$$

The symbolical equation (6), indicates by its form that there are no solutions expressible in descending powers of x, and infinite in one direction only—i. e. beginning with some finite exponent, and presenting a series of exponents thence descending. But the equation may be transformed so as to admit of a solution of this kind. For, assuming $u = \epsilon^{-qx} v$, we shall have

$$x \frac{d^2 v}{dx^2} + (a - 2qx) \frac{dv}{dx} - aqv = 0,$$

and of this the symbolical form will be found to be

$$D(D + a - 1)\,v - 2q\left(D + \frac{a}{2} - 1\right) \epsilon^\theta v = 0 \ldots\ldots\ldots (11);$$

whence, if v be developed in a series of the form $\Sigma v_m x^m$, the law of derivation of the coefficients will be

$$m(m + a - 1)\,v_m - 2q\left(m + \frac{a}{2} - 1\right) v_{m-1} = 0.$$

It follows from this that there will be two ascending and convergent series for v, and one descending and divergent series. The law of the latter series is by changing m into $m + 1$, more conveniently expressed in the form,

$$v_m = \frac{(m+1)(m+a)}{2q\left(m + \frac{a}{2}\right)} v_{m+1}.$$

Hence, the first exponent will be $-\frac{a}{2}$, and the ultimate value of u will be

$$u = C\epsilon^{-qx}x^{-\frac{a}{2}}\left\{1 + \frac{\frac{a}{2}\left(\frac{a}{2}-1\right)}{1.2qx} + \frac{\frac{a}{2}\left(\frac{a}{2}+1\right)\left(\frac{a}{2}-1\right)\left(\frac{a}{2}-2\right)}{1.2.4q^2x^2} + \&c.\right\} \quad (12).$$

If we assume $u = \epsilon^{qx}v$, and proceed as above, we shall obtain for v the symbolical equation,

$$D(D+a-1)v + 2q\left(D+\frac{a}{2}-1\right)\epsilon^{\theta}v = 0 \quad \text{......... (13)},$$

and as this differs from the previous equation for v, only by a change of sign affecting q, we at once deduce a second value of u, in the form

$$u = E\epsilon^{qx}x^{-\frac{a}{2}}\left\{1 - \frac{\frac{a}{2}\left(\frac{a}{2}-1\right)}{1.2\,qx} + \frac{\frac{a}{2}\left(\frac{a}{2}+1\right)\left(\frac{a}{2}-1\right)\left(\frac{a}{2}-2\right)}{1.2.4\,q^2x^2} - \&c.\right\} \quad (14),$$

the terms within the brackets being alternately positive and negative.

Both the descending series are finite when a is an even integer, and though for all other values of a they are infinite and ultimately divergent, yet if x be large they begin with being convergent, and may under certain circumstances be employed for numerical calculation.

Thus, we have obtained two solutions expressed in ascending series always convergent, and two solutions involving series expressed in descending powers of x, and ultimately divergent.

As concerns the convergent series for v, derivable from the transformed equations (11) and (13), we may remark that when multiplied by the developed exponentials, they will only reproduce the convergent series for u already obtained in (8).

One observation yet remains. We have seen that each of the assumptions $u = \epsilon^{qx}v$ and $u = \epsilon^{-qx}v$, transforms the proposed differential equation into another of which the solution in a descending series is finite when the given equation admits of finite integration. This species of transformation is frequently possible. To accomplish it we must assume $u = Qv$, the form of Q being determined by the solution of that differential equation upon which, by Props. 2 and 3, Chap. XVII., the solution of the proposed equation, when possible in finite terms, is dependent.

Solution of the Equation by Definite Integrals.

3. Comparing the proposed equation,

$$x\frac{d^2u}{dx^2} + a\frac{du}{dx} - q^2xu = 0 \quad\text{.................}\quad (15),$$

with the general form (1), we have

$$\phi\left(\frac{d}{dx}\right) = \frac{d^2}{dx^2} - q^2, \quad \psi\left(\frac{d}{dx}\right) = a\frac{d}{dx}.$$

Hence,

$$\phi(t) = t^2 - q^2, \quad \psi(t) = at;$$

$$\therefore \int\frac{\psi(t)\,dt}{\phi(t)} = \frac{a}{2}\log(t^2 - q^2).$$

Substituting these values in (4), we have

$$u = C\int\epsilon^{xt}(t^2 - q^2)^{\frac{a}{2}-1}dt \quad\text{..................}\quad (16),$$

while for the limits of integration, (5) gives

$$\epsilon^{xt}(t^2 - q^2)^{\frac{a}{2}} = 0.$$

Hence, *supposing a positive*, and confining our attention for the present to the factor $(t^2 - q^2)^{\frac{a}{2}}$, which alone determines t in perfect independence of x, we find $t = \pm q$. Thus,

$$u = C\int_{-q}^{q}\epsilon^{xt}(t^2 - q^2)^{\frac{a}{2}-1}dt,$$

Assuming then $t = q\cos\theta$, and changing the sign of the arbitrary constant,

$$u = C\int_0^\pi \epsilon^{qx\cos\theta}(\sin\theta)^{a-1}\,d\theta \text{ .... } (17),$$

and this, as its form suggests, and as we shall hereafter shew, is an expression for the particular integral represented by the first convergent series in the general value of u, given in (8).

To deduce another integral, let us in the symbolical equation (6), assume $u = \epsilon^{(1-a)\theta}v$. We find

$$v - \frac{q^2}{(D+1-a)D}\epsilon^{2\theta}v = 0 \text{ .... } (18).$$

Hence, a value of v may be determined from that of u by changing $a-1$ into $1-a$; i.e. by changing a into $2-a$. Thus we have, for the second particular integral,

$$u = C_2 x^{1-a}\int_0^\pi \epsilon^{qx\cos\theta}(\sin\theta\}^{1-a}\,d\theta,$$

provided that $2-a$ *be positive.*

Hence, *if* a *lie between* 0 *and* 2, we have for the complete integral,

$$u = C_1\int_0^\pi \epsilon^{qx\cos\theta}(\sin\theta)^{a-1}d\theta + C_2x^{1-a}\int_0^\pi \epsilon^{qx\cos\theta}(\sin\theta)^{1-a}d\theta \ldots (19).$$

If $a=1$, the two particular integrals in the above expression merge into one. To deduce the true form of the general integral, we may proceed thus,

$$u = \int_0^\pi \epsilon^{qx\cos\theta}\{C_1(\sin\theta)^{a-1} + C_2(x\sin\theta)^{1-a}\}\,d\theta,$$

$$= \int_0^\pi \epsilon^{qx\cos\theta}\left\{A(\sin\theta)^{a-1} + B\frac{(\sin\theta)^{a-1} - (x\sin\theta)^{1-a}}{a-1}\right\}d\theta,$$

on replacing C_1 and C_2 by two new arbitrary constants, A and B.

Now when $a=1$, we find by the usual mode of treating vanishing fractions,

$$\frac{(\sin\theta)^{a-1}-(x\sin\theta)^{1-a}}{a-1}=\log\{x(\sin\theta)^2\}.$$

Thus,

$$u=\int_0^\pi \epsilon^{qx\cos\theta}\left[A+B\log\{x(\sin\theta)^2\}\right]d\theta \;\ldots\ldots\ldots\; (20).$$

This is the complete integral of the equation,

$$x\frac{d^2u}{dx^2}+\frac{du}{dx}-q^2xu=0 \;\ldots\ldots\ldots\ldots\ldots\ldots\; (21),$$

and a similar form exists for all cases in which a is an odd integer.

4. We proceed to the cases in which a is fractional and does not lie between the limits 0 and 2. By the application of Props. 2 and 3, Chap. XVII., this case can be reduced to the case in which a does lie between the limits 0 and 2. First, suppose a negative; then we may assume $a=a'-2n$, where a' lies between 0 and 2, and n is a positive integer. In this case, the first term of (19) will need transformation. Now the symbolical equation (6), becomes

$$u-\frac{q^2}{D(D+a'-2n-1)}\epsilon^{2\theta}u=0.$$

Hence, if we assume

$$v-\frac{q^2}{D(D+a'-1)}\epsilon^{2\theta}v=0,$$

we shall have

$$u=(D+a'-1)(D+a'-3)\ldots(D+a'-2n+1)\,v$$

$$=\left(x\frac{d}{dx}+a'-1\right)\left(x\frac{d}{dx}+a'-3\right)\ldots\left(x\frac{d}{dx}+a'-2n+1\right)v \;\ldots\; (22),$$

in which

$$v=C_1\int_0^\pi \epsilon^{qx\cos\theta}(\sin\theta)^{a'-1}\,d\theta \;\ldots\ldots\ldots\ldots\ldots\; (23).$$

And this particular expression for u must replace the first term in the general value of u given in (19). The differentiations may obviously be performed under the integral sign.

As a particular illustration suppose n to lie between 0 and -2, then $n=1$, $a=a'-2$, whence

$$x\frac{d}{dx}+a'-1=x\frac{d}{dx}+a+1.$$

The particular value of u which must replace the first term in the general value (19), will therefore be

$$u=C_1\int_0^\pi\left(x\frac{d}{dx}+a+1\right)\epsilon^{qx\cos\theta}(\sin\theta)^{a+1}\,d\theta.$$

Effecting the differentiations, and substituting in (19), we have, for the general value of u,

$$u=C_1\int_0^\pi\epsilon^{qx\cos\theta}(qx\cos\theta+a+1)(\sin\theta)^{a+1}\,d\theta,$$

$$+\,C_2x^{1-a}\int_0^\pi\epsilon^{qx\cos\theta}(\sin\theta)^{1-a}\,d\theta.$$

Secondly, when $a>2$, the assumption

$$u=\epsilon^{(1-a)\theta}v,\quad \text{i.e. } u=x^{1-a}v,$$

in effect converts a into $2-a$. Compare (6) and (18). In effect, therefore, it converts a into a negative quantity, and reduces the present case to the preceding one.

It remains only to notice that when a is an even integer, the complete integral is expressible in finite terms. Chap. XVII. Art. 5.

Collecting these results together, we see that, according as a is an even integer, a fraction, or an odd integer, the complete integral is expressible in finite terms, or by definite integrals producing on development two algebraic series, or by definite integrals producing on development two series, one of which is multiplied by the factor $\log x$. We propose before going farther to verify these results.

Verification.

5. If in the solution (19), we develope the exponentials, and for brevity write

$$\int_0^\pi (\cos\theta)^m (\sin\theta)^{a-1}\, d\theta = A_m, \int_0^\pi (\cos\theta)^m (\sin\theta)^{1-a}\, d\theta = B_m \ldots (24),$$

we shall have

$$u = C_1 \Sigma \frac{A_m}{1\,.\,2\ldots m} q^m x^m + C_2 x^{1-a} \Sigma \frac{B_m}{1\,.\,2\ldots m} q^m x^m \ldots\ldots (25),$$

the summation denoted by Σ extending to all positive integral values of m, from $m = 0$ to $m = \infty$. Thus the general value of u is expressed by two series, whose equivalence to the series given in (8), it remains to establish.

Now, when m is odd, $A_m = 0$, $B_m = 0$, the positive and negative elements in each integral mutually destroying each other. Again, by a known formula of reduction,

$$\int (\cos\theta)^m (\sin\theta)^n d\theta = \frac{(\cos\theta)^{m-1} (\sin\theta)^{n+1}}{m+n}$$

$$+ \frac{m-1}{m+n} \int (\cos\theta)^{m-2} (\sin\theta)^n\, d\theta.$$

Supposing the limits 0 and π, the term free from the sign of integration vanishes at each limit when n is positive, and we have, changing n successively into $a-1$ and $1-a$,

$$A_m = \frac{m-1}{m+a-1} A_{m-2}, \quad B_m = \frac{m-1}{m+a-1} B_{m-2} \ldots\ldots (26).$$

Now let the coefficient of x^m in the first series in (25), be represented by u_m, then

$$u_m = C_1 \frac{A_m q^m}{1\,.\,2\ldots m}, \quad u_{m-2} = C_1 \frac{A_{m-2} q^{m-2}}{1\,.\,2\ldots m-2};$$

$$\therefore \frac{u_m}{u_{m-2}} = \frac{q^2 A_m}{m(m-1) A_{m-2}} = \frac{q^2}{m(m+a-1)} \text{ by (26).}$$

Now this is the law of the coefficients assigned in (7). And just in the same way may the second series in (25) be verified. Thus the development of the general solution (19), produces the two convergent series of the solution in Art. 2.

The verification of the solution (20), though somewhat more difficult, may be effected on the same principles.

Developing the exponential, and assuming

$$\int_0^\pi (\cos\theta)^m d\theta = E_m, \int_0^\pi (\cos\theta)^m (\log\sin\theta)\, d\theta = F_m,$$

we shall have

$$u = \Sigma\left(\frac{AE_m}{1.2\ldots m} + \frac{2BF_m}{1.2\ldots m}\right) q_m x^m + \log x \Sigma \frac{BE_m}{1.2\ldots m} q^m x^m \ldots (27),$$

the summation extending to all even integral values of m, from $m = 0$ to $m = \infty$.

Now it may be shewn that

$$E_m = \frac{m-1}{m} E_{m-2},\ F_m = \frac{m-1}{m} F_{m-2} - \frac{1}{m} E_m \ \ldots\ldots (28),$$

and it will be found that these relations establish, for the coefficients of the series involved in (27), the same laws of successive derivation as are assigned in (10).

The verification of the solution (22), involves no difficulty.

Solution by Definite Integrals resumed.

6. In Art. 3, we found for the equation of the limits,

$$\epsilon^{xt} (t^2 - q^2)^{\frac{a}{2}} = 0 \ \ldots\ldots\ldots\ldots\ldots\ldots (29),$$

from which, in order to determine the limits in perfect independence of x, we rejected the factor ϵ^{xt}. In the discussion of the same problem in the great work of Petzval*, now in course of publication, that factor is retained, giving, according

* *Integration der Linearen Differentialgleichungen mit constanten und veränderlichen coefficienten.*

as x is positive or negative, the additional limit $-\infty$ or ∞. And thus the following solutions are arrived at, viz.:

$$u = C_1 \int_{-q}^{q} \epsilon^{xt} (t^2 - q^2)^{\frac{a}{2}-1} dt + C_2 \int_{-\infty}^{-q} \epsilon^{xt} (t^2 - q^2)^{\frac{a}{2}-1} dt \ldots\ldots (30),$$

when x is positive, and

$$u = C_1 \int_{-q}^{q} \epsilon^{xt} (t^2 - q^2)^{\frac{a}{2}-1} dt + C_2 \int_{q}^{\infty} (t^2 - q^2)^{\frac{a}{2}-1} dt \ldots\ldots\ldots (31),$$

when x is negative. It will be observed that it is in their second terms that the above expressions for u differ from the expression given in (19), and the question arises, what do those second terms really represent? We propose here to consider this question.

Supposing x positive, we have to examine the term

$$C_2 \int_{-\infty}^{-q} \epsilon^{xt} (t^2 - q^2)^{\frac{a}{2}-1} dt.$$

Now this expression, on assuming $t = -q(1+\theta)$, so as to make the limits of integration 0 and ∞, and performing reductions affecting only the arbitrary constant, becomes

$$C \int_0^{\infty} \epsilon^{-qx(1+\theta)} (2\theta + \theta^2)^{\frac{a}{2}-1} d\theta,$$

$$\text{or,} \quad C\epsilon^{-qx} \int_0^{\infty} \epsilon^{-qx\theta} (2\theta + \theta^2)^{\frac{a}{2}-1} d\theta \ldots\ldots\ldots\ldots (32).$$

It is easy to see that this cannot produce either of the particular integrals represented by *ascending* developments in (8). For, if we develope the exponential under the sign of integration, the coefficient of x^m in the factor represented by the definite integral, will be

$$\frac{(-q)^m}{1 . 2 \ldots m} \int_0^{\infty} \theta^m (2\theta + \theta^2)^{\frac{a}{2}-1} d\theta.$$

But, m and a being positive, it is manifest that the expression is infinite.

We may, however, expand the definite integral in *descending* powers of x. Developing the binomial in ascending powers of θ, and integrating by the well known theorem

$$\int_0^\infty \epsilon^{-qx\theta}\,\theta^{\lambda-1}\,d\theta = \frac{\Gamma(\lambda)}{(qx)^\lambda},$$

(32) assumes the form

$$C\epsilon^{-qx}\left\{\frac{2^{\frac{a}{2}-1}\Gamma\left(\frac{a}{2}\right)}{(qx)^{\frac{a}{2}}} + \frac{\left(\frac{a}{2}-1\right)2^{\frac{a}{2}-1}\Gamma\left(\frac{a}{2}+1\right)}{(qx)^{\frac{a}{2}+1}} + \&\text{c.}\right\},$$

Now observing that $\Gamma\left(\frac{a}{2}+1\right) = \frac{a}{2}\Gamma\left(\frac{a}{2}\right)$ &c., substituting and merging the common factors in the arbitrary constant, we have

$$C\epsilon^{-qx}x^{-\frac{a}{2}}\left\{1 + \frac{\left(\frac{a}{2}-1\right)\left(\frac{a}{2}\right)}{2qx} + \frac{\left(\frac{a}{2}-1\right)\left(\frac{a}{2}-2\right)\frac{a}{2}\left(\frac{a}{2}+1\right)}{1\,.\,2\,\ldots\,(2qx)^2} + \&\text{c.}\right\}\ldots(33),$$

which agrees with (12). Exactly in the same way Petzval's second integral for the case in which x is negative, represents the other descending and divergent series (14).

7. We thus see the true nature of the distinction between Petzval's form of solution, and those obtained in Art. 4. The latter represent the two converging and ascending series derived immediately from the differential equation. The former represents one of those series accompanied by the divergent series derived from a transformed differential equation*.

* Spitzer, in a recent Memoir in Crelle's *Journal* (Vol. LIV. p. 280), shews that when the coefficients of the differential equation

$$(a_2+b_2x)\frac{d^2y}{dx^2} + (a_1+b_1x)\frac{dy}{dx} + (a_0+b_0x)\,y = 0$$

satisfy the condition $a_1b_2 - a_2b_1 = b_2^2$, the solution will be

$$y = \int \epsilon^{ux}\,V\,(A + B\log\{(a_2+b_2x)\,U_1\})\,du,$$

where

$$U_1 = b_2u^2 + b_1u + b_0,\quad V = \frac{1}{U_1}\,\epsilon^{\int\frac{a_2u^2+a_1u+a_0}{U_1}du},$$

It is known that in the employment of divergent series an important distinction exists between the cases in which the terms of the series are ultimately all positive, and alternately positive and negative. In the latter case we are, according to a known law, permitted to employ that portion of the series which is convergent for the calculation of its entire value. Now, a being positive, the series (12) assumes this character when x is positive, the series (14) when x is negative. But these are precisely the cases in which these series are represented by Petzval's integrals.

When, for the calculation of an element dependent on the solution of a differential equation, ascending and descending series are both employed (the former for small, the latter for large values of the independent variable), it is necessary to determine the connexion of the constants. For this purpose the expressions of the series by definite integrals may be of importance. On this, and on other points connected with this subject, the reader is referred to two most instructive Memoirs by Prof. Stokes*, in which some of the equations of this chapter are applied to physical problems.

Partial Differential Equations.

8. Some of the most interesting applications of the above method occur in the solution of partial differential equations. The following is an example.

Ex. Required the most general solution of the equation

$$\frac{d^2u}{dx^2}+\frac{d^2u}{dy^2}+\frac{d^2u}{dz^2}=0,$$

and the limits are given by

$$\epsilon^{ux}\, U_1 V=0.$$

The deduction of this as a *limiting case of the general solution* may serve as an exercise to the student. It will be proper to assume $a_2+b_2x=v$ as the independent variable.

Spitzer expresses surprise that Petzval has not arrived at the above solution. We see however that it has no proper place in Petzval's actual scheme.

* *On the Numerical Calculation of a Class of Definite Integrals and Infinite Series.* Cambridge *Philosophical Transactions,* Vol. IX. Part I. p. 166.

On the Effect of the Internal Friction of Fluids on the Motion of Pendulums. Ibid. Part II. p. 8.

which can be expressed in terms of z and r, supposing

$$r = \sqrt{(x^2 + y^2)}.$$

This equation, with its supposed condition, presents itself in the problem of determining the attraction of a solid of revolution on an external point, and in the problem of the motion of an incompressible fluid, disturbed by the motion of a solid of revolution in the direction of the axis of revolution z.

The transformed equation is easily found to be

$$r\frac{d^2u}{dr^2} + \frac{du}{dr} + r\frac{d^2u}{dz^2} = 0 \quad\text{.................}(34).$$

Now the solution of the equation

$$r\frac{d^2u}{dr^2} + \frac{du}{dr} + q^2ru = 0,$$

is

$$u = \int_0^\pi \epsilon^{qr\cos\theta\sqrt{(-1)}}\,[A + B\log\{r(\sin\theta)^2\}]\,d\theta.$$

Hence, replacing q by $\frac{d}{dz}$, and A and B by arbitrary functions of z, we have, for the solution of (34),

$$u = \int_0^\pi \epsilon^{r\cos\theta\frac{d}{dz}\sqrt{(-1)}}\,[\phi(z) + \psi(z)\log\{r(\sin\theta)^2\}]\,d\theta,$$

or, by the symbolical form of Taylor's theorem,

$$u = \int_0^\pi \phi\{z + r\cos\theta\sqrt{(-1)}\}\,d\theta$$

$$+ \int_0^\pi \psi\{z + r\cos\theta\sqrt{(-1)}\}\log\{r(\sin\theta)^2\}\,d\theta \quad\text{......}(35).$$

Such is the complete integral.

In all physical problems involving partial differential equations the determination of the arbitrary functions so as to satisfy given initial conditions is a matter of great importance, and sometimes, where discontinuity presents itself, of great difficulty. But though some general principles might be stated, the subject is best studied in the concrete application.

In applying the above solution to the problem of attraction it is required to determine the arbitrary functions so that when $r=0$ we should have $u=F(z)$. Now, since, when $r=0$, $\log r$ is infinite, it is necessary to suppose $\psi(z)=0$. We have then

$$F(z)=\int_0^\pi \phi(z)\,d\theta=\pi\phi(z).$$

Thus the solution under the proposed limitation becomes

$$u=\frac{1}{\pi}\int_0^\pi F\{z+r\cos\theta\sqrt{(-1)}\}\,d\theta.$$

Parseval's Theorem.

9. Equations whose symbolical form is binomial generally admit of solution by definite integrals. Pfaff's equation has thus been treated by Euler. (Lacroix, Tom. III. p. 529.) The very beautiful theorem of Parseval which makes the limit of the series $AA'+BB'+CC'+$ &c. dependent upon the limits of the series $A+Bu+Cu^2+$ &c., and

$$A'+\frac{B'}{u}+\frac{C'}{u^2}+\text{\&c.}$$

should be noticed.

Suppose that, for all values of u, real and imaginary,

$$A+Bu+Cu^2\ldots=\phi(u),$$

$$A'+\frac{B'}{u}+\frac{C'}{u^2}\ldots=\psi(u).$$

Then, multiplying the equations together,

$$AA'+BB'+CC'+\Sigma\left(\alpha_m u^m+\frac{\beta_m}{u^m}\right)=\phi(u)\,\psi(u).$$

Assume, in succession, $u=\epsilon^{\theta\sqrt{(-1)}}$ and $u=\epsilon^{-\theta\sqrt{(-1)}}$, and add the results.

We find

$$2(AA' + BB' + CC') + 2\Sigma(\alpha_m \cos m\theta) + 2\Sigma(\beta_m \cos m\theta)$$
$$= \phi\{\epsilon^{\theta\sqrt{(-1)}}\}\,\psi\{\epsilon^{\theta\sqrt{(-1)}}\} + \phi\{\epsilon^{-\theta\sqrt{(-1)}}\}\,\psi\{\epsilon^{-\theta\sqrt{(-1)}}\}.$$

Now multiply by $d\theta$, integrate between the limits 0 and π, observing that $\int_0^\pi (\cos m\theta)\,d\theta = 0$, and divide the result by 2π, then

$$AA' + BB' \ldots = \frac{1}{2\pi}\int_0^\pi [\phi\{\epsilon^{\theta\sqrt{(-1)}}\}\,\psi\{\epsilon^{\theta\sqrt{(-1)}}\}$$
$$+ \phi\{\epsilon^{-\theta\sqrt{(-1)}}\}\,\psi\{\epsilon^{-\theta\sqrt{(-1)}}\}]\,d\theta \ldots\ldots (36),$$

which is the theorem in question.

Solution of Differential Equations, by Fourier's Theorem.

10. As Fourier's theorem affords the only general method known for the solution of partial differential equations with more than two independent variables (and such are the equations upon which many of the most important problems of mathematical physics depend), we deem it proper to explain at least the principle of this application, referring the reader for a fuller account of it to two memoirs by Cauchy*.

As a particular example, let us consider the equation

$$\frac{d^2u}{dt^2} - h^2\left(\frac{d^2u}{dx^2} + \frac{d^2u}{dy^2} + \frac{d^2u}{dz^2}\right) = 0 \ldots\ldots\ldots\ldots\ldots\ldots (37).$$

Let $u = \phi(x, y, z, t)$

represent any solution of this equation. By a well-known form of Fourier's theorem,

$$\phi(x) = \frac{1}{2\pi}\int_{-\infty}^{\infty}\int_{-\infty}^{\infty} da\,d\lambda\,\epsilon^{(a-x)\lambda\sqrt{(-1)}}\,\phi(a),$$

* *Sur l'Intégration d'Equations Linéaires. Exercices d'Analyse et de Physique Mathématique,* Tom. I. p. 53.

Sur la Transformation et la Réduction des Intégrales Générales d'un Système d'Equations Linéaires aux différences partielles. Ibid. p. 178.

successive applications of which enable us to give to u the form

$$u = \frac{1}{8\pi^3} \iiiiiint_{-\infty}^{\infty} \epsilon^{A\sqrt{(-1)}} \phi(a, b, c, t)\, da\,db\,dc\,d\lambda\,d\mu\,d\nu \;\ldots\ldots\; (38),$$

where $A = (a - x)\lambda + (b - y)\mu + (c - z)\nu$.

Substituting this expression in (37), and observing that from the form given to A we have

$$\left(\frac{d^2}{dx^2} + \frac{d^2}{dy^2} + \frac{d^2}{dz^2}\right) \epsilon^{A\sqrt{(-1)}} = \epsilon^{A\sqrt{(-1)}} (-\lambda^2 - \mu^2 - \nu^2),$$

we have

$$\frac{1}{8\pi^3} \iiiiiint_{-\infty}^{\infty} \epsilon^{A\sqrt{(-1}} \left\{\frac{d^2}{dt^2} + h^2 (\lambda^2 + \mu^2 + \nu^2)\right\} \phi\, da\,db\,dc\,d\lambda\,d\mu\,d\nu = 0,$$

ϕ being put for $\phi(a, b, c, t)$. This equation will be satisfied if ϕ be determined so as to satisfy the equation,

$$\frac{d^2\phi}{dt^2} + h^2 (\lambda^2 + \mu^2 + \nu^2)\,\phi = 0.$$

Hence, integrating and introducing arbitrary functions of a, b, c in the place of arbitrary constants, we have the particular integrals,

$$\phi = \epsilon^{Bht\sqrt{(-1)}} \psi_1(a, b, c), \quad \phi = \epsilon^{-Bht\sqrt{-1}} \chi_1(a, b, c) \;\ldots\; (39),$$

where $B = (\lambda^2 + \mu^2 + \nu^2)^{\frac{1}{2}}$.

Substituting the first of these values in (38), and merging the factor $\frac{1}{8\pi^3}$ in the arbitrary function, we have

$$u = \iiiiiint_{-\infty}^{\infty} \epsilon^{(A+Bht)\sqrt{(-1)}} \psi_1(a, b, c)\, da\,db\,dc\,d\lambda\,d\mu\,d\nu \;\ldots.\; (40),$$

a particular integral of the proposed equation. It may easily be shewn that the employment of the second value of ϕ given in (39) would only lead to an equivalent result.

To complete the solution, we observe that if, representing $\frac{d^2}{dx^2}+\frac{d^2}{dy^2}+\frac{d^2}{dz^2}$ by H, we make $t=\epsilon^\theta$, so as to reduce the given equation to the symbolical form,

$$u-\frac{H}{D(D-1)}\epsilon^{2\theta}u=0,$$

then, by Propositions 2 and 3, Chap. XVII. the transformation $u=\epsilon^{-\theta}\frac{dv}{d\theta}=\frac{dv}{dt}$, will give

$$v-\frac{H}{D(D-1)}\epsilon^{2\theta}v=0,$$

which is of the same form as the equation for u. Hence, v admitting of expression in the form (40), we have on merely changing the arbitrary function,

$$u=\frac{d}{dt}\int\int\int\int\int\int_{-\infty}^{\infty}\epsilon^{(A+Bht)\sqrt{(-1)}}\psi_2(a,b,c)\,da\,db\,dc\,d\lambda\,d\mu\,dv \;\ldots\ldots\; (41).$$

The complete integral is thus expressed by the sum of the particular integrals (40) and (41). The sextuple integral by which the above particular values of u are expressed admits of reduction to a double integral leading to a form of solution originally obtained by Poisson. Cauchy effects this reduction by a trigonometrical transformation. It may be accomplished, and perhaps better, by other means; but this is a matter of detail which does not concern the principle of the solution. We may add, that when the function to be integrated becomes infinite within the limits, Cauchy's method of residues should be employed. The reduced integral in its trigonometrical form, together with Poisson's method of solution, which is entirely special, will be found in Gregory's Examples, p. 504.

Cauchy's method is directly applicable to equations with second members, and to systems of equations. The above example belongs to the general form

$$\frac{d^2u}{dt^2}=Hu,$$

where H is a function of $\frac{d}{dx}, \frac{d}{dy}, \frac{d}{dz}$. For all such equations the method furnishes directly a solution expressed by sextuple integrals, which are reducible to double integrals if H is homogeneous and of the second degree. In the above example the double integration proves to be, in effect, an integration extended over the surface of a sphere whose radius increases uniformly with the time. Integrals of this class are peculiarly appropriate for the expression of those physical effects which are propagated through an elastic medium, and leave no trace behind.

XVIII. MISCELLANEOUS EXERCISES.

1. The complete integral of the equation

$$\frac{d^2u}{dx^2} = \left\{h^2 + \frac{n(n+1)}{x^2}\right\} u,$$

is expressible in the form $u = A\epsilon^{hx} + B\epsilon^{-hx}$, A and B being series which are finite when n is an integer. (Tortolini, Vol. V. p. 161.)

2. The definite integral $\int_0^\pi \cos \{n(\theta - x\sin\theta)\}\, d\theta$, can be evaluated when $n = \pm\left(i + \frac{1}{2}\right)$, where i is a positive integer or 0. (Liouville, *Journal*, Tom. VI. p. 36.)

Representing the definite integral by u, it will be found that u satisfies an equation of the form $\frac{d^2u}{dx^2} = \left(A + \frac{B}{x^2}\right)u$.

The subject of the evaluation of definite integrals by the solution of differential equations has been treated with great generality by Mr Russel. (*Philosophical Transactions for* 1855.)

3. If $v = \alpha$ be the equation of a system of curves, v being a function of x and y which satisfies the equation $\frac{d^2v}{dx^2} + \frac{d^2v}{dy^2} = 0$, and if $u = \beta$ be the equation of the orthogonal trajectories of the system, then u may be found by the integration of an

exact differential equation of the first order, and when found will satisfy the equation $\frac{d^2u}{dx^2}+\frac{d^2u}{dy^2}=0$.

The above theorem is applied by Professor Thomson to the problem of determining the forms of the rings and brushes in the spectra produced by biaxal crystals. (*Cambridge Journal*, 2nd Series, Vol. I. p. 124.)

4. The normal at a point P of a plane curve meets the axis in G, and the locus of the middle point of PG is the parabola $y^2=lx$. Find the equation to the curve, supposing it to pass through the origin. (*Cambridge Problems.*)

5. The normal at any point of a surface passes through the line represented by $\frac{x}{l}=\frac{y}{m}=\frac{z}{n}$. Find the differential equation to the surface, and obtain the general integral. (*Ib.*)

6. Prove that the differential equation of the surfaces generated by a straight line which passes through the axis of z, and through a given curve which makes a constant angle with the axis of z, is

$$x\frac{dz}{dx}+y\frac{dz}{dy}=\sqrt{(x^2+y^2)}\cot\alpha. \quad (Ib.)$$

7. Integrate the above equation.

8. Express by a definite integral the series,

$$1-\frac{x^2}{2^2}+\frac{x^4}{2^2.4^2}-\frac{x^6}{2^2.4^2.6^2}+\&c.$$

Form the differential equation by Chap. XVII. Art. 11, and then apply Laplace's method, Chap. XVIII. The result is $u=\frac{2}{\pi}\int_0^{\frac{\pi}{2}}\cos(x\cos\theta)\,d\theta$. (Stokes, *Cambridge Transactions*, Vol. IX. p. 182.)

9. Hence express the series in a form suitable for calculation when x is large.

Proceeding according to the directions of Chap. XVIII. the complete integral of the differential equation expressed by descending series will be

$$u=x^{\frac{1}{2}}\{(A\cos x+B\sin x)R+(A\sin x-B\cos x)S\},$$

where $$R=1-\frac{1^2.3^2}{1.2(8x)^2}+\frac{1^2.3^2.5^2.7^2}{1.2.3.4(8x)^4}-\&c.$$

$$S = \frac{1^2}{1.8x} - \frac{1^2.3^2.5^2}{1.2.3(8x)^3} + \&c.$$

The values of A and B for the particular integral in question will be $A = B = \pi^{-\frac{1}{2}}$. These are deduced from the consideration that, when x tends to infinity, we have, in the limit,

$$\frac{2}{\pi}\int_0^{\frac{\pi}{2}} \cos(x\cos\theta)\,d\theta = \pi x^{-\frac{1}{2}}(\cos x + \sin x). \quad (\textit{Ibid.})$$

The above series occurs in several physical problems.

10. The complete integral of the equation,

$$x\frac{d^2y}{dx^2} + (a + bx)\frac{dy}{dx} + (f + gx + hx^2)\,y = 0,$$

may be expressed by a finite formula involving general differentiation. (Attributed to Liouville.)

Assume $y = z\epsilon^{\alpha x + \frac{\beta x^2}{2}}$; then, by a proper determination of α and β, the equation may be reduced to the form

$$x\frac{d^2z}{dx^2} + (a' + b'x)\frac{dz}{dx} + f'z = 0.$$

The symbolical equation obtained by assuming $x = \epsilon^{\theta}$ will be binomial, and the integration in the required form may be effected by Prop. 3, Chap. XVII.

11. Equations of the form

$$x^2\frac{d^2y}{dx^2} + (A_1 + B_1x^m)\,x\frac{dy}{dx} + (A_0 + B_0x^m + C_0x^{2m})\,u = 0,$$

may be reduced to the form,

$$t\phi\left(\frac{d}{dt}\right)z + \psi\left(\frac{d}{dt}\right)z = 0 \;\ldots\ldots\ldots\ldots\ldots\ldots\; (m),$$

considered in Chap. XVIII.

Assume $x^m = t$, $y = t^k z$; the determination of k will be found to depend on the equation $k(k-1)m^2 + k\{m(m-1) + mA_1\} + A_0 = 0$.

Petzval, *Linearen Differentialgleichungen*, Pt. 1st, p. 105. Riccati's equation is included in the above.

12. Equations of the form

$$(a_0 + b_0\log x)\,x^2\frac{d^2u}{dx^2} + (a_1 + b_1\log x)\,x\frac{du}{dx} + (a_2 + b_2\log x)\,u = 0$$

are reducible to the form (m). (*Ib.* p. 112.)

13. The complete integral of the equation

$$\frac{d^n y}{dx^n} = (\alpha + \beta x)\, y,$$

is $y = \int_0^\infty dt\, \epsilon^{-\frac{t^{n+1}}{n+1}} \left(C\epsilon^{tx} + C_1 \rho \epsilon^{\rho tx} \ldots + C_n \rho^n \epsilon^{\rho^n tx}\right)$,

where ρ is a primitive root of $\rho^{n+1} = 1$, and $C, C_1, C_2 \ldots C_n$, satisfy the condition $C + C_1 + C_2 \ldots + C_n = 0$, but are otherwise arbitrary. (Jacobi, *Crelle's Journal*, Vol. x. p. 279.)

14. The determination of the orthogonal trajectory of any system of straight lines on a plane, involving in their general equation one variable parameter, can be determined by the solution of an exact differential equation between x and y.

This interesting proposition, together with the following demonstration, was communicated to the author by Professor Donkin, with whose permission it is published.

The equation of the given system can always be expressed in the form $x \sin\theta - y\cos\theta = \phi(\theta)$, or, putting $\cos\theta = u$, $\sin\theta = v$,

$$vx - uy - F(u, v) = 0 \ldots\ldots\ldots\ldots\ldots (1),$$

$$u^2 + v^2 - 1 = 0 \ldots\ldots\ldots\ldots\ldots (2).$$

The equation of the trajectory will then be

$$udx + vdy = 0 \ldots\ldots\ldots\ldots\ldots (3),$$

u and v being determined from (1) and (2) as functions of x and y.

Now, if we represent the first members of (1) and (2) by F and Φ respectively, then, in order that (3) may be an exact differential equation, we must have, in virtue of (37) Chap. XIV.

$$\frac{dF}{dx}\frac{d\Phi}{du} - \frac{dF}{du}\cdot\frac{d\Phi}{dx} + \frac{dF}{dy}\frac{d\Phi}{dv} - \frac{dF}{dv}\frac{d\Phi}{dy} = 0 \ldots\ldots\ldots (4),$$

and this will be found to be identically satisfied. Hence (3) is an exact differential equation, as was to be shewn. The proposition applies generally to the problem of involutes. Thus, the tangents to a circle being represented by

$$vx - uy = a, \quad u^2 + v^2 = 1,$$

the equation (3) will become

$$\frac{\{x\sqrt{(x^2+y^2-a^2)} - ay\}\, dx + \{y\sqrt{(x^2+y^2-a^2)} + ax\}\, dy}{x^2 + y^2} = 0.$$

This is exact, and determines, on integration, the system of possible involutes.

15. To determine the connexion of the integrals of any system of simultaneous differential equations expressible in the form

$$\left.\begin{aligned}\frac{dx}{dt}=\frac{dF}{du},\quad \frac{dy}{dt}=\frac{dF}{dv}\\ \frac{du}{dt}=-\frac{dF}{dx},\quad \frac{dv}{dt}=-\frac{dF}{dy}\end{aligned}\right\}\ldots\ldots\ldots\ldots\ldots(1).$$

where F is a given function of x, y, u and v.

The complete solution will evidently consist of four equations determining x, y, u, v as functions of t, and four arbitrary constants.

Suppose that there exists an integral of the form $\Phi = c$, where Φ is a function of x, y, u, v, not involving t. Then, differentiating, we have

$$\frac{d\Phi}{dx}\frac{dx}{dt}+\frac{d\Phi}{dy}\frac{dy}{dt}+\frac{d\Phi}{du}\frac{du}{dt}+\frac{d\Phi}{dv}\frac{dv}{dt}=0,$$

or, substituting for $\frac{dx}{dt}$, $\frac{dy}{dt}$, &c. the values given in (1),

$$\frac{d\Phi}{dx}\frac{dF}{du}+\frac{d\Phi}{dy}\frac{dF}{dv}-\frac{d\Phi}{du}\frac{dF}{dx}-\frac{d\Phi}{dv}\frac{dF}{dy}=0\ldots\ldots\ldots\ldots\ldots(2).$$

Now this equation is identically satisfied if $\Phi = F$. Hence one integral will be $F = a$, where a is an arbitrary constant.

Suppose now that another integral not involving t can be found. Then representing it by $\Phi = b$, and observing that (2) is identical with the equation (4) in the last problem, it is seen that if, from the two equations $F = a$, $\Phi = b$, we determine u and v as functions of x, y, a, b, the expression $udx + vdy$ will be an exact differential. Hence, if $\int(udx + vdy) = \chi$, we have

$$u=\frac{d\chi}{dx},\quad v=\frac{d\chi}{dy}\ldots\ldots\ldots\ldots\ldots\ldots\ldots(3).$$

Now differentiating the integral $F = a$ with respect to a, and regarding u, v, as functions of x, y, a, b, we have

$$\frac{dF}{du}\frac{du}{da}+\frac{dF}{dv}\frac{dv}{da}=1,$$

or, putting for $\frac{dF}{du}$, $\frac{dF}{dv}$ their values given in (1), and for u, v their values given in (3),

$$\frac{d^2\chi}{dadx}\frac{dx}{dt}+\frac{d^2\chi}{dady}\frac{dy}{dt}=1,$$

or

$$\frac{d}{dx}\left(\frac{d\chi}{da}\right)dx+\frac{d}{dy}\left(\frac{d\chi}{da}\right)dy=dt,$$

whence, integrating,

$$\frac{d\chi}{da} = t + c \quad \text{.........................(4)},$$

c being an arbitrary constant. Since the form of χ is known, this constitutes a third integral.

Lastly, differentiating $F = a$ with respect to b and proceeding as above, we find

$$\frac{d\chi}{db} = e \quad \text{...................... (5)},$$

e being an arbitrary constant. And this is the fourth integral.

The above is a simple illustration of the methods of Theoretical Dynamics referred to in Chap. XIV. Thus the equations for the motion of a body attracted towards fixed centres (all in one plane) are

$$\frac{d^2x}{dt^2} = -\frac{dR}{dx}, \quad \frac{d^2y}{dt^2} = -\frac{dR}{dy},$$

R being a function of x, y, and the co-ordinates of the fixed centres. These equations may be expressed in the form

$$\frac{dx}{dt} = u, \quad \frac{dy}{dt} = v,$$

$$\frac{du}{dt} = -\frac{dR}{dx}, \quad \frac{dv}{dt} = -\frac{dR}{dy}.$$

Now, if we represent the function $\frac{1}{2}(u^2 + v^2) + R$ by F, the above equations assume the general form (1).

It was intimated in Chap. XIV. that the solution of the equations of Dynamics is finally dependent on the obtaining of the complete primitive of a non-linear partial differential equation of the first order; and this was previously shewn to depend on the integration of an *exact* differential equation the coefficients of which were determined by the solution of a *linear* partial differential equation of the first order. Now all this agrees with what has been exemplified above. For the last two integrals, (4) and (5) are derived, by mere differentiation, from χ, while χ is found by the integration of an *exact* differential equation whose coefficients, u and v, are obtained from equations which satisfy the *linear* partial differential equation (2).

The student is especially referred to the original memoirs by Sir W. R. Hamilton (*On a General Method in Dynamics. Philosophical Transactions*, 1834—5), to various memoirs by Jacobi contained in his collected works or scattered through Crelle's *Journal*, and to the recent memoirs of Prof. Donkin (*On a Class of Differential Equations including those of Dynamics. Philosophical Transactions*, 1854—5). Liouville's *Journal* is rich in valuable memoirs on the subject.

ANSWERS.

The following table does not contain answers to all the questions proposed in the Exercises, but to a selected number of them, thought amply sufficient for ordinary requirements.

CHAPTER I.

2. (1) $y = px + \sqrt{(1+p^2)}$. $\left(\text{Here, } p = \dfrac{dy}{dx}\right)$.

(2) $p - ay = \epsilon^{ax}$. (3) $(1+x^2)p + y = \tan^{-1} x$.

(4) $xp + y = y^2 \log x$. (5) $yp^2 + 2xp = y$.

(6) $y = xp + \phi(p)$.

3. (1) and (2) $\dfrac{d^2y}{dx^2} + m^2 y = 0$. (3) $x^3 \dfrac{d^2y}{dx^2} = \left(y - x\dfrac{dy}{dx}\right)^2$.

5. $y^2 - a^2x^2 = b$. 6. (1) $(x-a)^2 + (y-b)^2 = 1$.

(2) $bx - ay = ab(xy - 1)$.

8. $x - \dfrac{m}{p^2} = a$, $y - \dfrac{2m}{p} = b$, $x - \dfrac{m}{p^2} = f\left(y - \dfrac{2m}{p}\right)$.

9. $(y - c)^2 = 4c'x$.

CHAPTER II.

1. (1) $\log xy + x - y = c$. (2) $\log \dfrac{x}{y} - \dfrac{y+x}{xy} = c$.

(3) $(1+x^2)(1+y^2) = cx^2$.

(4) $\dfrac{x}{\sqrt{(1+x^2)}} - \dfrac{1}{2}\log(1+y^2) - \log\{y + \sqrt{(1+y^2)}\} = c$.

(5) $\cos y = c \cos x$. (6) $\tan x \tan y = c$.

2. Yes. 3. (1) $y = c\epsilon^{-\frac{x}{y}}$. (2) $y = c\epsilon^{-\sqrt{\left(\frac{x}{y}\right)}}$.

(3) $x^2 = c^2 + 2cy$. (4) $x = c\epsilon^{-\sin\frac{y}{x}}$. (5) $(y+x)^2(y-2x)^3 = c$.

4. (1) $x^2-xy+y^2+x-y=c$. (2) $(y-x+1)^2(y+x-1)^5=c$.

5. $y=Cx^a+\dfrac{x}{1-a}-\dfrac{1}{a}$. 6. (2) $y=ax+cx\sqrt{(1-x^2)}$.

(3) Read x for 2 in second member,

$$y=c\epsilon^{-\frac{x}{\sqrt{(1-x^2)}}}+\frac{x}{\sqrt{(1-x^2)}}.$$

9. (1) $z=\{c\sqrt{(1-x^2)}-a\}^{-1}$. (2) $z=\left(c\epsilon^{ax}-\left(\dfrac{x+1}{a}+\dfrac{1}{a^2}\right)\right)$

(3) $z=\{c\epsilon^{2x^2}+\dfrac{a}{2}(2x^2+1)\}^{-\frac{1}{2}}$. (5) $y=(cx+\log x+1)^{-1}$.

CHAPTER III.

1. $x^4+6x^2y^2+y^4=C$. 2. $x^2-y^2=cx$. 3. $x^2-y^2=cy^3$.

4. $\dfrac{x^2+y^2}{2}+\tan^{-1}\dfrac{y}{x}=c$. 5. $x+y\epsilon^{\frac{x}{y}}=c$.

6. $\epsilon^x(x^2+y^2)=c$. 7. $\sin(nx+my)+\cos(mx+ny)=c$.

9. $\sqrt{(1+x^2+y^2)}+\tan^{-1}\dfrac{x}{y}=c$, $\log xy+\sin^{-1}\dfrac{x}{y}+\epsilon^{\frac{x}{y}}=c$.

10. Assuming $\dfrac{y}{x^a}=v$, we have $\displaystyle\int\frac{dv}{c-bv^2}+\frac{x^a}{a}=c$.

CHAPTER IV.

2. $\dfrac{1}{y^3-x^2y}$. 3. $\dfrac{1}{y^4}f\left(\dfrac{x^2-y^2}{y^3}\right)$.

5. (1) Integrating factor, $\dfrac{1}{x\sqrt{(x^2+y^2)}}$. Solution, $x^2=c^2+2cy$.

(2) Integrating factor, $\dfrac{1}{2x^2+3xy+y^2}$.
Solution, $(y+x)^2(y+2x)^3=c$.

(4) $y = c\sqrt{\left(1+\frac{2y}{x}\right)}$. (5) $xy\cos\frac{y}{x} = c$.

6. $y = cx$. 7. (1) $\frac{1}{xy(xy+1)}$. (2) $\frac{1}{x^3y^3+x^2y^2}$.

CHAPTER V.

1. (1) ϵ^x. (2) $\frac{1}{x^2}$. 2. y^{-4}. 3. $\epsilon^{\frac{x}{y}}$, and $\frac{1}{y^2}$.

4. (2) $y^{-3}\epsilon^{-\frac{ax}{y}}$. (3) $y^{-1}\epsilon^{\frac{x}{y}}$. (4) $(1+y^2-x^2)^{-2}$.
(5) $(x^3+y)^{-2}$. (6) $(x+y+xy)^{-2}$. (7) $(x+y^2)^{-3}$.

5. $\epsilon^x(x^2+y^2)^{-\frac{3}{2}}$.

9. When $\frac{1-n}{n^2}Q = \frac{d}{dx}\cdot\frac{P}{Q}$. Then $f(x) = -\frac{nP}{Q}$.

CHAPTER VI.

Equations 1 to 5 must be reduced to the form

$x\frac{dy}{dx} - ay + by^2 = cx^{2a}$, of which the solution is

$$y=x^a\sqrt{\left(\frac{b}{c}\right)}\frac{C\epsilon^{\frac{2\sqrt{(bc)}x^a}{a}}+1}{C\epsilon^{\frac{2\sqrt{(bc)}x^a}{a}}-1}, \text{ or } y=x^a\sqrt{\left(\frac{b}{c}\right)}\tan\left\{C-\frac{\sqrt{(-bc)}\,x^a}{a}\right\},$$

according as b and c are like or unlike in sign. In 1 we find $i=1$, and the solution by (A) is $y = a + \frac{x^{-2a}}{y_1}$, where y_1 is given by changing, in the first of the above solutions, a into $-a$, b into 1, c into 1, n into $-2a$. In 2, $i=2$. Apply (A). In 3 Apply (D).

7. $\sqrt{(\beta^2-4\alpha\gamma)} + n(i+\frac{1}{2})$, i being any integer positive, negative, or 0.

9. $x\frac{dy}{dx} - (2Ab+1)\,y + by^2 = cx^{m+2}$, where A is a root of the equation $bA^2 + A - h = 0$.

10. Compare with p. 95.

CHAPTER VII.

1. $(y-2x-c)(y-3x-c)=0$.

2. $(y-a\log x-c)(y+a\log x-c)=0$.

5. Eliminate p by means of $a\log p + 2bp + c = x$.

6. By $y = \frac{ap^2}{2} + \frac{2bp^3}{3} + c$.

8. By $y = \frac{p^2}{2} + \frac{p}{2}\sqrt{(1+p^2)} - \frac{1}{2}\log\{p+\sqrt{(1+p^2)}\} + c$.

12. Comp. Prim. $y = cx + c - c^2$. Sing. Sol. $y = \frac{(x+1)^2}{2}$.

13. Complete Primitive, $y = cx + \sqrt{(b^2 - a^2c^2)}$.

Singular Solution, $\frac{x^2}{a^2} + \frac{y^2}{b^2} = 1$. 14. $x^2 + y^2 = cx$.

16. Eliminate p by $x = \frac{p}{\sqrt{(1+p^2)}}\left[C + m\log\{p+\sqrt{(1+p^2)}\}\right]$.

17. By $x = \frac{p}{\sqrt{(1+p^2)}}\left(c + \frac{a}{p} + a\tan^{-1}p\right)$.

19. $(x-a)^2 + \{y - f(a)\}^2 = 1$. 21. $xf(a) - ya = af(a)(xy-1)$.

CHAPTER VIII.

4. Sing. Sol. $x = a$. 6. Diff. equation, $p = \frac{1}{2\sqrt{(x-a)}}$.

10. (1) $xy=1$. (2) $\left(\frac{x}{m}\right)^{\frac{1}{2}} \pm \left(\frac{y}{n}\right)^{\frac{1}{2}} = 1$. (3) $y = \frac{(x-1)^2}{4}$.

11. Particular Integral. 13. $y = \epsilon^{c(x^2+1)}$.

16. (1) Envelope species, $y = \frac{-x^4}{4}$.

(2) Envelope species, $y^2 = 4x^2$.

(3) Not of envelope species, $y = x^n$.

17. Singular solution, $\sqrt{x} + \sqrt{y} = a$.

18. Singular solution, $x^{\frac{2}{3}} + y^{\frac{2}{3}} = a^{\frac{2}{3}}$.

19. $x = \cos^{-1} y^{\frac{1}{2}} + (y - y^2)^{\frac{1}{2}}$.

CHAPTER IX.

1. $y = c\epsilon^{3x} + c'\epsilon^{4x}$. 2. $y = c\epsilon^{3x} + c'\epsilon^{4x} + \frac{x+7}{12}$.

3. $y = \epsilon^x (c_0 + c_1 x + c_2 x^2 + c_3 x^3)$.

4. $y = (c_1 + c_2 x) \cos x + (c_3 + c_4 x) \sin x$.

5. $y = c\epsilon^{-x} + (c_1 + c_2 x) \epsilon^{2x}$.

6. $y = c_1 \cos x + c_2 \sin x + (c_3 + c_4 x) \epsilon^x + 1$.

7. $y = (c_1 + c_2 x) \epsilon^{kx} + \frac{\epsilon^x}{(k-1)^2}$. 8. $y = \left(c_1 + c_2 x + \frac{x^2}{2}\right) \epsilon^x$.

9. $y = cx^3 + \frac{c'}{x}$, 10. $y = c(x+a)^2 + c'(x+a)^3$.

11. $y = \epsilon^{\frac{bx^2}{2}} \{c \cos (x \sqrt{b}) + c' \sin (x \sqrt{b})\}$.

12. $y = c\epsilon^{q \sin^{-1} x} + c_1 \epsilon^{-q \sin^{-1} x}$.

14. Add x^2 to the previous value of y.

CHAPTER X.

1. $y = \frac{x^3}{6} - \sin x + c + c'x$.

3. $y = \int \sqrt{(a^2 + x^2)}\, dx + c \log x + c'$. 5. $y = \frac{cx^2}{3} + \frac{c'}{x}$.

7. $x + c = (a^2 - y^2)^{\frac{1}{2}}$. 8. $y = \frac{cx^2}{2} + f(c)\, x + c'$.

9. $x = \frac{1}{c} \log \{cy + f(c)\} + c'$. 11. $y = x \log \left(\frac{x}{c + c'x} \right)$.

14. $y = \epsilon^{-\int P dx} \left(\int \epsilon^{\int P dx} C dx + C' \right)$. 19. $y = cx$.

20. $y = -a + \frac{1}{2} \left(a\epsilon^{\frac{x}{a}} + a\epsilon^{-\frac{x}{a}} \right)$. 22. $x + c + (c_1^{\frac{2}{3}} - y^{\frac{2}{3}})^{\frac{3}{2}} = 0$.

23. $y = c \log \{x + c + \sqrt{(x^2 + 2cx)}\} + c'$.

26. $t = \int \frac{dx}{\left(C\epsilon^{-\frac{k}{x^2}} + \frac{\mu}{k} \right)^{\frac{1}{2}}} + C'$.

28. $x^2 \frac{d^2y}{dx^2} - x \frac{dy}{dx} + xy^2 = c$. 31. $\left(\frac{dy}{dx} \right)^2 - x^2 = 0$.

32. $(y - c)^2 - \frac{x^2}{4} = 0$. 33. $y = \frac{1}{2a} \left(\frac{x^3}{3} + a^2 x + b \right)$.

CHAPTER XI.

1. $x = cy^n$. 2. $x + c = \frac{1}{n} \log \{ny + \sqrt{(n^2 y^2 - 1)}\}$.

4. $2cx + c' = b \left(\frac{c^2 y^{\frac{b-a}{b}}}{b - a} - \frac{y^{\frac{b+a}{b}}}{b + a} \right)$.

6. Let $y^2 = 2cx - x^2$ represent the circles, then the trajectory is $x^2 = 2cy - y^2$.

7. $y^2 + x^2 - c = 2a^2 (\log x)^2$. 8. $r = c\epsilon^{m\theta}$.

10. $4ay + c = 2ax \sqrt{(4a^2 x^2 - 1)} - \log \{2ax + \sqrt{(4a^2 x^2 - 1)}\}$.

CHAPTER XII.

1. $(x - a)(y - b)(z - c) = C$.

2. $x^2+2y^2-6xy-2xz+z^2=C.$ 3. $yz+zx+xy=c.$

5. $\epsilon^x(y+z)=c.$ 6. $\frac{a}{x}+\frac{b}{y}+\frac{c}{z}=C.$

7. $\frac{y+z}{x}+\frac{z+x}{y}=C.$ 8. $\epsilon^{x^2}(x+y+z^2)=C.$

9. In first term read $2xz$ for $2yz$, then

$$x^2+xy^2-w+x^2z=c.$$

10. No.

CHAPTER XIII.

1. $x=c\epsilon^{-\frac{7t}{2}}-\frac{y}{2},\quad y=(ct+c_1)\,\epsilon^{-\frac{7t}{2}}.$

2. $y=\epsilon^{-6t}(c\cos t+c'\sin t),$

$$x=\frac{\epsilon^{-6t}}{2}\{(c+c')\sin t+(c-c')\cos t\}.$$

5. $x+y=c\epsilon^{-4t}+\frac{\epsilon^t}{5}+\frac{\epsilon^{2t}}{6},$

$$x-2y=c'\epsilon^{-7t}+\frac{\epsilon^t}{8}-\frac{2\epsilon^{2t}}{9}.$$

7. $x=\frac{1}{7}+4c_1\epsilon^{2t}+4c_2\epsilon^{-2t}-3c_3\epsilon^{t\sqrt{7}}-c_4\epsilon^{-t\sqrt{7}},$

$$y=\frac{9}{14}+c_1\epsilon^{2t}+c_2\epsilon^{-2t}-c_3\epsilon^{t\sqrt{7}}-c_4\epsilon^{-t\sqrt{7}}.$$

CHAPTER XIV.

2. $z=y\sin^{-1}\frac{x}{y}+\phi(y).$ 3. $\epsilon^{-\frac{z}{y}}(x+y+z)=\phi(y).$

5. $z=\frac{x}{a}+\phi(ay-bx)$. 6. $z=\epsilon^{\frac{y}{a}}\phi(x-y)$.

7. $z=x+y\phi(x+y)$. 8. $z=\frac{y^2}{3x}+\phi(xy)$.

9. $\frac{y-b}{z-c}=\phi\left(\frac{x-a}{z-c}\right)$. 10. $z^2=xy+\phi\left(\frac{x}{y}\right)$.

11. $x^2+y^2+z^2=z\phi\left(\frac{y}{z}\right)$.

13. $x+\sqrt{(x^2+y^2+z^2)}=x^{1-a}\phi\left(\frac{y}{x}\right)$. 15. $z=c\sqrt{(x^2+y^2)}$.

16. $(a-1)z+\frac{xy}{t}=x^a\phi\left(\frac{y}{x},\frac{t}{x}\right)$.

18. Complete Primitive, $z=ax+by+ab$. 19. $z=-xy$.

20. $z=ax+\frac{y}{a}+b$. 21. $z=axe^y+\frac{a^2}{2}\epsilon^{2y}+b$.

23. $z=xy+y\sqrt{(x^2-a^2)}+b$ and $z=\frac{x-y}{a}+\frac{ay}{x-y}+b$.

CHAPTER XV.

2. $z=x\phi\left(\frac{y}{x}\right)+\psi\left(\frac{y}{x}\right)$. 3. $y=x\phi(z)+\psi(z)$.

4. $x=f(y)+\phi(z)$. 5. $x=F(z)+f(x+y+z)$.

8. $z=\phi(x+ay)+y\sqrt{(-1-a^2)}$.

CHAPTER XVI.

1. $y=c+\left(c_1+c_2x+\frac{x^2}{2}\right)\epsilon^x$. 3. $y=c\epsilon^{2x}+c'\epsilon^{3x}-x\epsilon^{2x}$.

2. $\dfrac{\epsilon^{mx}}{m^2 - 5m + 6} + c\epsilon^{2x} + c'\epsilon^{3x}.$

4. $u = \epsilon^{3x}\left(\dfrac{x^2}{2} + \dfrac{3}{2}x + \dfrac{9}{16}\right) + c\epsilon^{4x} + c'\epsilon^{5x}.$

5. $u = \dfrac{3m \sin mx - (m^2 - 2)\cos mx}{9m^2 + (m^2 - 2)^2} + c_1\epsilon^{-x} + c_2\epsilon^{-2x}.$

8. $u = x^n\phi\left(\dfrac{y}{x}\right) + x\psi\left(\dfrac{y}{x}\right).$

9. $u = \dfrac{(x^2 + y^2)^{\frac{n}{2}}}{n(n-1)} + x\phi\left(\dfrac{y}{x}\right) + \psi\left(\dfrac{y}{x}\right).$

10. $u = \cos n \log x\phi\left(\dfrac{y}{x}, \dfrac{z}{x}\right) + \sin(n \log x)\,\psi\left(\dfrac{y}{x}, \dfrac{z}{x}\right).$

11. Assume $\dfrac{d}{dx} + X = \pi.$

CHAPTER XVII.

1. $u = c\epsilon^x(x-1) - c'\epsilon^x(x+1).$

2. $u = \left(x^2\dfrac{d^2}{dx^2} + x\dfrac{d}{dx}\right)\dfrac{c + c'\log x}{1 - x}.$

7. $u = x^i\left(\dfrac{d}{dx}\right)^i \dfrac{c_1 + c_2\int x^{b-1}(1+qx)^{a-b}dx}{x^b(1+qx)^{a-b+1}}.$

9. $z = \left(x\dfrac{d}{dx} - 1\right)\{\phi(y+x) + \psi(y-x)\}.$

12. $\left(x\dfrac{d}{dx}\right)^n \epsilon^x.$

THE END.

Cambridge:
PRINTED BY C. J. CLAY, M.A.
AT THE UNIVERSITY PRESS.

APPENDIX.

NOTES TO CHAPTER XIV.

Art. 5. To complete the theory of the linear partial differential equation $Pp + Qq = R$ it ought to be shewn that the solution $u = f(v)$, or as it may be expressed,

$$F(u, v) = 0 \qquad \text{.............................} \quad (1),$$

includes every possible solution.

Let $\chi(x, y, z) = 0$, or for simplicity $\chi = 0$, represent *any* particular solution. Differentiating, we have

$$\frac{d\chi}{dx} + \frac{d\chi}{dz} p = 0, \quad \frac{d\chi}{dy} + \frac{d\chi}{dz} q = 0,$$

and substituting the values of p and q hence derived in the given equation

$$P \frac{d\chi}{dx} + Q \frac{d\chi}{dy} + R \frac{d\chi}{dz} = 0.$$

Similar equations being obtained from the particular integrals $u = a$, $v = b$, we have, on eliminating P, Q, R,

$$\frac{d\chi}{dx}\left(\frac{du}{dy}\frac{dv}{dz} - \frac{du}{dz}\frac{dv}{dy}\right) + \frac{d\chi}{dy}\left(\frac{du}{dz}\frac{dv}{dx} - \frac{du}{dx}\frac{dv}{dz}\right) + \frac{d\chi}{dz}\left(\frac{du}{dx}\frac{dv}{dy} - \frac{du}{dy}\frac{dv}{dx}\right) = 0 \quad \text{.........} \quad (2).$$

Now suppose the forms of u and v to be

$$u = \phi(x, y, z), \quad v = \psi(x, y, z) \quad \text{..............} \quad (3),$$

$\phi(x, y, z)$ and $\psi(x, y, z)$ being given functions. From these two equations some two of the quantities x, y, z may be determined as functions of the other and of u and v. Suppose x and y thus determined as functions of z, u, and v; then by substitution $\chi(x, y, z)$ becomes a function of z, u, and v, and we may write

$$\chi(x, y, z) = \chi_1(z, u, v).$$

Hence we find

$$\frac{d\chi}{dx} = \frac{d\chi_1}{du}\frac{du}{dx} + \frac{d\chi_1}{dv}\frac{dv}{dx},$$

$$\frac{d\chi}{dy} = \frac{d\chi_1}{du}\frac{du}{dy} + \frac{d\chi_1}{dv}\frac{dv}{dy},$$

$$\frac{d\chi}{dz} = \frac{d\chi_1}{du}\frac{du}{dz} + \frac{d\chi_1}{dv}\frac{dv}{dz} + \frac{d\chi_1}{dz}.$$

Substituting these in (2) and reducing, we have

$$\frac{d\chi_1}{dz}\left(\frac{du}{dx}\frac{dv}{dy} - \frac{du}{dy}\frac{dv}{dx}\right) = 0 \;\ldots\ldots\ldots\ldots\;(4).$$

But, were the second factor of the first member equal to 0, u would be a *definite* function of v and z (Chap. IV. Art. 3) and the equations (3) could not determine x and y as by hypothesis they do. We have then $\frac{d\chi_1}{dz} = 0$, whence χ_1 does not involve z. Thus, χ being expressible as a function of u and v, the equation $\chi = 0$ is included in the general form (1).

The student will easily extend this proof to the case in which the number of variables is n.

ART. 14. The most important form of the problem of this Article is the following, and the reader is requested to substitute it for the one in the text, sufficient account not being there taken of the conditions among the constants.

Required a value of z as a function of $x_1, x_2, \ldots x_n$ which shall satisfy the partial differential equation

$$F(x_1, x_2 \ldots x_n, z, p_1, p_2 \ldots p_n) = 0 \;\ldots\ldots\ldots\ldots\;(1),$$

and shall, when $x_n = 0$, assume a *given* form,

$$z = \phi(x_1, x_2 \ldots x_{n-1}) \;\ldots\ldots\ldots\ldots\ldots\ldots\;(2).$$

Representing the second member of (2) by ϕ, and $\frac{d\phi}{dx_1}$ by ϕ_1, &c., we shall have, when $x_n = 0$,

$$p_1 = \phi_1,\ p_2 = \phi_2,\ \ldots\ p_{n-1} = \phi_{n-1}, \ldots\ldots\ldots\ldots\ldots\ldots (3),$$

for, in seeking the forms which $\frac{dz}{dx_1}, \frac{dz}{dx_2} \ldots \frac{dz}{dx_{n-1}}$ assume when $x_n = 0$, we are permitted to make $x_n = 0$ in the general value of z before differentiating.

Now the auxiliary system of the linear equation, (45) in the text, yields $2n$ integrals connecting $x_1 \ldots x_n$, z, $p_1 \ldots p_n$ with $2n$ arbitrary constants. But since one of the integrals is $F = c$, and since to make this agree with (1) we must have $c = 0$, the $2n$ integrals will effectively contain $2n - 1$ arbitrary constants. This however being the number of the variables contained in (2), (3), namely of the variables $x_1 \ldots x_{n-1}$, z, $p_1 \ldots p_{n-1}$, we may express, and so replace, these arbitrary constants by initial values of the above variables corresponding to $x_n = 0$.

Let $\xi_1, \ldots \xi_{n-1}, \zeta, \pi_1, \ldots \pi_{n-1}$ be the new constants in question; then, substituting these for the variables whose initial values they represent in the n equations (2), (3), we obtain n conditions connecting the above constants.

Thus we have finally $3n$ equations, consisting of $2n$ integrals with n equations of condition connecting the $2n - 1$ constants which those integrals contain. From these $3n$ equations we can eliminate the above $2n - 1$ constants together with the n quantities $p_1, p_2 \ldots p_n$. The result will be a final relation between z, x_1, $x_2 \ldots x_n$, which will be the solution sought.

If we regard the function $\phi(x_1, x_2 \ldots x_{n-1})$ as arbitrary, the above solution will constitute a *general* primitive; but if we give to it a particular form involving n arbitrary constants, we shall obtain a *complete* primitive. (Cauchy, *Exercices*, Vol. II. p. 238.)

CORRECTIONS.

The author is indebted to Mr Todhunter, and to other members of the University of Cambridge, for most of the following corrections, and desires to record his thanks.

Wherever the correction is such as to leave no doubt as to its application, it has been thought unnecessary to reprint the error of the text.

Chapter I.

P. 5, (8): $\left(\frac{dy}{dx}\right)^2 + x\frac{dy}{dx} - y = 0$. p. 15, l. 25: $n-r+1$. p. 21, Ex. 2, (3): $ce^{-\tan^{-1}x}$. dele Ex. 5. p. 477, Ans. 3, (3): $-$ after $=$. Ans. 9: $(y-c)^2$.

Chapter II.

P. 30, l. 6: $=0$. p. 34, l. 8: $\frac{dx'}{x'}$. p. 36, l. 11: $\therefore$. p. 41, Ex. 1, (4): dy before $=$. p. 42, Ex. 6, (3): $x+$ for $2+$. p. 43, Ex. 6: $\int \epsilon^{\int P dx}$. Ex. 7: $-$ for $+$. p. 477, Ans. 3, (5): $+$ for $-$. p. 478, Ans. 6, (3): $\sqrt{1-x^2}$. Ans. 9, (2):

$$z = \left(ce^{\frac{3ax}{2}} - \frac{x+1}{a} - \frac{2}{3a^2}\right)^{\frac{1}{3}}.$$

Chapter III.

P. 44, l. 16: member. p. 478, Ans. 9: $\sin^{-1}\sqrt{x^2+y^2}$ for $\log xy$.

Chapter IV.

P. 52, l. 9, 18: $=0$. p. 61, l. 16: $=0$. p. 64, l. 9: Ny. l. 18 and 20: $=0$. p. 67, Ex. 1: $2(xy-1)$ for $(2xy-1)$.

CHAPTER V.

P. 73, Ex. 1 : $=0$. p. 74, l. 2 : $\frac{x}{y}$ for xy. p. 78, l. 11: $3n$ for $3x$. p. 80, l. 3 : omit 'concluding.' l. 4 : for 5, read 6. l. 19 : x^{m+1}. (17): $=$ for $=-$. p. 81, l. 15 : $+v\psi(v)$ for $-v\psi(v)$. p. 85, l. 9 : XXIV. p. 86, l. 3 : $+A''BC'$. p. 89, Ex. 4 (3): $(y-x)\,dy$. p. 479, Ans. 9 : $\frac{n-1}{n^2}$.

CHAPTER VI.

P. 98, l. 15 : (16) for (18). p. 111, Ex. 5 : dele dx. Ex. 6 : b for 6. p. 112, Ex. 13 : $b-a$ for $y-x$. p. 479, Ans. 7 : $=0$.

CHAPTER VII.

P. 117 & 118: for μ_1, μ_2 read $\mu_1 dx$, $\mu_2 dx$. p. 118, (17): $\frac{dV_1}{dx}$, $\frac{dV_2}{dx}$, $\frac{dV_3}{dx}$. p. 119, l. 13 : formed. p. 120, l. 21 : parallel to. l. 22 : the axis of y. p. 125, l. 11 & 14 : (2) for (3). p. 128, l. 5 : $+4ax$. l. 14 : $-a\log$ for $+a\log$. l. 15 : p for y. (1): x^n for x^4. p. 138, Ex. 18 : $x+f\{$. Ex. 19 : $\sqrt{1+\left(\frac{dy}{dx}\right)^2}$; $x-$ for $x+$. p. 480, Ans. 12: $\frac{(x+1)^2}{4}$. Ans. 13 : $-\frac{x^2}{a^2}$. Ans. 21: $ax-f(a)y=$.

CHAPTER VIII.

P. 146, (7): $\frac{-df(x,c)}{dc} \div \frac{df(x,c)}{dx}$. p. 150, last line : $-$ for $+$. p. 151, l. 8 : $y^2=4mx$. p. 156, l. 4 : X. p. 164, l. 4 : X. p. 171, (3) : $xy^{\frac{1}{2}}$ for $4xy^{\frac{1}{2}}$. p. 172, l. 9 : $\frac{dp}{dy}=\frac{2}{x}$. p. 183, l. 16 : $=mnp$. p. 184, Ex. 14 : omit second sentence. Ex. 16, (1): $+$ for $-$. Ex. 16, (3): y for z. p. 185, l. 7 : y^2. p. 186, l. 2 : $\frac{d\phi}{dc}$. l. 5 : $c-3x-y$. p. 481, Ans. 13 : Sing. Sol. $y=0$; comp. Prim. $y=\epsilon^{c(x^2-1)}$. Ans. 19 : $\cos^{-1}y^{\frac{1}{2}}+(y-y^2)^{\frac{1}{2}}$.

CHAPTER IX.

P. 187, (1): $\frac{d^2y}{dx^2}$. l. 19 : Art. 2. p. 197, l. 2 : r times. l. 21 and 23 : $\cos nx$; $\sin nx$. p. 205, l. 18 : is -1. p. 206, l. 4 : $C\epsilon^{a \sin^{-1} x}$. p. 481, Ans. 2 : $\frac{12x+7}{144}$.

CHAPTER X.

P. 209, l. 18 : + for =. p. 215, l. 10 : θ by $\log x$. p. 217, l. 2 & 4 : z for x. l. 11 : $+c'$ or. p. 218, last line : $-bu^2$. p. 221, (36): $y^2\left(\frac{dy}{dx}\right)^2$. p. 223, l. 18 : q^2y. p. 229, l. 12 : n^{th}. l. 13, et seq.: (47), (48), (49) for (46), (47), (48). p. 231, Ex. 12 : x^3+ for x^2+. p. 233, l. 7 : $(y^2+x^2)^2$. Ex. 31, 1st term : $\left(\frac{d^2y}{dx^2}\right)^2$. p. 482, Ans. 3 : $x=\frac{a^{\frac{1}{4}}}{2}\int\frac{dy}{\sqrt{c+\sqrt{y}}}+c'$. Ans. 7: $(y^2-a^2)^{\frac{1}{2}}$. Ans. 11 : $x\log\frac{x}{c+c'x}$. Ans. 32 : $(y-c)^2-\frac{x^4}{4}=0$.

CHAPTER XI.

P. 235, last line : x, y for x', y'. p. 236, Ex. 1 : 2 for z. p. 238, l. 11 : $X+C=$. p. 246, (a): h^2 for c^2. p. 247, l. 20 : (4) and (5). l. 22 : (6). p. 249 : omit first four lines. p. 261, l. 17: $(m^2+1)^{\frac{1}{2}}$. p. 265, Ex. 14 : $\left(\frac{y}{a}\right)^{\frac{2m}{1-m}}$. p. 266, l. 3 : $y\sqrt{1+\left(\frac{dy}{dx}\right)^2}$. p. 482, Ans. 6: $x^2=2c'y-y^2$. Ans. 7 : $2a^2\log x$.

CHAPTER XII.

P. 274, l. 27 : $(y+a)^2$. p. 277, l. 9 and 11 : $v+1$. l. 13 : $y+z$. last two lines : μ for u. p. 279, l. 20 : $=0$. p. 280, l. 16 : (c). l. 20 : Art. 4. p. 282, (27): $P($. p. 286 : dele Ex. 4. Ex. 9 : xz for yz.

CHAPTER XIII.

P. 290, l. 3 : (3). p. 293, (12): dele – each time. p. 294, l. 4 and 5 : $\frac{d^2x_1}{dx^2}$, $\frac{d^{n-1}x_1}{dx^{n-1}}$. p. 297, l. 9: m. p. 300: z'. p. 302, (20): $\frac{dt}{T}$. l. 21 : of t. p. 305, (a): $(ab' - a'b)\,x$. p. 306, l. 17: $\frac{dy}{dt}$, $\frac{d^2y}{dt^2}$. p. 309, (i): B). p. 483, Ex. 1 : $c\epsilon^{-\frac{7t}{2}}$

CHAPTER XIV.

P. 320, last line but one : q for g. p. 325, (14) et seq. : v_{n-1} for v_n. p. 327, l. 3 : $\frac{1}{3}$ for $\frac{1}{4}$. p. 333, l. 1 : f for p. l. 7 : dx, dy for da, db. p. 325, l. 1: $\psi(c)$. p. 340, l. 3 : $\frac{-z^3dp}{p}$. p. 344, last line : F for f. p. 346, (44) : dx_i for dx_t. p. 350, l. 4 : ψ for ϕ. last line : $\phi'(a) + \frac{1}{c^2}$. p. 484, Ans. 7 : $(x+y)$. Ans. 10 : z^2. Ans. 11 : $\phi\left(\frac{y}{z}\right)$.

CHAPTER XV.

P. 352, last line : s for 0. p. 359, (31) : $f_1(v_1)$; $f_2(v_2)$. p. 363, l. 24 : $\frac{4pdx}{2y-a}$. p. 364, l. 11 : a for c. p. 366, (49): dX; dY. p. 368, l. 3 and 6 : z for Z.

CHAPTER XVI.

P. 378, (18): o for X. p. 382, (33) : r for 2. p. 383, l. 3 : $\frac{1}{2n\sqrt{-1}}$. l. 7 : $\sin nx$. l. 9: for the last $-\sqrt{-1}$ read $+\sqrt{-1}$. (34): $u =$. p. 384, l. 23 : member. p. 388, l. 3 : omit last index. p. 391, l. 1 : Φ_1 and Φ_2. p. 392, (2): N_2 for A_2. p. 394, l. 5 : member. p. 398, (9): $y_3 = a_3$. p. 484, Ans. 3 : $y =$. p. 485, Ans. 4: $\frac{7}{4}$ for $\frac{9}{16}$. Ans. 10 : $\cos n \log x$.

CHAPTER XVII.

P. 402, l. 6 : X. p. 404, l. 13 : form. p. 405, l. 16 : with $\{f_0(D)\}^{-1}$. p. 430, l. 7 : (b). p. 445, last line : $\dfrac{F''(x)}{1.2}$. p. 448, Ex. 8 : $\dfrac{4c}{a}$ for $4c$.

CHAPTER XVIII.

P. 452, (3) : $\{\phi(t)\, T\}$. p. 456, (14) : q^2x^2 for qx^2. p. 467, l. 23 : & 468, l. 2 : &c. for CC'.

MISCELLANEOUS EXERCISES.

P. 472, Ex. 6 : curve, and which.

CAMBRIDGE: PRINTED BY C. J. CLAY, M.A. AT THE UNIVERSITY PRESS.